Aspects in
GAS CHROMATOGRAPHY

Papers based on main lectures,
held on the Sixth Symposium on Gas Chromatography 1968 in Berlin,

Sponsored by
Deutsche Akademie der Wissenschaften zu Berlin

Edited by H. G. STRUPPE

A K A D E M I E - V E R L A G · B E R L I N
1 9 7 1

Erschienen im Akademie-Verlag GmbH, DDR-108 Berlin, Leipziger Straße 3—4

Lizenznummer: 202 · 100/578/71
Offsetdruck und buchbinderische Weiterverarbeitung: VEB Druckerei „Thomas Müntzer", 582 Bad Langensalza/DDR
EDV-Nummer: 761 481 9
Printed in German Democratic Republic

CONTENTS

Foreword

The German Academy of Sciences at Berlin by its Sub-Commission for Gas Chromatography has held six symposia on gas chromatography in co-operation with the chemical and scientific instrument industries. These symposia tock place in Leipzig October 1958, in Boehlen near Leipzig October 1959, in Schkopau (district of Merseburg) May 1961, in Leuna (district of Merseburg) May 1963, and in Berlin May 1965 and May 1968, respectively.

These conference have shown an ever increasing international participation. The Sixth Symposium had registered 350 members from the German Democratic Republic as well as 150 participants from Austria, Australia, Belgium, Bulgaria, CSSR, Danemark, Democratic Republic Vietnam, France, German Federal Republic, Great Britain, Hungaria, Italy, Netherlands, Poland, Roumania, Sweden. Switzerland, USA, USSR, and Yugoslavia. During the symposium 18 main lectures were read as well as 45 original contributions, reprinted in full text in the volume of preprints, and other remarkable discussions. Because of the considerable volume of the materials publisher and editor have agreed to publish the papers not reprinted in the preprints, i. e. the main lectures, in the present form. By an edition in English language may be expected that the readers, corresponding to the international character of the conference, would not be restricted to the German speaking area. Besides the extensively reprinted review articles, an appendix shows the English summaries of the papers given in German in the Preprints. On interest either the complete preprints or copies (microfilms) of the English translation of single papers may be ordered from the Sub-Commission for Gas Chromatography. Reproduction costs will be charged by the bookdealers.

The Editor acknowledges the support granted by Preston Technical Abstracts Co., Evanston for proofreading of the English translations and wishes to express his gratitude to all authors for their surplus efforts in preparing the English version.

Leipzig, Spring 1970 Hans Georg Struppe

State of Development and Application of Radiation Ionisation Detectors

H.-K. BOTHE

German Academy of Sciences in Berlin, Institut of Applied Radioactivity,
Leipzig

1. Introduction

Considering the fact that in the future an analyst has to perform an increasing number of analyses with a decreasing number of personnel, it seems usefull to divide the numerous detectors for gas chromatography in three classes according to their states of development;

1. Standard detectors, whose properties are well known and which are so easy to use that a laboratory assistant is able to install and to operate them without any assistance. To date only the thermal conductivity and the flame ionisation detectors belong in this class.
2. Detectors offered by most of the manufacturers, but whose installation requires the knowledge and experience of a graduat chemist or physicist.
3. Detectors which are developed at various institutes for special purposes but are not in common use.

This report is limited to four radiation ionisation detectors which belong to the second class; The cross-section detector, the electron capture detector, the argon ionisation detector, and the helium detector. I'll point out what improvements should be done if these detectors once should belong to the standard detectors of class one mentioned above.

As stated before, the use of a standard detector should be so easy that a laboratory assistant can operate it by following simple instructions. To achieve this, the following requirements are taken into account by the designers and manufacturers:

1. Little interferences in routine as well as non-routine operation.
2. Little and simple attendance.

3. Good reproducibility of the properties of the various designs made by different manufacturers in order to obtaine optimal parameters and response factors from the literature.

4. The response beeing dependent on as few parameters as possible.

2. The Ionisation Cross-Section Detector

Historicaly the oldest radiation ionisation detector is the cross-section detector first described by OTVOS and POMPEIO [1] and BOER respectively [2] but it falls into oblivion by the sensational successes of the argon ionisation detectors. Since a few years a micro cross-section detector is offered by some firms which is described first by LOVELOCK, SHOEMAKE and ZLATKIS [3 4]. To this publication is to refer the common misconception, that the response of the detector may be enhanced by diminution the volume of the ionisation chamber.

Table 1. Properties of cross-section detectors according to LOVELOCK, SHOEMAKE and ZLATKIS

Ionisation chamber volume		250 µl	80 µl	8 µl
Limits of detection	(Vol.-%)	$3{,}2 \cdot 10^{-4}$	$2 \cdot 10^{-4}$	$2 \cdot 10^{-4}$
	(g/sec)	$1{,}6 \cdot 10^{-9}$	$3{,}2 \cdot 10^{-10}$	$3{,}2 \cdot 10^{-11}$
Flow rate		?	?	16 µl/sec
Current	(Amp)	$2{,}5 \cdot 10^{-8}$	$5 \cdot 10^{-9}$	$1{,}3 \cdot 10^{-9}$
Linear dynamic range		$1{,}5 \cdot 10^{5}$	$3 \cdot 10^{5}$	$3 \cdot 10^{5}$

In Table 1, there are some characteristic data of three cross-section detectors with 250, 80 and 8 µl volume respectively, which were published by LOVELOCK, SHOEMAKE and ZLATKIS [4]. It is obvious that the lowest detectable concentration by volume is nearly independent of the volume of the detector. The increase of the response with decreasing volume found by LOVELOCK and coworkers is only recognizable in the data with the dimension g/s. Obviously the authors have reduced the flow rate simultanously with the volume. In this manner it is possible to lower the limit of detection of a concentration sensitive detector at will.

The diminuation of the volume of the ionisation chamber does not result in an enhancing of its response but it makes it useful for capillary gas chromatography in which it has a large response by reason of the little flow rate. Fig. 1 shows a micro cross-section detector developed by LOVELOCK, SHOEMAKE and ZLATKIS [3]. Inside a narrow tube of 1,6 mm diameter there is a stainless steel foil with a surface layer of tritium on titanium. The distance between this radioactive source and the inner electrode ist about 0,4 mm. The ionisation chamber is formed by this narrow cylindrical slit of a few millimeters in length through which the gas stream passes axially.

A rather interesting property of the cross-section detector already pointed out by LOVELOCK and coworkers [4] has been found little consideration till now. It is the possibility of the exact calculation of the mass of a substance of known ctructure using peak area without any experimental calibration. The formula for this is given by

$$m = \frac{v \cdot p}{6{,}24 \cdot 10^4 \cdot T \cdot i_0} \cdot \frac{M \cdot Q_T}{Q - Q_T} \cdot F \quad [g]$$

where

v = Flow rate of the carrier gas

T = temperature

p = pressure

i_0 = current intensity

Q_T = ionisation cross section of carrier gas

Q = ionisation cross section of the cubstance

M = relative molecular mass of the substance.

Because the other parameters are easy measurable the accuracy of the calculation is essentially dependent on the accuracy of the measurement of the ionisation cross-sections. Already 12 years ago OTVOS and STEVENSON [5] computed the relative atomic ionisation cross-sections of the most chemical elements on a quantum mechanic. basis. They demonstrated that the molecular cross-sections of a number of organic compounds may be computed from the cross-sections of the atoms with rather a good accuracy. This method of evaluation was taken by BOER [2] for the compution of the response-factors for the cross-section detector with good success. However, it is necessary to take into consideration that the quantum mechanical evaluation is exact only when the colliding electrons have a large enough energy. This is obviously not the case when using tritium sources for the ionisation of the gas.

Table 2. Relative atomic ionisation cross sections

Atoms	Calculated by OTVOS and STEVENSON [5]	MANN [6]	Evaluated by LOVELOCK and coworkers [4]
H	1	1	1
C	4,16	4,23	3,69
O	3,29	3,67	4,56

Table 2 shows the relative atomic ionisation cross-sections of carbon and oxygen as related to hydrogen = 1. They were evaluated by OTVOS and STEVENSON [5] and - more exactly - by MANN [6], respectively. The empirical investigation by LOVELOCK and coworkers [4] based on the response of the tritium cross. section detector, Fig. 1.

The considerable differences between the calculated and the empirical values are evident.

We 7 investigated the specific response of the detector, Fig. 2 which is similar in its construction to the argon detectors produced by W. GIEDE K. G., Berlin, and E. ZIMMERMANN K. G., Leipzig. It has a krypton-85 source, typ AI, by Isocommerz GmbH., Berlin. This source forms the inner electrode of a cylindrical ionisation chamber. The emitted radiation has an energy considerably larger than that of a tritium source. For more than 40 organic compounds investigated the specific response of this detector did not differ from the values computed by using the atomic cross-sections, evaluated by OTVOS and STEVENSON or MANN in excess of 7%, resp. 6%. This deviation could be attributed to the influence of the chemical linkage on the cross-section of the atoms not taken into consideration by OTVOS and STEVENSON and MANN. In case of homologous compounds in which the method of the calculation of the cross-sections was tested until now, this influence is not so evident as in case of compounds which contain the atoms in different type of linkage. The differences between the evaluated values and the values found experimentally, could be reduced below 2% by using special empirical evaluated cross-sections for the several types of linkage (Table 3).

Table 3

Linkage	Ionisation cross-section
C - H	1,95
C - C (σ)	2,35
C - C (π)	1,90
O - H	2,81
C - O (σ)	3,20
C - O (π)	1,94
H - H	1,3

These linkage cross-sections are choosen in such a manner that the cross-section of pentane would have the value 32,8 in accordance with the value found by the several other modes of evaluation used until now [7]. Using this linkage cross-sections which are now beeing investigated for other types of linkage by us, it is possible to evaluate the response of a cross-section detector for a given substance with an accuracy better than 2%.

However, a detector that functions as a true cross-section detector is the prerequisite to do this. In such a cross-section detector the range of almost all of the emitted ionising particles should be essentially larger than their way through the ionisation chamber.

For this reason and also to secure an absolute safety against any contamination by radioactive substances we used krypton-85 source to ionise the gas (Fig. 2) [23].

The use of ionisation cross-section detectors in gas chromatography is recommended for three purposes;

1. Exact quantitative analyse with packed columns. The linear dynamic range reaches almost 100 % of the probe in the carrier.
2. Calibration of other detectors especially if the components of interest are not available in pure form. The two detectors should be connected in line. The quantity of each component may be evaluated from the peak areas produced by the cross-section detector using equation 1.
3. Capillary gas chromatography especially for such substances which are not detected by the FID. For this purposes the volume of the ionisation chamber may not be larger than about 10 μl; whereas for the operation with packed columns a volume of 100 μl to 1 ml should be sufficient.

The ultimate performance of a cross-section detector is achieved only, if the temperature and the pressure in the ionisation chamber are well stabilised, or their influences on the ionisation current are compensated. It is best done by connecting two identical detectors in compensation in an isothermal block similar to that of a catharometer.

The properties and the physical laws of the cross-section detector are so well known that there are only little difficulties to produce a detector of this type which may fulfil the requirements mentioned above. For example, argon- or helium-detectors with sources emitting a more energetic radiation than tritium may be used as cross-section detectors when hydrogen used for carrier gas.

3. The Electron-Capture Detector

The selective detection of substances by the electron capture detector is based on the essentially stronger electron affinity of these substances in comparison with the carrier gas. The electron affinity and consequently the response of the electron capture detector are specific to the character of the substances. For this reason it is suggested to use the electron capture detector for the identification of unknown components. Up to day this didn't acquire a greater practical importance because the properties of the electron capture detector are still not too well reproducible.

The most important use of the electron capture detector till now is the residue analysis of halogenated pesticides, because this type of compounds has such a high electron affinity that it is possible to detect them besides other hydrocarbons in picogram quantities. In the commercially produced electron capture detector the negative ions are formed by the electron capture process and they are detected by recombination with positive ions which results in the reduction of the ionization current. It is possible to enhance the detector's sensitivity and to eliminate some disturbances by pulsing the voltage, but this yields two additional parameters with the length of the impulses (ca. 1 μs) and with their time intervals (ca. 1000 μs). Generally it is possible to use the same detector in both kinds of operation. Fig. 3 shows the three different kinds of electron capture detectors, produced commercially at present. Detector shown in Fig. 3a is a parallel plate detector first described by LOVELOCK and LIPSKY [8]. In the cylindrical ionisation chamber the two electrodes are plane opposite one to another. The radioactive source generally tritium on zirkon or titan fixed on the catode at the exit of the detector. The main disadvantage of this kind of construction is that the carrier gas passes laterally around the source.

The relative large distance between the electrodes is striking if we consider the fact that the ionisation of the gas happens only within the narrow sphere ($<$ 1 mm) immediately at the surface of the tritium source. The electrons and negative ions will be drawn by the electric field through the whole volume to the anode, but a recombination is possible only in the sphere at the surface of the source because only there are the positive ions necessary for this process. Nevertheless, the region between the sphere of ionisation and anode has a positive influence on the response of the electron capture detector. Because the absence of positive ions in this region a volume charge is formed, decreasing the strength of field in the ionisation sphere. Negative ions will increase this charge due to their lower mobilities. Because of this the rate of recombination increases in the sphere of ionisation. Some properties of the electron capture detector found by several authors which has not yet been explained are probably due to this effect.

The use of tritium sources has proved to be very disadvantageous when an electron capture detector is used for the detection of high boiling pesticides because at temperatures above perhaps 150 $^{\circ}$C tritium escapes from the source to a degree which is not allowed by the authorities for radiation protection [9]. In addition to this already the smallest contamination of the surface of the source will absorb a considerable amount of the emitted radiation and decrease the current. Nickel-63 is a nuclide which has almost as small energy of decay as tritium but it is solid and chemically stable. For this reason it has an increasing importance especially for electron capture detectors. However the following two facts hamper the general use of this nuclide in place of tritium.

1. The starting product for producing nickel-63 is the isotop Ni-62 which is in the natural nickel only in a amount of 4 % and it has to be enriched first up to 100 %. Because of this the costs of production will be considerably increased.
2. The layer of a Ni-63 source from which an essential amount of the radiation is escaping not thicker than 2 to 3 mg/cm^2. The largest commercial available specific activities may be 0,1 to 1 Curie per gram.

For this reason only sources with an effective activity of not more than 3 mC/cm^2 are producable. To get a strong ionisation current, a surface of the source as large as possible is necessary without making the volume of the ionisation chamber larger than about 1 cm^3. The strength of the generated current of the strongest nickel-63 sources in the best position in an ionisation chamber of acceptable dimensions is considerable smaller than that of a tritium source. In the last years electron capture detectors are produced with a source which is the outer electrode in a cylindrical ionisation chamber [10,

11], Fig. 3b. In this construction a relatively large volum exists without any ionisation between the anode and the ionisation sphere near the source. In this Volume the electric charge will be enhanced by negativ ions. This effect will be further increased by shiffing the catode with the source in direction of the gas flow [12] (fig. 3c).

By such measures it is possible to influence the response in a positive manner, but at the same time the number of the parameters will be increased and the difficulties in explaining the different properties of the detectors are increasing. A considerable part of this difficulties may be eliminated if the negative ions are detected directly and not by their recombination with positive ions [14]. Fig. 4 shows a detector which operates in such a manner and contains a krypton-85 source, typ AI available by Isocommerz G.m G.m.b.H., Berlin.

The source is the anode of a cylindrical ionisation chamber. The uncaptured electrons will be drawn away by an ac or dc voltage whereas the ions because of their little mobilities will be rinsed out of the ionization chamber and transferred into a second chamber by the gas flow where the negative ions will be drawn to a second anode which is the probe of the electrometer that records the negative ions current.

In pure nitrogen or argon the strength of this current is less than 10^{-12} amps. The lowest detectable amount for this detector is about 1 picogram lindane or dieldrine.

The main problem in using electron capture detectors is the dependence of the response on temperature and pressure and the short linear dynamic range.

The temperature dependence is dependent on the nature of the substances analysed and may be both strongy positive or negative but also relatively small in other cases. Fortunately the latter is the case with the pesticides. For this reason the temperature of electron capture detectors must be carefully regulated for common use.

WENTWORTH and CHEN [15] tried to interpret the temperature dependence of the electron capture process theoretically. A part of their conclusions found contradictions [16].

The response of the electron capture detector is linear only if the number of electrons captured per second is small as compared to the total number of electrons generated per second (standing current), respectively, if the deviation of the ionisation current by the component is small as compared to the remaining current of free electrons. Peaks that exceede about 10% of the standing current have peak areas which are not proportional the amount of the component that caused the peaks. If a greater linear dynamic range is required, the signals of the detector may be corrected electronically before the signal is registered by the recorder or integrator [11] Electron capture detectors are also used for the detection of other halogenated compounds like thyroid hormones. Anaestetics may be detected in blood or in tissues with such a high sensitivity that only a few μg of material are required from which the anaestetics will be extracted by heptane [13]. Substances such as steroids which have relatively low electron affinity may also be detected with a high sensitivity after halogenation.

In the qualitative analysis the electron capture detector simplifies the identification especially of those substances with a extremly high or low electron affinity. An absolute identification by means of the specific response of the detector is not yet possible because the reproducibility of the response is still unreliable. If the electron capture detector is to be a reliable standard detector the designers and manufacturers have to aspire after the following:

1. A more simple function by clear physical conditions especially in view of the gradients of the electric field, the gas flow and the drift of electrons and ions.
2. The use of other sources than tritium for ionization.
3. Standardization of the carrier gas. A gas which is easily to get in good and reproducible quality, is prefered e. g. nitrogen.
4. Explanation of the electron capture mechanism in diluted mixtures with inert gases.
5. Extention of the linear dynamic range.
6. Absolute gastightness of the whole instrument and elimination of materials and liquid phases which may contaminate the carrier gas.
7. Rigorous thermostating of the detector.

4. The Argon Ionisation and the Helium Detector

Among the ionisation detectors the argon ionisation detector is the most problematic detector. Nevertheless, until a few years ago this detector had won a great importance, because it enables to detect organic components up to 10^{-13} g/sec. But with the improvement of the FID (hydrogen flame ionization detector) it lost its importance gradually ever since it became apparent that the physical basis of its operation is more complex than it was assumed in the beginning. At the present argon detectors should be used for such purposes only for which a FID may not be applied (e. g. not burnable substances or danger of explosion).

The main shortcomings of the argon detector in comparison with the FID can be characterized as follows:

1. While the FID does not detect inorganic gases and also is not disturbed in its operation by them, just small concentrations of inorganic gases alter the ionisation current as well as the response of the argon detector.
2. The liquid phase bleeding disturbs only the zero line of the FID whereas it may contaminate an argon detector to such a degree that its operation is disturbed.

3. The argon detector is overloaded by small amounts of sample (10^{-7} g/sec) and its response to larger amounts is nò longer linear.

These disadvantages of the argon detector however, become less significant when trace amounts of such substances are to be detected which can not be detected by the FID. These are in the first place the inorganic gases and vapours. Three years ago we reported [18] three different methods for the detection of inorganic gases with an argon detector. The lowest detectable amount was about 10^{-11} g/sec. Argon, argon plus 0,1 % propan, or neon were used as carrier gases. With all the three types of carrier gas the inorganic components diminished the ionisation current by quenching the metastable excited atoms of the noble gases. These methods which also had been developed in various forms by several authors should be considered as a temporary solution only, because during the last years the helium detector reached a remarkable improvement. It can detect organic as well as inorganic substances with the same sensiticity of almost 10^{-13} g/sec [19].

The operation of the helium detector is the same as that of the argon detector: The radioactive source ionises the gas flowing through the small ionisation chamber and the free electrons generated by the radiation are accelerated by the electric field. The accelerated electrons excite the helium atoms by inelastic collisions. Molecules with a ionisation energy smaller than the excitation energy of the helium atoms will be ionised by the transfer of this excitation energy. This is the case for all organic and inorganic compounds with the exception of neon.

In his first publication, LOVELOCK already pointed out the possibility of the detection of inorganic compounds by argon detector using helium carrier [20]. However, it was not until the last few years that the availability of ultra pure helium gas made this possible, because the previously described detection mechanism is operative only in very pure helium gas.

The details of the operation mechanism of the helium detector are by no means fully elucidated yet.

To date, HARTMANN and DIMICK [19] had the best success in developing the helium detector, shown in Fig. 5. The ionisation chamber is formed by the narrow slit between a round plate tritium source of 250 mC and a similar plate catode. The advantage of this construction is the homogeneous field in the ionisation chamber. In most of the argon detectors, especially in those described by LOVELOCK, the field in the ionisation chamber is remarkably inhomogeneous. This results in different properties and a very complex function of the detectors in the different zones of the ionisation chambers.

In the helium detector developed by HARTMANN and DIMICK the gas flow passes the tritium source laterally and then directed radially through the ionisation chamber to the exit in the center of the cathode. This has a disadvantage, because the linear velocity of the gas flow increases from the border to the center of the ionisation chamber. It is known that the response of an argon or helium detector increases with the linear velocity of the gas flow [21]. A cylindrical ionisation chamber in which the gas flows axially is more favourable.

At the present the helium detector is the most sensitive detector for inorganic substances including water. The smallest contaminations in the carrier gas, either present in it initially, or, more likely, comming in through small leaks or from the desorption of the column or other materials may disturb the function of the detector in a much larger degree than commonly found with other detectors. Therefore, the injection systems, valves and the different materials comming in contact with the carrier gas which were normally suitable may not be used for trace analyses with a helium detector.

For this reason it may not be advisable to use helium or argon detectors as an interchangeable part of common gas chromatographs, but rather develope special gas chromatographs with these detectors which are fitted with equipment for the purification of the carrier gas and other performances required for an optimal use of the detector for trace analysis.

In the construction of the helium, and argon detectors a homogenous electric field as well as a homogenous gas flow through the ionisation chamber should be secured more than in the past, because only this way will it be possible to make the properties of the detectors e. g. their specific responses, explicable on an exact physical basis.

5. Conclusion

The three radiation detectors: cross-section detector, electron capture detector, and helium detector are essential for many special applications of gas chromatography, but further improvements are required to develope them into reliable standard detectors useable for everyone.

1. The cross-section detector could be improved in rather short time if sources with a more energetic radiation than that of tritium are used. The accuracy of computing the substance specific response factors may be increased by using linkage cross-sections instead of the atomic cross section. The cross-section detector is especially suited for quantitative analyses when no pure substances are availeble for calibration.

2. The electron capture detectors should be simplified in their functions and operations in order to make their responses dependent on as few parameters as possible. Tritium sources should be avoided. The detectors should thermally controlled. The electron capture detector is especially suited for the selective and sensitive detection of compounds containing iodine, bromine, or more than two chlorine atoms as well as metall organic compounds or other compounds with a large electron affinity. After improving the reproducibility of the substance specific response of the detector it will also be applicable for the identification of unknown components.

3. The helium detector, and also the argon detector should be used only with gas chromatographs which guarantee the purity of the carrier gas. The electric field and the gas flow in the ionisation chamber should be homogenous as much as possible. The helium detector is the most sensitive detector for both inorganic and organic substances. The trace analysis of gases is the main field of its use.

The problem of the radioactive sources of all radiation ionisation detectors should be solved in a manner that anybody can clean the détector in a normal laboratory without any danger. By this reason tritium sources should be eliminated. Ni-63 sources - apart from their high prices - are not usefull without objections because in cleaning them the dissipating of radioactive material may be possible.

With the use of Kr-85 sources no radioactive contamination is possible [23].

References

1. Pompeo, D. J., and Otvos, J. W.: U. S. Patent 2641710 (1953).
2. Boer, H. in D. H. Desty: Vapour Phase Chromatography S. 142, Butterworth London 1958.
3. Lovelock, J. E., Shoemake, G. R., and Zlatkis: Anal. Chem. 35, 460 (1963).
4. Lovelock, J. E., Shoemake, G. R., and Zlatkis, A.: Anal. Chem. 36, 1410 (1964).
5. Otvos, J. W., and Stevenson, D. P.: J. Amer. Chem. Soc. 78, 546 (1956).
6. Mann, J. B.: J. Chem. Phys. 46, 1646 (1967).
7. Große, J., and Bothe, H.-K.: Z. f. Naturforsch., 23a 1583 (1968).
8. Lovelock, J. E., and Lipsky, J.: J. Amer. Chem. Sbc. 82, 431 (1960).
9. Kahn, L., and Goldberg, M. C.: J. Gas Chromatography 3, 287 (1965).
10. Maggs, R. J., and Swan, D. F. K.: Column (W. G. Pye) 1, 3, 2 (1966).
11. Simmonds, P. G., Fenimore, D. C., Pettit, B. C., Lovelock, J. E., and Zlatkis, A.: Anal. Chem. 39, 1428 (1967).
12. Hartmann, C. H., and Oaks, D. M.: Aerograph. Res. Notes Winter 1965.
13. Morgenstern, S., Bothe, H.-K., and Popp, P.: Der Anaestesist 15, 194 (1966).
14. Bothe, H.-K., Hilbig, H., and Popp, P.; in Struppe, H. G.: Gas-Gromatographie 1965, Akademie-Verlag, Berlin 1966, p. 525.
15. Wentworth, W. E., and Chen, E.: J. Gas Chromatogr. 5, 170 (1967).
16. Christophorou, L. G., and coworker: J. Chem. Phys. 45, 536 (1966).
17. Jaakonmaki, P. I., and Stouffer, J. E.: Advances in Gas Chromatogr. 1967 Ed. A. Zlatkis, p. 149. Preston Techn. Abstracts, Evanston 1967.
18. Leonhardt, J., and Bothe, H.-K.: Gas-Chromatographie 1965, p. 307, Akademie-Verlag, Berlin 1966.
19. Hartmann, C. H., and Dimick, K. P.: J. Gas Chromatogr. 163 (1966).
20. Lovelock, J. E.: J. Chromatogr. 1, 35 (1958).
21. Leonhardt, J.: Dissertation Math.-Nat. Fakultät der Karl-Marx-Universität, Leipzig, 1966.
22. Bothe, H.-K.: Isotopentechnik 1, 163 (1960/61).
23. Prösch, U., Vormum, G., and Zöpfel, H. J.: Kernenergie 7, 26 (1964)!

Conditions of Analytical and Preparative Separations

Some new Column Types in High-Speed Gas Chromatography

I. HALÁSZ and E. HEINE

Institut für Physikalische Chemie der Universität Frankfurt am Main

1. Introduction

In gas chromatography it is wanted to resolve the compounds completely as fast as possible and with the apparative expenditure as low as possible. The optimal parameters are determined by the object of the separation. Only analytical separations will be discussed in this paper. The problems of prepscale gas chromatography and that of trace analysis will be bypassed.

Constant temperature and constant pressure drop on the column are assumed. The problems of programmed temperature and programmed flow rate are only shortly discussed.

Up to now seven different column types are described in the literature [1, 2]. However, most of the routine gas chromatographic analysis are made on conventional packed columns as described by JAMES and MARTIN in 1952 [3]. This column type is frequently used for separations, where the application of other column types should be much more advantageous. The reason for this strange situation is, that it is simple to build conventional packed columns. Furthermore the concentration and mass flow rate of the compounds at the end of these columns are relatively high, consequently insensitive, simple and robust detector can be applied.

In general it is believed, that the manufacture of the other column types is sophisticated and that their life time is short. To destroy this assumption the simple preparation of some column types will be described here.

2. Speed of analysis

At constant flow rate and temperature the time required for the separation of two compounds, the necessary-pressure drop and -column length are simply calculable [1, 4]. All the parameters used in these calculations are to infer from the chromatogram itself as shown in the „List of Symbols" at the end of this paper. The assuptions applied by the derivation of these equations are good approximations for the routine work. However, these equation can only be used if the ascending branch of the H vs. $\bar{u}$ curve is linear. Without discussing the details of these equations [4] it has to be mentioned, that - because of pragmatical reasons - the number of effective plates N and not that of the theoretical plates n are to be used in these equation. The height equivalent to an effective plate H and that to a theoretical plate h will be used alternatively, where $H = h\ [(1 + k')/k']$.

As a consequence of these equations, the speed of analysis may be characterised by the number of effective plates produced by unit time (N/t). Furthermore it follows, that the time of analysis is proportional with $H/\bar{u}$ and the necessary pressure drop is a function of $H\bar{u}$.

3. Column types

From the point of view of aerodynamics the chromatographic columns are classified [1, 4] in Table 1. In the second but the last line of this table the known column types are to be seen. In the last line the kind of the stationary phases available for the given column type is shown, i. e. an active solid (GSC), an adsorption layer (GAC) or a liquid stationary phase (GLC).

It is a well known experimental fact that the minimum of the height equivalent to a theoretical plate h in a well prepared analytical column is more ore less independent of the column type and h_{min} usually lies between 0.5 - 1.0 mm. On the other hand it is known from the theory and practice, that best results are achieved [5] if $k' = 1 - 5$ (less commonly $k' = 0.5 - 10$). Although it is unexpected, the minimum heights equivalents to an effective plate (H_{min}) are very similar in different column types.

As it has been shown [1, 4], only the phase ratio a, the permeability K and the slope of the ascending branch in the H vs. $\bar{u}$ curve are essential in the characteristic of column type. The partition coefficient k is defined by the sample, the stationary phase and its temperature. Consequently, because of the wanted $k' = 1 - 5$, the phase ratio $a = k/k'$ is of importance. As pointed out above the wanted phase ratio ascertain the optimal column type. It should be mentioned, that overlap of the range of phase ratios of the neighboring column types is possible.

The maximum pressure drop is limited by the pressure regulating and injection system of the gas chromatographic equipment. The greater the permeability K the longer columns may be used with a given pressure drop and the velocity of the carrier gas will be higher. Long columns are necessary to resolve substances with small relative retentions. Note: the analytical problem determines the optimal column type. Choose always the column with the possible highest permeability in which the phase ratio is attainable yielding optimum k' for the given partition coefficients.

The flatter the ascending branch of the H (or h) vs. $\bar{u}$ curve the shorter the possible time of analysis which is proportional with $H/\bar{u}$. It seems to be an experimental experience,

Table 1

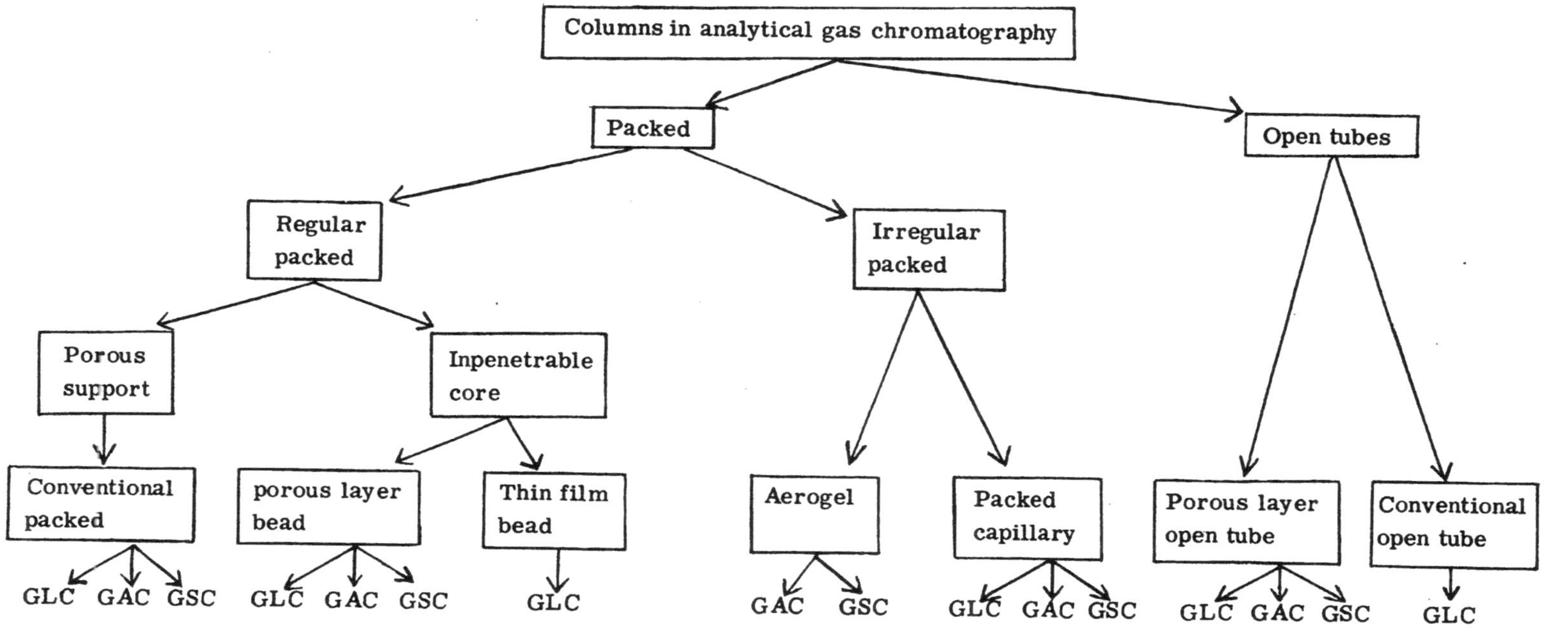

Columns in analytical gas chromatography
Packed
Open tubes
Regular packed
Irregular packed
Porous support
Inpenetrable core
Conventional packed
porous layer bead
Thin film bead
Aerogel
Packed capillary
Porous layer open tube
Conventional open tube
GLC GAC GSC
GLC GAC GSC
GLC
GAC GSC
GLC GAC GSC
GLC GAC GSC
GLC

Table 2. Characteristic column properties

1	2	3	4	5	6	7	8
Column type	Phase ratio a	Temperatur of analysis T	Permeability K	Loadability s	Detector sensitivity	Conc. of sample c	N/t sec^{-1}
Conventional packed	↓	↑	↓	↑	↓	↑	5
Porous layer bead							100
Thin film bead							30
Aerogel							100
Packed capillary							100
Porous layer open tube							100
Conventional open tube							70

that in column types with high permeability the ascending branch of the H vs. $\bar{u}$ curve ist flat.
The alteration of the characteristic properties of the different column types [2] are shown in Table 2. The parameters are growing in the direction if the arrows. The following characteristic properties were chosen: phase ratio $\underline{a}$, temperature of analysis T, Permeability K, maximum ample size entering the column (s) or loadability, the required sensitivity of the detector for the given column type, the concentration of the sample in the carrier gas at the end of the column (important for prepscale work and if combining with other analytical methods) and the average speed of analysis characterised by the available number of effective plates per second. It seems to be, that the difficulties in the preparation of the column are growing from the conventional packed columns into the direction of the open tubes.

As to be seen from Table 2 the choice of the optimal column type is always a compromise. This compromise is determined by the purposes to be achieved, e. g. the given separation problem, experimental limits (temperature, pressure drop, etc.), speed of analysis, high loadability, coupling with other analytical methods, etc.

In the followings the preparation of some less common column types and their analytical advantages should be discussed. Mainly experiences of our group will be reviewed.

4. Porous layer beads

Porous layer beads (PLB) are prepared in coating a core impenetrable to gas with a porous solid layer. Mostly nonporous glass, metal or plastic beads are used as impenetrable cores. The solid porous layer used to be prepared of highly dispersed (in the particle range of 1 micron or less) substances as metal oxides (for example Fe_2O_3, Al_2O_3, ZrO_2, etc.), organic dyes, or graphited carbon black with specific surface areas of 30-150 m^2/g. These supports are poreless, consequently the speed of the mass transfer in the stationary phase is high. Because of the high interfacial tension of these highly dispersed solids the powdered particles are apparently held to the surface of the beads by van der Waals forces. The adhesion forces between core and porous layer become greater if the active powder is mixed with a liquid stationary phase, i. e. gasadsorption-layer GAG. If powdered inactive supports are used (Sterchamol, Chromosorb, etc.) sometimes it is necessary to add adhesives (for example Epicote resin). The preparation of PLB is extremely simple and is summed up in an other paper in this book [6].
Beads, powder and if wanted the liquid stationary phase (and) or adhesive are suspended in a low boiling solvent in a mortar by stirring [7, 8]. The solvent is evaporated under an infrared lamp. Afterwards the porous layer adheres on the beads. To achieve a more regular packing in the column the porous layer beads may be rolled on a granulation plate. Before and after rolling the coated beads are sieved to separate the PLB from the small remaining excess of powder. Thereafter the column is packed as usual with PLB. The loadability of the PLB is usually smaller than that of the conventional packed columns, but it is much more higher than that of the beads coated only with liquid stationary phase. It is a surprise, that in the region of GAC porous layer beads columns could have

been prepared with a loadability adequate to that of the conventional packed columns [9].

With the help of the technology described above active solids with insufficient mechanical properties (for example graphited carbon black) can be used as stationary phases. These solids can not be used in routine work utilizing conventional packed columns.

Short times of analysis and low temperatures of the separation are the main advantages of the PLB columns. The excellent mechanical properties are especially wanted in programmed temperature and (or) programmed flow rate work [10]. A further advantage of this packing material is the possibility to pack columns extremely reproducable. This is a consequence of the eveness of the particles. For the same reason it can be advantageously used for prepscale columns.

Some advantages of the PLB, especially in GAC, are demonstrated in the followings. In Fig. 1 the h vs. $\bar{u}$ curve for n-heptane (k' = 2.9) in shown on glass beads coated with 6.24 % b. w. ZrO_2 and 0.16 % triethylene glycol [11]. The minimum of this curve appears at time averaged carrier gas velocity $\bar{u}$ = 20 cm/sec. This value is roughly with a factor 5 higher than usual with conventional packed columns. The minimum of the h and H values is extremely small, h_{min} = 0.043 cm an H_{min} = 0.078 cm. The minimum of the curve is very broad indeed. The optimal $H/\bar{u}$ value [4] is about 4.2 msec, i. e. the speed of analysis is extremely high. It should be mentioned, that the h vs. $\bar{u}$ curve shown in Fig. 1 is not exceptional, but rather typical for columns packed with PLB [7, 10].

A further advantage of the column described in Fig. 1 is its versatility. Hydrocarbons, organic compound containing oxygen and halogenated aliphatic hydrocarbons are resolved well on the same column [11]. It should be pointed out, however, that the peak broadening is, among other, a function of the quality of the sample too.

Remarkable separation is shown in Fig. 2, where aromatic hydrocarbons are resolved on glass beads coated with 5.63 % Fe_2O_3 powder and 0,57 % triethylene glycol [11]. The relative retention of m- and p-xylene is poor, these compounds remain unresolved. The speed of analysis is unusual high. The following N/t values per second are achieved: 85 for benzene (k'= 1.0); 219 for toluene (k' = 2,0); 294 for m- and p-xylene (k' = 4.0) and 332 for o-xylene (k'= 4.7).

Preparation and some more analytical application of beads coated with metal oxides and a liquid stationary phase (GAC) with and without flow program are described [6 - 11]. Special methods for preparing porous layer beads and their analytical application are summed up [1].

5. Packed capillary columns

Packed capillary (PC) columns are irregular packed columns with 0,2 to 0,5 particle/ columns diameter ratios and with a packing density substantially below that of conventional packed columns (preferrably with an inner diameter of less than 0.5 mm) [12 - 16]. PC columns have only 25 to 35 %, sometimes 50 % [17] of the packing density of the conventional packed columns. Only few chamels exist for the carrier gas because of the irregular packing and the continous change of free gas cross section (up to 1 : 100) prevents a parallel flow patern. Consequently a mechanical mixing takes place in the gas phase. Further consequences of the low packing density are:

1) the permeability of PC columns is roughly by a factor 10 greater than that of conventional packed columns with the same particle size,
2) the phase ratio of PC columns is higher than that of conventional packed columns,
3) the loadability of the PC columns is relatively small, but high compared with open tubes.

In some recent publications the name packed capillary column were applied to regular packed columns (i. e. particle/column diameter ratio less than about 0.2) with inner diameter less than 1 mm [18-20]. Excellent result were achieved with these columns, which were an improvement over those with classical packed column. Because of the regular packing of these columns, however, it seems to be better to speak of micro packed columns, in contrary to the irregular packed PC columns.

Packed capillary columns are produced simply be drawing out thick walled glass tubes with an i. d. of 1-2 mm [21], which are initially loosley filled with a granular solid of narrow mesh range [12, 14, 16, 17]. Glass drawing mashines are now commercially available. Supports with poor mechanical properties - which never could be used in conventional packed columns - as for example pelleted graphited carbon black [22] - can be employed.

During the drawing process temperatures are reached, which are 100-300 ^{o}C higher than the softening point of the glass used. If it is wanted to coat the support with an organic liquid phase, this is done in the column in situ after the drawing [16]. The in situ coating is a well known technique for conventional packed columns [23]. All the experimental details for the preparation of PC columns are summed up [16]. Some routine and skill is wanted if preparing PC columns. In our view, however, this procedure is not much more sophisticated, than the preparation of conventional packed columns.

The shape of the h vs. u curves of PC columns is very convenient. The minimum appears at high linear velocities ($\bar{u}$ = 15-35 cm/sec) with about h_{min} = 0.02-0.05 cm. The minimum is broad, the mass transfer in the gas phase is in some cases smaller than 1 msec. The PC columns can be used for GSC, GAC or GLC, Especially in GSC and GAC - where the mass transfer in the stationary phase is quicker, than that in the mobile phase and where the bleeding and noise of the column is negligible - diffecult separation problems can be solved at high speed with CP columns.

Because of the relatively high phase ratio of the PC columns best results are achieved if the partition coefficients of the samples are high, as usual in GSC and GAC.

Because of the reasons mentioned above the separation of low boiling substances on active solids is exceptionally recomandable. In Fig. 3 the express separation of methane, ethane, ethylene and propane on alumina in PC column in less than 3 seconds [15] is shown. Although the numbers of effective plates generated per second are only 92-250, the separation is quick because the relative retention are unusual high, i. e. about 1.9. This separation demonstrates the often forgotten truism, that the time of a separation is a function as well of the N/t values as of the relative retentions. The high permeability of the column and the low viscosity of the dry hydrogen carrier gas allow exceptional high linear velocities ($\bar{u}$ = 80 cm/sec) with usual pressure drop. As a consequence of the base line separation and of the negliable noise level the absolute errors of the reproducibility in the quantitative evaluation of Fig. 3 are smaller than 0.1 % [15]. In process monitoring the discontionous gas chromatographic method becomes quasi continous if the time of analysis is short enough.

Difficult separation of C_1 - C_5 hydrocarbons on alumina [24] is shown in the Figure 4. The nitrogen carrier gas was moistened over $Na_2SO_4 \cdot 1OH_2O$ at room temperature. The moistening of the carrier gas with cristal hydrates is a simple, insensitive and reproducible procedure [16]. The carrier gas velocity is „only" $\bar{u}$ = 9.6 cm/sec. The base line separation of 16 peaks is done in less than 5 minutes. In this analysis the two pairs of substances; n- and trans-buten-2, and propylene - butadiene-1,2 remain unresolved.

Some 20 compounds are resolved in Fig. 5 in 7.5 minutes [24]. The stationary phase is alumina coated with 16 % b. w. NaCl. The wanted amount of NaCl is solved in water and the active alumina is mixed with the concentrated solution. Thereafter the water is evaporated with the help of an infrared lamp. The stationary phase is activated for 16 hours at 450 °C. Dry nitrogen carrier gas is used (GSC). By comparing the Figures 4 and 5 it is to be seen, that not only the relative retentions, but in some cases the order of retention is changed. The stationary phase used in Fig. 5 is extremely stable and is to recommend for routine work.

Another analytical application of PC columns filled with graphited carbon black may be of some interest. Here [22] the order of the retentions of C_1 to C_8 hydrocarbons is that of the number of carbon atoms in the molecule. The C_8 hydrocarbons with a quaternary carbon atom fall in the C_7 range. The relative retentions are more or less independent of wether the graphited carbon black is uncoated or coated with small quantities (less than 1 %) of polar or apolar liquid stationary phase.

Of course not only hydrocarbons, but also polar substances can be separated in PC columns. Some more analytical applications in GLC with PC columns containing chromosorb with squalane [14] or firebrick with tri-2,4-xylenyl-phosphate [17] are described. Special methods for the preparation and analytical application of packed capillary and micro packed columns are summed up [16].

6. Aerogel columns

This column type was first described in 1966 [25]. Although the particles appear to be closely packed, the packing is in fact highly dispersed and only 2 to 4 % of the total column volume is occupied by the stationary phase. Up to now only highly dispersed (340 m^2/g) spherical, poreless silica (Aerosil 2491/380, Degussa, Frankfurt/Main) was used as stationary phase in Aerogel columns. The preparation of Aerogel columns is like that of packed capillary columns, but more skill is wanted. Aerosil is moistened with CCl_4 until a gelatinous product is obtained. After evaporation of the solvent the residue is ground in a mortar. The mortar is „sieved" on a special way [25]. This product is filled in a thick walled glass tube (with 2.2 mm i. d. and 6 mm o. d.). The filled tube is drawn out to a capillary of 0.4 mm i. d. More details are described in the literature [25]. No one has yet succeded in constructing Aerogel columns with an inner diameter larger than 1 mm, or with on other support than highly dispersed silica. However, this support can be coated with liquid stationary phases or coated with highly dispersed active solids (for example ferric oxide). The analytical application of Aerogel columns is limited.

The permeability of the Aerogel columns is comparable with that of conventional packed columns although the particle size of the support is by a factor 100 greater in the conventional packed columns. Small particle size is in Aerogel columns compensated by high porosity. The small particle size results in small peak broadening. In Fig. 6 an h vs. $\bar{u}/j = u_0$ is shown [26]. Further dates for the same column are given in Table 3.

Table 3. [26] Characteristic Properties for an Aerogel Column

Parameters same as in Fig. 6

Necessary parameters for separation (R = 1.5) of propane and propylene at 25 °C (S = 112, k′= 1.46)

Δp at	u_0 cm/sec	$\bar{u}$ cm/sec	H cm x 100	$H/\bar{u}$ msec	t_{ne} sec	L_{ne} cm	Δp_{ne} at
0.5	8.4	6.6	8.91	13.4	2.19	9.98	0.20
1.0	22.3	14.3	4.77	3.34	0.546	5.34	0.23
1.5	37.8	20.3	4.03	1.99	0.325	4.51	0.27
2.0	59.6	27.5	3.97	1.44	0.235	4.46	0.36
2.5	86.2	34.7	3.52	1.01	0.165	3.94	0.40
3.0	111.	39.6	3.69	0,93	0.152	4.13	0.49
3.5	142.	45.4	4.54	1.00	0.164	5.08	0.70
4.0	186.	53.9	6.13	1.14	0.186	6.87	1.14
4.5	218.	57.8	6.27	1.09	0.178	7.02	1.26

The mass transfer on the surface of the solid support is much more quicker, than that in the mobile phase. Consequently the h vs. u_0 curve yields a linear ascending branch, as to be seen in Fig. 6. The shape of this curve is similar to those measured in packed capillary columns, but the corresponding h values are in the Aerogel columns smaller (h_{min} = 0.0124 cm in Fig. 6 at $\bar{u}$ = 34,7 cm/sec). Therefore short columns with usual pressure drops can be applicated and the time of analysis is short. In Table 3 the necessary parameters [4] are calculated for the separation of propane/propylene (r_{12} = 2.3 at 25 °C) on the same column for different pressure drops. As to be seen only a few centimeter of column length is wanted. The calculated parameters were realised. Baseline separation of methane, propane and propylene was achieved in 1 second. This and other analytical applications are described [25].

7. Porous Layer open tubes

Porous layer open tubes (PLOT) are prepared in depositing on the wall a coated or uncoated porous solid layer of support. The free cross section of the tube has to be symmetrical, possibly circular. From the point of view of aerodynamics and that of the mass transfer a smooth surface of the layer is wanted. The porous layer is made of active poreless and highly dispersed support. The layer adheres on the glass, plastics or metal wall of the tube, as discussed earlier in the chapter on porous layer beads. The thickness of the layer is variable within broad limits [27-30].

The difficulties in preparation of PLOT are usually overestimated. Such layer can be produced by special chemical reactions in situ in the tube. For example the inside of glass tubes may be etched to porous silica layer [31-36] or a layer of alumina may be produced on the wall of an aluminium tube [37-38].

A general method for producing a porous layer of active or inactive, coated or uncoated support on the inner wall of an open tube was described in 1963 [7, 39, 40].

It is more or less simple to produce PLOT made of highly dispersed metal oxides (e. g. Fe_2O_3. ZrO_2, Al_2O_3, etc.), with a particle size of about 1 micrometer. These materials adhere excellently on the inner wall of the tube. According to experience the adhesive forces grow with adding a liquid stationary phase to the active powder. To produce PLOT all the compounds of the stationary phase are suspended in a low boiling solvent with the aid of a high-speed stirrer. The suspension has to be stable at pressures as high as 150 atm. Therefore dispersing agents with high density are prefered. Halogenated hydrocarbons are excellent solvents here, because of their polarities, high densities and low boiling points. It has to be pointed out, that the most difficult problem in this technique is the choice of the dispersing agent. This can determine the structure of the layer and consequently the peak broadening on this stationary phase [6].

A capillary tube is filled with the suspension, care being taken that no air bubbles remain. Some 10 to 150 atm of inert gas pressure is required for this procedure. After closing one end of the capillary it is introduced slowly and with constant speed through a heated slit into an air thermostat, beginning with the open end. The temperature of the slit is 50-100 ^{o}C, that of the oven some centigrade higher than the atmospheric boiling point of the dispersing agent. After the careful evaporation of the solvent on the inner wall of the open tube a thin layer of support impregnated with the liquid phase remains. Subsequently the PLOT is flushed for some hours with the carrier gas at temperatures 50 ^{o}C higher than that of the analysis [40].

It is more difficult to produce PLOT with highly dispersed inert support [7, 27-30, 40]. The adhesion forces between powder and the wall of the tube are here much smaller. Sometimes the use of an „Adhesive" (glue) is necessary. Such adhesives are for example Epicote resins or „Baymal" [4]. Care has to be taken using such adhesives. Although their concentration is only 0.1 - 5 % of the support, the relative retentions may be changed compared with the support alone. The advantage of PLOT with inert support without adhesive is, that liquid stationary phases of any kind can be used. The reproducable preparations of PLOT in GSC are descriped, e. g. graphited carbon black [39], fibrilar Bochmite [41], molecular sieve [42] and colloidal silica [43-44]. In GAC, PLOT columns have employed ferric oxide coated with triethylene glycol [40] or with Carbowax [6] and diatomaceous earth coated with squalane [27, 29].

Because of the porous structure of the support the surface area of the PLOT is by a factor 6-10 or more greater than that of the conventional packed columns [34, 40]. Consequently greater amounts of partitioning liquids may be used than in conventional open tubes with the same „film thickness". With growing amount of stationary phase the speed of the separation decreases [29], while the loadability increases. The injectable sample size in PLOT columns is in special cases nearly as high as in conventional packed columns [28, 29]. Because of the lower phase ratio PLOT with inner diameter of 0.5 mm and more are usual. The permeability of an open tube with i. d. = 0.5 mm is roughly by a factor 1000 better than that of conventional packed columns with a particle size of 0.1 mm. Extremely long PLOT can be used with small pressure drop and high linear velocity of the carrier gas. The PLOT are mechanically stable and can be used for a long time in routine work. Because of the adhesive forces the vapour pressure of the partitioning liquid is in PLOT smaller (up to a factor 10^4 [10, 45]), than in conventional open tubes.

High permeability, good mechanical properties, negliable bleeding and the convenient shape of the H vs. $\bar{u}$ curves in PLOT are the big advantages of this column type in programmed temperature [28] and in programmed flow rate [6, 29] work.

The measurement of physical constants with the help of gas chromatography [46] is convenient and speed using PLOT type columns [40].

In Fig. 7 H vs. $\bar{u}$ curves in PLOT of graphited carbon black are shown as recalculated from Fig. 41 in [7]. The minimum appears at unusual high time averaged linear velocities ($\bar{u}$ = 35 cm/sec). The minimum is broad, the slope of the ascending branch is very small. It should be pointed out again, that in this curve the hight equivalents to an effective plate are given. The calculated C terms for the H vs. $\bar{u}$ curves are 15 msec for k'= 1.42 and 8.3 msec for k' = 4.07, i. e. this curve represents rather an average than a good result. On the same stationary phase in PLOT 465 effective plates per second were achieved [2, 39]. Some more analytical applications are shown in the literature cited in this chapter.

8. Conclusion

In this paper the characteristic properties of some speedy column types were discussed, e. g. permeability, phase ratio, shape of the H vs. $\bar{u}$ curve (speed of analysis). The properties of the stationary phases (relative retentions, linearity of the distribution coefficients, speed of the mass transfer in the stationary phase, temperature of analysis, loadability, etc.) were not discussed. It should be pointed out, however, that the properties of the column construction and the properties of the stationary phase determine together the compromise in the choice of the optimal column for a given separation problem.

Acknowledgement

We express our gratitude to the Max-Buchner-Forschungsstiftung and to the Deutsche Forschungsgemeinschaft for financial furtherance of the works referred in this paper.

List of symbols

a = V_m/V_s = k/k' = phase ratio of the column

at = 1 kilogram weight (kilopond) per sq. cm.

h = $L\,(w/4t_R)^2$ = L/n = height equivalent to a theoretical plate

H = $L\,(w/4t'_R)^2$ = L/N = $h\,[(1+k')/k']^2$ = height equivalent to an effective plate

i. d. = inner diameter

j = $\bar{u}/u_o$ = James-Martin pressure correcting factor

j' = pressure correcting factor in the equation: $\bar{u} = K\Delta p j'/\eta L$ [4]

K = $-\eta u d_p/dx$ = specific gas permeability

k = partition coefficient

k' = kV_2/V_m = k/a = $(t_R - t_o)/t_o$ = capacity ratio of the column

L = column length

L_{ne} = HS = necessary column length

n = $[4\,t_R/w]^2$ = number of theoretical plates

N = $[4\,t'_R/w]^2$ = $n\,[k'/(1+k')]^2$ = number of effective plates

Δp = pressure drop on the column

Δp_{ne} = $H\bar{u}S/Kj'$ = necessary pressure drop across the column [4]

r_{12} = k_2/k_1 = k'_2/k'_1 = t'_{R2}/t'_{R1} = relative retention

R = $2(t_{R2} - t_{R1})/(w_1 + w_2)$ = resolution

s = sample size entering the column

S = $[4\,R_{12}/(r_{12} - 1)]^2$ = separation factor = N_{ne}

t = time of analysis

t_c = column temperature

t_{ne} = $H(1+k')S/\bar{u}$ = necessary time of analysis

t_o = gas hold up time

t_R = retention time of peak measured from the start

t'_R = $t_R - t_o$ = net retention time

$\bar{u}$ = L/t_o = time averaged mean carrier gas velocity

u_o = $\bar{u}/j$ = outlet velocity of carrier gas

V_m = volume of the gas (mobile) phase in the column

V_s = volume of the partitioning stationary phase in the column

w = peak width at base line, segment of peak base intercepted by tangents to the inflection point

x = distance from inlet end of column

η = carrier gas viscosity

μ = micrometer

References

1. Halász, I., and Heine, E., in: Advances in Analytical Chemistry and Instrumentation Vol. 7 (Ed.: J. H. Purnell), Interscience Publ., New York 1968, in print.
2. Halász, I.: Z. anal. Chem., in print.
3. James, A. T., and Martin, A. J. P.: Biochem. J. 50,679 (1952)
4. Halász, I., Hartmann, K., and Heine,E., in: Gas Chromatography 1964, Brighton (Ed.: A. Goldup), The Institute of Petroleum, London 1965, p. 38.
5. Purnell, J. H., and Quinn, C. P., in: Gas Chromatography 1960, Edinburgh (Ed.: R. P. W. Scott), Butterworths, London 1960, p. 184.
6. Halász, I., and Holdinghausen, F., in: this book, p.
7. Horvath, C., and Thesis, Ph. D.: Universität Frankfurt/Main, Germany, 1963.
8. Halász, I., and Horváth, C.: Anal. Chem. 36, 1178 (1964).
9. Gross, K.-D.: unpublished results.
10. Halász, I., and Holdinghausen, F.: J. Gas Chrom. 5, 385 (1967).
11. Halász, I., and Horváth, C.: Anal. Chem. 36, 2226 (1964).
12. Halász, L., and Heine, E.: Nature 194, 971 (1962).
13. Heine, E., and Thesis, PH. D.: Universität Frankfurt/Main, 1963.
14. Halász, I., and Heine, E.: Analy. Chem. 37, 495 (1965).
15. Halász, I., and Heine, E.: Chem. Ing. Techn. 37, 61 (1965).
16. Halász, I., and Heine, E., in: Advances in Chromatography, Vol. 4 (Eds.: J. C. Giddings and R. A. Keller), Dekker, New York 1967, p. 207.
17. Landault, C., and Guiochon, G., in: Gas Chromatography 1964 Brighton (Ed.: A. Goldup), The Institute of Petroleum, London 1965, p.121.
18. Virus, W.: J. Chromatog. 12, 406 (1963).
19. Carter, H. V.; Nature 197, 684 (1963).
20. Berezkin, W. C., Swjatoschenko, A. T., and Klemmentjewskaja, L. N., in: Gas Chromatographie 1965, Berlin (Ed.: H. G. Struppe), Akademie-Verlag, Berlin 1965, preprints, p. 7

21. Desty, D. H., Haresnape, J. N., and Whyman, B. H. F.: Anal. Chem. 32, 302 (1960).
22. Schneider, W., Bruderreck, H., and Halász, I.: Anal. Chem. 36, 1533 (1964).
23. Averill, W.: J. Gas Chromatog. 1, 34 (1963).
24. Brudereck, H., and Schneider, W.: unpublished results.
25. Halász, I., and Gerlach, H. O.: Anal. Chem. 38, 281 (1966).
26. Gerlach, H. O.: Diplomarbeit, Universität Frankfurt/Main 1964.
27. Ettre, L. S., Purcell, J. E., and Norem, S. D.: J. Gas Chromatog. 3, 181 (1965).
28. Purcell, J. E., and Ettre, L. S.: J. Gas Chromatogr. 4, 23 (1966).
29. Ettre, L. S., Purcell, J. E., and Billeb, K.: J. Chromatog. 24, 335 (1966).
30. Ettre, L. S., in: „Open Tubular Columns in Gas Chromatography", Plenum, New York 1965.
31. Leibnitz, W., and Mohnke, M.: Chem. Techn. 14, 753 (1962).
32. Mohnke, M., and Saffert, W., in: Gas Chromatography 1962 Hamburg (Ed.: M. van Swaay) Butterworth, London 1962, p. 216.
33. Zhdanov, S. P., Kalmanovskii, V. I., Kiselev, A. V., Fiks, M. M., and Yashin, Y. I.: Zhur. Fiz. Khim. 36, 1118 (1962) or Chem. Soc. Translation 595 (1962).
34. Bruner, F. A., and Cartoni, G. P.: Anal. Chem. 36, 1522 (1964).
35. Bruner, F. A., and Cartoni, G. P.: J. Cnromatog. 18, 390 (1965)
36. Bruner, F. A., Cartoni, G. P., and Liberti, A.: Anal. Chem. 38, 298 (1966).
37. Petitjean, D. L., Leftoult, C. F.: J. Gas Chromatog. 1, 18 (1963).
38. Desty, D. H., in: Advances of Chromatography I (Eds.: J. C. Giddings and R. A. Keller) Dekker, New York 1965, p. 199.
39. Halász, I., and Horváth, C.: Nature 197, 71 (1963).
40. Halász, I., and Horváth, C.: Anal. Chem. 35, 499 (1963).
41. Kirkland, J. J.: Anal. Chem. 35, 1295 (1963).
42. Purcell, J. E.: Nature 201, 1321 (1964).
43. Schwartz, R. D., Brasseux, D. J., and Shoemake, G. R.: Anal. Chem. 35, 497 (1963).
44. Schwartz, R. D., Brasseux, D. J., and Mathews, R. G.: Anal. Chem. 38, 303 (1966).
45. Kabaker, G., and Thesis, Ph. D.: Universität Frankfurt/Main 1965.
46. Purnell, J. H.: Endeavour 23, 142 (1964).

Trace Analysis by Gas Chromatography

J. JANÁK

Institute of Instrumental Analytical Chemistry, Czechoslovak Academy of Science, Brno

1. Introduction

Let us first clarify the term trace analysis. What is meant by trace analysis by gas chromatography? (i) The analysis of small quantities of a substance, (ii) the analysis of a low concentration of it.

The first approach constitutes a broader viewpoint of the problem. It deals with methods of handling and detection of substances in micro and submicro quantities. It is characteristic of these methods that the total quantity of sample under analysis is small, but the sample comprises one or several components in high concentrations. To set a limit, quantities in the range 10^{-3} to 10^{-4} g. can be cited below this limit, the amount of a substance to be treated is always a trace amount. Gas chromatographic analysis makes use of a great variety of sensitive detectors, permitting the determination of 10^{-6} to 10^{-9} g., less often 10^{-10} to 10^{-12} g., and occasionally even still smaller quantities of substances; thus it is obviously an important method applicable to trace analysis.

The second approach constitutes an analytical definition, where the term trace analysis characterizes the determination, of substances present in concentrations less than 0.1 % = 1000 p. p. m. The limit of direct trace analysis by GC is, today, about 0.1 p. p. m. The object of investigation is the determination of concentrations 1 p. p. b. and lower ones.

The difference between the two approaches can be illustrated by example. Analysis of mixtures by capillary gas chromatography (GC) requires very small quantities of samples.

Capillary gas chromatography is thus a convenient technique for the analysis of small samples. If, however, capillary gas chromatography is used for the determination of a component present in the sample as a trace admixture only, the analysis will be a failure. In such a case, the use of a packed column will prove to be more convenient, since the chromatographic process will require such quantity of sample that the detection limit for the particular component will be attained. This is the reason why capillary gas chromatography will prove to be an unsuitable method for analyzing samples for components present in low concentration in spite of its suitability for analysis of small sample quantities.

A classification [1] dividing the method of trace analysis by gas chromatography into three main groups was found to be useful. It turned out that a number of analytical procedures have been developed since that time without affecting the above classification. Since then only the limits of sensitivity have changed (Table 1).

Table 1. Classification of gas chromatographic methods for trace analysis.

	Method	Application	Sensitivity limit p.p.m.+)
Direct:	a) Use of more sensitive detectors	general	0.01 - 0.001
	b) Method of large samples	special	2.5 - 10
	c) Conversion of substance into ones easier to determine	special	-
Indirect:	1. Enrichment of the trace component		
	a) Chromatographic sorption	general	0.1 - 0.001
	b) Desorption by thermal gradient	general	0.1 - 0.01
	c) Condensation	special	1
	2. Elimination of the main component		
	a) Chemical reaction	special	10
	b) Sorption	special	1
	c) Condensation	special	10
	d) Extraction	special	10

+) The limit of sensitivity is understood to be generally obtainable sensitivity; higher or lower values are possible in some special cases.

For a wider literature search see [2]. The best results can be achieved by adopting a combination of several modes of operation which can be termed „generally applicable". Let us pursue the individual modes of operation while adopting the newly described methods and auxiliaries only.

2. Sensitive and selective detectors

A comparison of some features of the detectors most often used in trace analysis has been made in Table 2.

The parameters of interest are marked by capitals. Of widest applicability are those detectors that have most of their data underlined: the flame ionization detector, Lovelock's detectors, and special detectors important in trace analysis such as the electron capture detector and the socalled sodium thermionic detector.

Table 2. Comparison of some features of detectors used in trace analysis.

Detector	Dynamic linearity	Background current	Minimal detectable quantity g/s
Katharometer	10^4	-	10^{-9} (C_3)
Argon	10^5	10^{-8}	10^{-13} (C_3)
Electron capture	10^3	10^{-9}	10^{-14} (CCl_4)
Electron mobility			
indirect	10^2	10^{-8}	10^{-9} (CO_2)
direct	10^3	10^{-10}	10^{-11} (CO_2)
Flame ionization	10^6	10^{-12}	10^{-12} (C_3)
Sodium thermionic	10^3	10^{-9}	10^{-13} (Parathion)
Photoionization	10^5	10^{-10}	10^{-13} (C_3)

Of these, the rewest and least known is the thermionic detector [3]. A diagram of it is shown in Figure 1. It differs from the classical flame ionization detector in that the cone of the hydrogen flame is burning at the surface of an insert made of pressed sodium sulfate or other alkali metal salt. The alkali metal ions influence the lifetime of the hydroxonium ion, which is responsible for ionization of C-C bonds, i. e., for the response and sensitivity of the FID to organic compounds. The sensitivity of response to a C-C bond drops rapidly if alkali metal ions are present in the flame. On the contrary, they will enhance the response in presence to compounds containing phosphorus or halogens. Figure 2 illustrates the selectivity of the thermionic detector to certain compounds as compared with that of the flame ionization detector.

Table 3 summarizes the values of ionization efficiences showing relative sensitivities of the flame ionization and thermionic detectors.

Table 3. Comparison of the ionization efficiency of the FID and TID

Compound	effective carbons	ionization effciency C/mole FID[a)]	ionization effciency C/mole TID	FID C/mole / TID C/mole
n-Hexyl alcohol	5.4	$3.4.10^{-1}$	$6.3.10^{-2}$	0.18
p-Dichlorobenzene	6.0[b)]	$4.7.10^{-1}$	3.4	7.2
0,0-Diisopropyl methyl phophonate	7.0[c)]	$5.0.10^{-1}$	$1.5.10^2$	300
Naphtalene	10.0	$4.1.10^{-1}$	no signal	0.00

[a)] Maximum ionization effciency of FID without the TID for the platinum nozzle (internal diameter 0.6 mm) amounts to 0.164 C/gramatom of carbon for corrected flow rates of N_2 = 3.19 ml/sec, H_2 = 1.85 ml/sec;
[b)] Effect of substitution with chlorine neglected; [c)] Approximate value (determined by us).

On this basis the thermionic detector is applicable to the analysis of phosphorus pesticides and of additives.

A thorough study of this detector carried out by us [4] has shown that the thermionic detector response, in terms of ionization efficiency vs. background current, changes until it reaches a maximum and than decreases again (Fig. 3). The maximum and the point at which the peak will become inverted (below the baseline) are not restricted to phosphorus or halogens only; and they assume, for various elements, different positions in the plot.

Certain alterations in design can also make this detection type selective for nitrogen and sulphur and other elements. By varying the background current it is possible to find conditions in which compounds containing one of the above named elements will give a negative signal. It is obvious that such data can be used to advantage for analyses specific for different compounds, above all for heterocyclic and metalloorganics.

Table 4. Sensitivity limits of a helium detector (Varian Aerograph).

Substance	Response factor C/mole	Ionization efficiency %	Sensitivity limit p.p.b
H_2	630	0.6	3
N_2	800	0.8	2
Ar	860	0.9	2
O_2	1100	1.1	.5
CO	1200	1.2	.5
CH_4	1600	1.7	.6
SF_6	1850	1.9	.2
C_2H_6	2700	2.8	.1
C_2H_4	3200	3.3	.2
C_3H_8	3900	4.0	.2
C_4H_{10}	4300	4.5	.2
CO_2	5400	5.6	.1
H_2S	5000	5.2	.1+)
NO	5400	5.6	.1+)
H_2O	5400	5.6	.1+)
N_2O	5400	5.6	.1+)
SO_2	5400	5.6	.1+)
NO_2	5400	5.6	.1+)

+) Estimated values assuming noninterference with column. 100% = 96,500 Coulombs per mole.

Our survey of detectors, to be complete, must include several other detector types. The argon detector in its „helium version" (Figure 4), i. e., using extremely pure helium carrier gas, can be used in trace analysis for permanent ok and other gases where the argon detector fails and other modes of detection lack the necessary sensitivity. Table 4 gives the figures for detection limits of a Varian Aerograph helium detector. The flame emission detector also deserves mentioning.

Figure 5 illustrates the Mikrotek flamephotometric detector. This detector is intended for the determination of trace amounts of sulfur dioxide in air and other sulfur-containing substances in gas chromatographic effluents.

Similarly, traces of pesticides can be determined by means of

Transformation of the trace admixture to a more easily detectable substance.

In spite of its uniqueness, in view of its dependence on a particular type of compound, this approach (preparation of derivatives) is often made use of in trace analysis. Several typical examples of this approach are given in Table 5.

Table 5. Some examples of chemical transformation of trace substances to more easily detectable derivatives

Substance	Derivative	Method of detection
Carbon monooxide	methane	flame ionization
Water	hydrogen	thermal conductivity
Steroids Amino acids	trifluoroacetate	electron capture
Heptachlor (X-pesticides	HCl	microcoulometry
Nucleotides Nucleotides	monophosphate	thermionic flame ionization

3. Concentration methods

The most common method is based on the sorption of the trace component in an active bed. The simplest method is to let the mixture in question pass through a sorbent bed, where a frontal chromatographic process will take place. This procedure is justified from the viewpoint of analysis only if the difference in volatilities of the trace component and other constituents is large enough. The trace component can be either more or less volatile than the major component of the mixture.

In such cases we speak of component enrichment by frontal chromatography (Figure 6). This can be achieved both in a gas-liquid and in a gas-solid system. The procedure suffers from one restriction, i. e., the degree of enrichment is determined by the partition coefficient of the most volatile component to be enriched. The amount of a mixture which can be forced through a sorption bed, therefore, is mostly given by the component which can be trapped in the bed quantitatively [5].

Of wider scope is the use of the equilibrium state in frontal chromatography [6] (Fig. 7).

It follows from the theory of phase equilibria that the amount of enriched component is proportional to its partition coefficient. This offers interesting possibilities which can be shown by means of an example. A mixture of benzene and nitrobenzene vapors containing a very high proportion of the former (400 : 10 mg/m^3) was subjected to analysis using a large sample (Figure 8a). There is no sign of nitrobenzene in the chromatogram, since its absolute quantity did not suffice to produce a signal. If subjected to the fronta chromatographic equilibration method, the mixture gives a complete record (Figure 8b). It should be noted that the high percentage of benzene is not reflected by an excessively large peak, since enrichment of nitrobenzene by phase equilibration is much greater, i.e., proportional to its partition coefficient. On the same grounds the peak of water vapor is missing in the chromatogram, in spite of the fact that the water concentration in the sa sample was 2500 mg/m^3; since the partition coefficient of water in SE-30 equals zero, no enrichment will occur. It is obvious that the enrichment effect can be varied in favor of a given type of compound by making use of a stationary liquid varying in polarity.

A very interesting example of the use of frontal chromatographic phase equilibration of the use of frontal chromatographic phase equilibration is the procedure known as reverse chromatography [7] or vacancy chromatography [8].

If the gas passing through the column is a gas containing the trace component, then, upon injecting a sample of the pure carrier gas (Figure 9B and 9D), the disturbance of the phase equilibrium will travel along the column at the same velocity at which the trace admixture is eluted, the only difference being that a negative peak will appear (compare Figure 9A and 9C). This method is also often used in trace analysis.

It is neither possible nor of consequence to include in this very short survey every possible type and application of trace analysis by gas chromatography. A special critical review is available [2]. This presentation should only serve to draw attention to the possibilities offered by properly combining the best features of different chromatographic techniques.

References

1. Janak, J.: in „Chromatography". Ed. Heftmann, E., 2nd Edition, Reinhold Pub. Co., New York, 1967, p. 774.
2. Svojanovsky, V., Krejči, M., Tesařik, K., Janák, J.: Trace Analysis by Means of Gas Chromatography, Chromatog. Rev., 8, 116 (1967).
3. Karmen, A., Giuffrida, L.: Nature, 201, 1204 (1964).
4. Dressler, M., Janák, J.: Collection Czech. Chem. Commun., in press.
5. Cropper, F. R., Kaminsky, S.: Anal. Chem., 35, 735 (1963).
6. Novak, J., Vasak, V., Janák, J.: Anal. Chem., 37, 660 (1965).
7. Reilly, C. N., Hildebrand, G. P.: Ashley Jr., X. J. W.: Anal. Chem. 34, 1198 (1962).
8. Zhukhovitskii, A. A., Turkeltaub, N. M.: Dokl. Akad. Nauk SSSR, 143, 646 (1962).

Elementary Analysis Using Principles of Gas Chromatography

H. KELKER

Farbwerke Hoechst AG. vormals Meister Lucius & Brüning, Frankfurt (Main)-Hoechst

The determination of the most important elements of organic chemistry is generally comprised under the name „elementary analysis". It is typical of these elements (above all, of C, H, N, O, S, Cl) that they can be converted to volatile compounds with relatively little experimental difficulties. In this field, oxidizing pyrolysis has for more than 100 years been the preferable method. The trend to the analysis of minimum quantities led - via the classic technique of microanalysis with 2 to 5 mg of substance - to the sphere of „ultramicrotechniques" using tenths or hundredths of a milligram, on the one hand, and back to the especially important field of semi-microtechnique with quantities between 20 and 100 mg, on the other hand. Some examining laboratories do their work almost exclusively on the micro-scale, while others use micro- and semimicroanalysis in about equal parts.

As an introduction, and for better understanding of the situation, I would mention the scheme of elementary micro- and semi-microanalysis (Fig. 1). The apparatus consists of the gas-purifying section and the reaction and absorption section proper. The two sections are very similar in composition and contain approximately 12 zones (marked by numbers in the picture) in which a reaction or auxiliary reaction of importance to the result of the analysis takes place [1]. Let me also say a few words about timing. Before the beginning of a series of CH-determinations, and after every prolonged interruption, the small absorption tubes for H_2O and CO_2 have to be tared; they have to be attached, conditioned by means of a „preliminary combustion", and finally weighed. This takes a total of about 30 minutes; only then can a test analysis be carried out and - in case of a positive result - the CH-analyses started. 21 minutes are estimated per analysis; this time includes weighing of the small tubes and weighing of the sample (see Fig. 2). About 20 analyses can therefore be carried out within an 8 hours' working day.

These preliminary remarks serve to indicate the complexity of the system and to point out that elementary microanalysis is a difficult trade which for decades seemed to defy any entirely rational, physico-chemical approach and thereby the chance of technical development. It is also justly pointed out in what an extremely high degree, e. g. personal errors enter into the total analytical error. Generally, values for C, H and N are accepted if in duplicate determinations they differ not more than 0.3 %. If therefore the standard deviation is calculated, according to DEAN and DIXON [2],

$$\partial = 0.89 \cdot 0.3 = 0.27$$

is obtained. Accordingly, for the permissible scatter Δx, with a confidence coefficient of 95 %,

$$\Delta x = 1.96 \cdot 0.27 = 0.54.$$

Thus, among 20 values we shall have to expect one that deviates from the accurate value by more than $\pm$ 0.54. Accordingly, as confidence interval $\Delta \bar{x}$ of the duplicate determination,

$$\Delta \bar{x} = \frac{2}{\sqrt{2}} \cdot 0.27 = 0.38$$

is obtained, i. e. among 20 mean values only one deviates by mor than $\pm$ 0.38%. It is therefore a matter of course that percentage data in the case of elementary analyses should only be permitted within the framework of these Δx-values. However, somewhat more favorable values are by all means attainable; thus, W. SIMON [3] gives 0 = 0.15 to 0.20 for H and C.

The diversity of the samples makes itself, of course, also felt particularly strongly in the discussion of errors.

The above notes serve to characterize the situation which at the end of the 1950s, after the success of the methods of gas chromatography, led to new studies, while the application of gas chromatography in microanalysis was discussed. We started extending the alkoxyl group determination by splitting up the CH_3J-, C_2H_5J-, etc. components; subjecting the Dumas nitrogen to closer investigation by gas chromatography (CH_4-content), and determining traces of water by means of the acetylene generated from CaC_2. After this the publications by DUSWALT and BRAND [4] and SUNDBERG and MARESH [5] brought the first methods of determining C and H by gas chromatography. According to Duswalt, 2 - 6 mg of substance are subjected to combustion in an oxygen stream, and H_2O formed is converted to C_2H_2 with $CaC_2 \cdot CO_2$ and C_2H_2 are taken out at the temperature of liquid nitrogen, and after completion of the combustion these components reach the thermal conductivity cell by means of a helium stream (flowing system).

This showed the possibility of further development which was due to the desire for faster, simpler, safer and cheaper methods of analysis. A point that is thought to be particularly awkward by everybody is the differential weighing of the absorption vessels. The weight of absorption tubes used in microanalysis is about 10-12 g. The weight difference to be measured is therefore only about 1/1000 of the total weight, and it is evident that gas chromatography particularly in this respect was able to make a contribution to simpler and more sensitive detection. In order not to have to omit these essential viewpoints of a development, we have given our subject a somewhat broader sense, including also those methods of elementary analysis among „gas chromatography" which have taken over one of its detection principles. If the subject were interpreted still more generally, it ought to comprise also the physical detection methods successfully tested in recent years, which are only listed here:

Automatic weighing, gravimetric direct detection [6, 7]
Mass spectroscopy [8, 9]
Manometry, volumetric analysis [3, 10, 11, 12, 12, 14, 15, 16, 17]
Titrimetric methods [18, 19]
Infrared detection (non-dispersive) [20]
Measurement of electrical conductivity [21, 22, 23, 24, 25, 26, 27, 28, 29]
Coulometry [30, 31, 32]
Measurement of thermal conductivity
a) stationary system [33]
b) flowing system [4, 5, 34, 35]

However, great influence was exercised on the present state of technology by improvements of the process of combustion and by systematic investigations into it, such improvements originating from the sphere of elementary analysis itself. It is impossible to mention all the names, but the following should be listed:

VECERA [36]
INGRAM, BELCHER [37, 38, 39, 40]
KAINZ, MONAR [41, 42]
LINDNER [43]
KÖRBL and HORACEK, MIZUKAMI and JEKHI [44, 45, 46, 47]

The decisive breakthrough occurred when WALISCH (in Saarbrücken) had developed his SIMULTANEOUS DETERMINATION of C, H an N [34]. It should also be mentioned that, about the same time, a method for direct oxygen determination had been found (GÖTZ, BOBER 1960 [35]) which - with some improvements - was incorporated, as a variant of like importance, in the standard program of CHN-determination a few years later. At the same time, articles by W. SIMON [3] appeared. He was working in Zurich at a static CHN-method which was in contrast to the dynamic detection principles so far know and which seemed promising in respect of accuracy. The first commercial apparatuses appeared on the market. By 1963 it was generally acknowledged that CHN-analysis had undergone a decisive change or, rather, supplementation by the methods of gas chromatography.

The operation of weighing the sample is a bottleneck in the attempt to devise a CHN-method that is more efficient and, if possible, automated. In the other hand, this operation has the advantage that the sample may be weighed while the preceding analysis is still going on, or the weighing may even be done at any time and in any place whatever. Only newly constructed types of the electrobalance (Cahn, Sartorius), and the well-known types made by Mettler (UM 6, UM 7) and Bunge (UM 25), enabled the technique of gas chromatography to be used as an ultramicrotechnique (as had been demanded). The widely used electrobalance is therefore shwon here in a schematic representation (Fig. 3). In the weighing of an object an accuracy of at least $\pm$ 0.5 μg is required, and this requirement is met. Because of the electric output, this type of balance seems ideal for connection to a transducer and thus for automation.

When the possibilities of automation are mentioned, and provided that automation is supposed to mean considerably more than just automatic performance of independent partial steps, we may state that the present means of technology enable all individual steps up to the digital or analogous result of the analysis to be realized in percentage by weight. It is only in the process of weighing the sample that it will probably never be possible to dispense with the human hand and the human judgment because the samples are mostly solids. This problem is far less serious for gases and liquids.

As far as the apparatus is concerned, the types may differ in the following points:

a) Combustion section (O_2 / catalysis / dynamic / stationary)
b) Separation of combustion products
   1. chromatographic
   2. adsorptive / freezing out
c) detection
   1. detection directly or after conversion
   2. static or dynamic detection
d) Combination of balance and apparatus
e) Degree of automation.

A. To begin with, we would mention an example of the genuine chromatographic method as it is realized in the F & M 185 [48, 49] and its predecessor F & M 180. The use of pure oxygen is dispensed with; combustion is carried out in a stationary manner with an oxygen catalyst, e. g. Ag_2O/MnO_2 or WO_3/MnO_2; the mixture is passed over a copper catalyst ($NO_2 \rightarrow N_2$) at a relatively low temperature, and the reaction products are separated chromatographically (formerly with two columns and interposed H_2O-trap (see Fig. 4), now (Fig. 5) $N_2/CO_2/H_2O$ successively on one column). In order to avoid the difficulties of the H_2O-blank- or noise levels which are hard to control, the carrier gas stream is charged with H_2O on $CuCO_4 \cdot 5\,H_2O$. With the aid of an especially constant column thermostat, detector signals were obtained which were quite reproducible and which in many cases even made it possible to manage with the calibration by height. In connection with the electrobalance there is also the technique of „radio recording" which means that a signal corresponding to the weight of the sample is transferred to the recorder potentiometer, whereupon the potentiometer recorder produces the signal voltage of the detector in such a form that all signals are related to a sample weight of 1 mg, and the comparison between different samples is somewhat facilitated.

The principle of gas chromatography is also applied in the Fisher analyzer (see Fig. 6). Combustion is carried out according to the Dumas principle: H_2O-conversion (according to Sundberg and Maresh), intermediary adsorption of the reaction gases with low-temperature cooling(liquid N_2), chromatographic detection in the order $N_2/CO_2/C_2H_2$ by means of thermal conductivity cells (dynamic principle).

Another process (CHUMACHENKO and PAKHOMOVA [50]) also operates in a genuine chromatographic manner: combustion takes place on NiO_2 in a helium stream; separation takes place in the final step on active carbon ($N_2/CO_2/C_2H_2$), likewise after conversion of the water.

B. WALISCH's [34] method is realized in the commercially available apparatuses of Technicon [51], in the apparatus made by WEITKAMP and KORTE [52] and in the Heraeus apparatus [53] „CHN = O -Ultra" which originated in a development in our laboratory - jointly with F. EHRENBERGER and O. WEBER [54].

WALISCH undertook the risk of carrying out the combustion in a flowing system (Fig. 7a). A helium stream containing 3 % oxygen is still suitable for complete combustion of most substances. Catalysts such as those mentioned at the beginning may be used but (as is particularly pointed out by WEITKAMP) they must not cover the sample of the substance because otherwise the release of the combustion products to the helium stream is hindered. An excess of oxygen is taken out with Cu which at the same time reduces the nitric oxides. (The correct treatment of the nitric oxides is characteristic of and important for the simultaneous determination of CHN! In the classic PREGL-CH they are absorbed, as far as they are present, and the elementary nitrogen, no matter what its proportion to the total nitrogen present, leaves the apparatus.) Since the chemical conversion of water always presents difficulties, WALISCH carried out its elimination on silica gel (to be expelled later on by heating, and determined separately). The combustion products N_2 and CO_2 pass the adsorption trap for H_2O and are conveyed together to one branch of a thermal conductivity cell where the components CO_2 and N_2 produce a joint signal. After this, via a bypass of sufficient length, CO_2 is removed from the mixture (on soda lime), and nitrogen is conveyed alone to the thermal conductivity cell. The N_2-signal is always considerably smaller than that of $CO_2 + N_2$, so that the value for CO_2 may be FOUND, within the permissible, error, by substraction of the signals.

We hesitated for a long time to adopt this principle; we preferred to carry out a GC-separation CO_2/N_2 instead after the H_2O-absorption. But finally it turned out that the results were of the same value, and that the analysis is easier without the separation column. The reason for this is that the water must, of course, subsequently be passed over this column or around it. The special feature of the Walisch apparatus built by Technicon is a integration motor with servo amplifier whose speed is always proportional to the katharometer voltage. A linearization element is used (as is was used subsequently by KORTE) which is considered necessary to ensure proportionality between katharometer voltage and CO_2-concentration at maximum concentrations of 5-10 % CO_2 in the helium (velocity of flow: approx. 30 ml/min. The values of accuracy stated by Walisch are realistic; they read as follows: ∂_C = 0.5 % abs., ∂_N max. 1 % abs., ∂_H = 0.1 % abs.

The enormous improvement of accuracy and reproducibility of the hydrogen values as against the classic method is typical of all gas chromatography elementary analyzers. The high nitrogen error should not, actually, be taken too hard, as is justly pointed out by Walisch: For identification, i. e. distinction between 1, 2, 3, or 4 N-atoms (a larger number has practically not to be taken into consideration) the accuracy of 1 % is completely sufficient. A high CH-accuracy is of much greater value.

In the following we would like to discuss the „2nd generation" developed by the Shell Grundlagenforschung (Birlinghoven). The work done by WEITKAMP and KORTE is mainly automation. After detection, the signals are very ingeniously converted immediately to percentages after putting in the weight of the sample be means of potentiometers. Minor changes to the apparatus, such as separate heating zones and cooled introduction of samples, were built in. The operation cycle is theoretically 10 minutes. There is a relatively fast flow up to 100 ml/min.; therefore, as in Walisch's method, a linearization element is used with the thermal conductivity detector, where this element depresses the output signals for high content values by as much as 2-3%. A voltage-divider circuit has the effect that with increasing input voltage the output voltage decreases more than linearly. By timing for the entire unit, the first almost fully automatic apparatus was produced. Only the weighing of the sample, and the adjustment of the potentiometer at the operational amplifier must be done by hand.

In view of the extremely great trouble taken, and the long practical experience with this apparatus, one may rely without hesitation on the accuracy data given by WEITKAMP which are considerably improved in comparison with WALISCH's data and read as follows: $\pm$0.2% C; $\pm$0.2% N; $\pm$0.2% H.

The exact price of the system is not kown; it is estimated to be relatively high, as for all laboratory developments.

The third device of the WALISCH family is our apparatus, which is built under license by Heraeus, under the designation „CHN-O ultra". The design of such an instrument by an analytical laboratory is simply due to the fact that at the time when this work started there was no suitable apparatus on the market. On the basis of some experience with integration [55], and after successful use of the Cahn balance and convincing demonstrations of the apparatus by Walisch we decided to begin the work. The apparatus is shown in Fig. 7b (the principle) and Fig. 8 (general view); Fig. 9 demonstrates the course of the experiment. Absorption of water on silica gel was replaced by a cold trap, and the vertical furnace built by EHRENBERGER [56] jointly with HERAEUS was used. The Cahn balance was built in and connected with the integration unit in such a way that the voltage frequency converter, the heart of the electric arrangement, receives the sample weight signal analogously and passes it on via memory and indicating tube (i. e. digital voltmeter) to a printer (Fig. 10). After the tare has been put in and the gross weight has been taken over, the printer at once prints out the net weight of the sample. After the converter has been switched over to the detector thermal conductivity cells as used by WALISCH and KORTE but without linearization elements), the converter takes the output voltages of the detector (as a function of time), integrates by counting the pulses, and stores and transfers the sums of pulses to the printer. The putting in of sample weight and response factors (which is, in principle, possible without difficulty) and the calculation of the percentage were dispensed with.

The determination of oxygen is, after installation of the carbonreactor unit, easily possible and has proved good for two years past (Fig. 11).

Our limits of error (reproducibility):

$s_C = 0.3...0.4$
$s_N = 0.4...0.5$
$s_H = 0.1...0.3$

The accuracy is of the same magnitude!

The low values apply to easily combustible substances, the high values to substances of low combustibility.

Fig. 12 and Fig. 13 show some more results of experiments. In particular, we have not been able experimentally to detect any deviations from the linearity of the CO_2-calibration curve. However, our working conditions are: max. 30 ml/min.; by the given arrangement, a max. CO_2-content of 2% is ensured, and a normal duration of combustion is ensured by the conditions anyway. The limits of applicability are given by the type of substances to be subjected to combustion. An especially plausible example, namely the CH-analysis of a material of silicate nature - a highly refractory stone containing carbonaceous, graphitized residues - resulted in the following values, showing the typical difficulties:

	% C	% H	with AgCl % O	% residue on ignition	sum
classic	60.6/60.8	0.9/0.8	5.5/5.7	34.3	100%
GC (Heraeus)	56	0.7	-	-	

C. We shall now discuss the method of SIMON. The important differences as compared with the apparatus mentioned unter A and B (Fig. 14) are: Combustion is carried out in pure oxygen with any catalyst whatever or without catalyst. The catalyst in the tube is usually CuO; the nitric oxides are reduced, the excess of O_2 is removed. After the process of combustion, the tube is flushed with helium, and its contents are pressed into an expansion- and mixing vessel which is kept at a temperature of 70 °C. In some deviation from the original conception by SIMON, according to which this vessel had first to be evacuated, the PE 240 [57] (that version of the Simon method which is now ready for practical use) employs a pressure up to 2 gauge atmospheres. After mixing and assuming the required temperature, the gas mixture is allowed to flow into a long tube where it is at the same time brought to normal pressure. The tube is relatively long; its contents may be pressed into the katharometer section by the helium stream, as a long plug, without changing its composition. This apparatus enables samples up to 5 mg to be analyzed (good dilution in the mixing vessel!). Measurement of H_2O, CO_2 and N_2 takes place according to the subtraction principle shown in the picture, and statically. Conversion to percentage by weight here takes place - like in all other apparatus - after previous calibration. Fig. 15 shows the Perkin-Elmer apparatus; Fig. 16, the reaction zones.

By courtesy of Perkin-Elmer we have been able to test the apparatus, and we have achieved good results with it. The standard deviations are:

C: s = 0.2% abs.
H: s = 0.1...0.2% abs. — The accuracy is of
N: s = 0.1.../.2% abs. — the same magnitude!

The examination of our afore-mentioned sample of silicate nature was especially convincing:

	% C	% H
Classic	60.6/60.8	0.9/0.9
PE 240	$60.9_5/60.9_3/60.7$	$0.7_6/0.8_0/0.8_3$

Because of some technical difficulties we had to interrupt the experiments at an early stage. However, it may already be said with certainty that the apparatus deserves much attention. (Time for one determination: about 15 minutes.)

D. A few more apparatus will now be discussed. The arrangement devised by KAINZ and WACHBERGER [58] should be mentioned because in it the principle was realized that most of all suggests itself (cf. MILLER [59], MILLER and WINEFORDNER [60]): Combustion in oxygen which serves as a carrier gas at the same time. CO_2 and H_2O are determined, but the method is not suitable for N_2. One series of measurements deserves particular acknowledgment: % CO_2 in oxygen against the corresponding signal in the thermal conductivity detector. The higher the oxygen concentration, the greater the signal and its increase with increasing concentration. By this the theory of WALISCH is confirmed for O_2. But up to 20 % O_2, corresponding to 3.4 mv., there is linearity between concentration and voltage, so that calibration with constant factors of proportionality is sufficient.

Carlo Erba has devised a CHN-analyzer which is practically in the service stage. Unfortunately, we have not yet seen it in operation, and the firm's prospectus does not give any characteristic details. The function of Model 950 resembles that of F & M 185; as the most interesting variant, it contains an automatic sample introduction device.

Practical experiences

Many a dispute on analysis results ends with the assertion that the sample is nonuniform. Practical experience has shown that nonuniformity may indeed be very great and that it often depends, in a typical manner, on the origin. Substances isolated from chromatograms; evaporation residues; lacquer samples; biological material - all these are often very nonuniform, not only in respect of the number of components present in a homogeneous phase, but also in the sense of different composition. But I do not intend here to mention longknown truths from the field of microanalysis. Is should only be pointed out that semi-microtechnique, i. e. the analysis of samples above one milligram is now as before of importance, so that in practice there is both the need for semi-microanalysis and the necessity of ultramicroanalysis.

Finally, a few economic aspects should be discussed. A major CH-laboratory has to furnish approximately 50 000 C-, H- and N-analyses per year. 250 working days of 81/4 hours each may be available. So there would be 100 CH- and 100 N-determinations per working day. Since a CH-determination requires about double the expenditure of an N-analysis, one C-, one H- or one N-determination may justly be counted as „one elementary determination". In conventional microanalysis, the 300 elementary determinations demanded here require about 13 persons and 13 apparatuses, taking real values as a basis and making allowance for illness of personnel and breakdown of apparatus. On this basis, one elementary determination costs DM 7.--. The 50 000 analyses would therefore be DM 350 000.--, including all laboratory expenses and the overhead expenses.

If all these analyses could be carried out by means of CHN-automata, this would require 4 apparatuses and 4 employees where 20 CHN-determinations are demanded of each apparatus per day. This value has been deliberately fixed at a low figure; 26 CHN-determination are theoretically possible, without preparation and calibration which are taken into account separately. The price for this installation is:

Capital expenditure	DM 200 000.-- (4 x DM 50 000.--)
4 employees	DM 120 000.--
	DM 320 000.--

Full amortization of this price within one year would require the same expenses as for the 13 persons and 13 apparatuses otherwise needed. From the second year onwards, the saving will be approximately DM 230 000.-- per year.

Unfortunately, the calculation does not work out so smoothly in practice, at least not today. According to the present state of technology, 50 % of the analyses must still be left to the classic method. But, as is immediately seen, even then a considerable advantage will accrue.

References

1. Ehrenberger, F.: Rationelle Industrieverfahren zur elementaranalytischen Bestimmung von Kohlenstoff, Wasserstoff, Stickstoff und Sauerstoff in organischen Verbindungen mit dem Verbrennungsautomaten Typ Mikro U der W. C. Heraeus GmbH, Hanau. Firmenschrift der W. C. Heraeus GmbH, Hanau
2. Dean, R. B., and Dixon, W. I.: Anal. Chem. 23, 636 (1951)
3. Simon, H., and Müllhofer, G.: Z. analyt. Chem. 181, 85 (1961)
4. Duswalt, A. A., and Brand, W. W.: Anal. Chem. 32, 272 (1960)
5. Sundberg, O. E., and Maresh, Ch.: Anal. Chem. 32, 274 (1960)
6. Trutnovsky, H.: Z. analyt. Chem. 222, 254 (1966)
7. Cahn, L.: DECHEMA-Monographien Band 44, S. 45
8. Grosse, A. V., Hindin, S. G., and Kirshenbaum, A. D.: Anal. Chem. 21, 386 (1949)
9. Kirshenbaum, A. D., Hindin, S. G., and Grosse, A. V.: Nature 160, 187 (1947)
10. Schöniger, W.: Mikrochim. Acta (Wien) 1957, 545
11. Frazer, J. W.: Mikrochim. Acta (Wien) 1962, 993
12. Frazer, J. W., and Crawford, R.: Mikrochim. Acta (Wien) 1963, 1132
13. Christmann, D. R., Stuber, J. E., and Bothner-By, A. A.: Anal. Chem. 28, 1345 (1956)
14. Ihn, W., Herb, W., and Noack, A.: Mikrochim. Acta (Wien) 1963, 1132
15. Pfab, W., and Merz, W.: Z. analyt. Chem. 200, 385 (1964)
16. Koch, C. W., and Jones, E. E.: Mikrochim. Acta (Wien) 1963, 734

17. Lindner, J.: Z. analyt. Chem. 66, 305 (1925)
18. Anderson, R. C., Delabarre, Y., and Bothner-By, A. A.: Anal. Chem. 24, 1298 (1952)
19. Blom, L., and Kraus, M. H.: Z. analyt. Chem. 205, 50 (1964)
20. Kuck, J. A., Berry, J. W., Andreatch, A. J., and Lentz, P. A.: Anal. Chem. 34, 403 (1962)
21. Gel'man, N. E., and Sheveleva, N. S.: Zhur. Anal. Khim. 20, 719 (1965)
22. Gel'man, N. E., Presler, P. J., Grek, N. V., Sheveleva, N. S., and Mel'nikova, A. A.: Doklady Akad. Nauk SSSR 161, 107 (1965)
23. Malissa, H.: Mikrochim. Acta (Wien) 1960, 127
24. Salzer, F.: Z. analyt. Chem. 205, 66 (1964)
26. Salzer, F.: Microchem. Journ. 10, 27 (1966)
27. Greenfield, S.: Analyst 85, 486 (1960)
28. Greenfield, S., and Smith, R. A. D.: Analyst 88, 886 (1963)
29. Stuck, W.: Mikrochim. Acta (Wien) 1960, 421
30. Oelsen, W., and Graue, G.: Angew. Chem. 64, 24 (1952)
31. Keidel, F. A.: Anal. Chem. 31, 2043 (1959)
32. Olson, C. E., Houtman, R. L., and Struck, W. A., Microchem. Journ. 5, 611 (1961)
33. Clerk, J. T., Dohner, R., Sauter, W., and Simon, W.: Helv. Chim. Acta 46, 2369 (1963)
34. Walisch, W.: Ber. 94, 2314 (1961)
35. Götz, A., and Bober, H.: Z. analyt. Chem. 181, 92 (1961)
36. Vecera, M., Snobl, D., and Synek, L.: Z. analyt. Chem. 176, 377 (1960)
37. Belcher, R., and Spooner, C. E.: J. Chem. Soc. London 1943, 313
38. Belcher, R., and Ingram, G.: Anal. Chim. Acta 4, 118 (1950)
39. Ingram, G.; Analyst 86, 411 (1961)
40. Ingram, G.: Analyst 73, 548 (1948)
41. Kainz, G., and Horwabitsch, H.: Mikrochim. Acta (Wien) 1962, 7
42. Monar, I.: Mikrochim. Acta 1965, 213
43. Lindner, J.: Ber. 65, 1696 (1932)
44. Horacek, J., and Körbl, J.: Chem. Listy 51, 2132 (1957)
45. Horacek, J., and Körbl, J.: Mikrochim. Acta (Wien) 1959, 303
46. Horaček, J., Körbl, J., and Pechanec, V.: Mikrochim. Acta (Wien) 1960, 294
47. Mizukami, S., and Jekhi, T.: Microchem. Journ. 7, 493 (1963)
48. Derge, K.: Chem.-Ing.-Techn. 37, 718 (1965)
49. Derge, K.: Chem.-Ztg. 90, 283 (1966)
50. Chumachenko, M. N., and Pakhomova, J. E.: Dokl. Akad. Nauk SSSR 170, 125 (1966)
51. Hochenegger, M.: Sonderdruck Nr. 422 der Technicon GmbH
52. Weitkamp, H., and Korte, F.: Chem.-Ing.-Techn. 35, 429 (1963)
53. Ehrenberger, F., and Weber, O.: Mikrochim. Acta (Wien) 1967, 513
54. Ehrenberger, F., Kelker, H., and Weber, O.: Z. analyt. Chem. 222, 260 (1966)
55. Kelker, H., Rohleder, H., and Weber, O.: GC-Symp. Hamburg 1962
56. Ehrenberger, F.: Mikrochim. Acta (Wien) 1962, 265

57. Condon, R. D., Precioso, A. N. , and Ettre, L. S.: Gas Chromatography and Methods of Immediate Separation, Vol. 1, 385 (1966)

58. Kainz,G., and Wachberger, E.: Mikrochim. Acta (Wien) 1968, 395

59. Miller, C. D.: Microchem. Journ. 11, 366 (1966)

60. Miller, C. D., and Winefordner, J. D.: Microchem. Journ. 8, 334 (1964)

Advances in Gas-Solid Chromatography

A. V. KISELEV

Laboratory of Adsorption and Gas Chromatography, Chemistry Department, M. V. Lomonosov State University of Moscow, and Laboratory of Surface Chemistry, Institute of Physical Chemistry, USSR Academy of Sciences, Moscow.

1. Introduction

Nine years ago, at the symposia in the USSR [1] and GDR [2], I tried to draw the attention to the possibilities of the development of gas-solid chromatography (GSC) on the basis of the synthesis of new homogeneous, crystalline and amorphous adsorbents with a controllable geometry and chemical composition of the surface. The years that have passed since then have shown that GSC can occupy an important place in the molecular chromatography of a wide range of substances, from gases to high-boiling liquids and many solids. The syntheses of new adsorbents and the development of methods for their modification solve problems of regulating the specificity of the immobile phase in the column leading to two important advantages in GSC: a high thermal stability and a rapid exchange of molecules between the gas and the immobile phase, the adsorbent surface. During the recent years a number of crystaline non-porous adsorbents with homogeneous surfaces, such as graphitized thermal carbon black; disperse, non-porous crystals of various inorganic substances of particular salts, and porous crystals (for instance, zeolites) have been synthetized. Also, use has been made of relatively homogeneous porous bodies with high surface areas, such as Saran carbons, porous glasses and styrene-divinylbenzene copolymers (Chromosorb 102, certain Poropaks, Polysorb-1). Methods have been developed für obtaining macroporous glasses, silica gels, aerosilogels, alumina gels, and macroporous organic and organosilicon polymers. Various methods have been developed for modifying the surfaces of solid adsorbents chemically or by adsorption deposition of dense and thermally stable monolayers of organic or organosilicon substances, including polymer molecules, or by deposition of thick layers of high-boiling substances. Finally, the formation of layers of adsorbents on and in the walls of open capillary columns has been studied.

Apart from important analytical and preparative applications, gas-solid chromatography is a sensitive and simple physicochemical method of investigation of the surface areas and properties of solid surfaces.

In this brief survey I cannot dwell on all the aspects of gas-solid chromatography. Some problems have already been considered in the reviews [3, 4]. I will limit myself to examples of the possibilities for gas-solid (and partly liquid-solid) chromatography on adsorbents of various geometrical and chemical structures.

2. Gas-Chromatography on First-Type Adsorbents (Non-Specific)

The surfaces of such adsorbents as, for instance, graphitized thermal carbon blacks, carry no ions or active functional groups. The more highly dispersed grades of graphitized carbon blacks, Saran carbons, oxides whose surfaces are modified by sufficiently dense chemically grafted or adsorbed layers of alkanes and alkylsilanes, and polymers containing no active functional groups are also virtually non-specific.

The surface of graphitized thermal carbon black, is the most homogeneous, both geometrically and chemically, The specific surface area of the different graphitized thermal carbon black vary from 6 to 15 m^2/g. The non-specificity of the surface of graphitized thermal carbon black is easy to reveal by comparing the adsorption of molecules capable of specific interaction [3 - 9] with that of „reference molecules" similar in size and polarizability, which are only capable of non-specific interaction (n-alkanes, noblesgases) [3, 4, 6, 8, 9]. Such a comparison is made in Fig. 1 in which the values of the heats of adsorption of various substances are plotted as a function of the electron polarizability of molecules, which determines the interaction with a given adsorbent [3, 4, 8, 9]. The points for molecules belonging to group B (having π-bonds or lone electron pairs) and even to group D (having functional groups types OH and NH), i. e. for molecules capable of specific interaction for this adsorbent tends to be close to the straight line obtained for reference molecules belonging to group A (n. alkanes, noble gases), which are incapable of specific interaction. The slight larger values of the heats of adsorption for lower alcohols is partly due to the difficulties in determining the heats of adsorption of isolated alcohol molecules capable of association on the surface of graphitized carbon black. Thus, the presence of π-bonds, lone electron pairs and protonized hydrogen in the group B and D molecules [3, 4, 6, 9], which leads to considerable quadropole and dipole moments in these molecules, has virtually no effect on the energy of interaction with the non-specific adsorbent - graphitized thermal carbon black. The interaction of group B and D molecules with this adsorbent, as in the case of group A molecules, is basically determinde by dispersion forces.

Examples of the application of graphitized thermal carbon black to the separation of various mixtures have been given in reviews [3, 4, 10] and at the previous symposia [6, 11-14]. There are also cited in the communications of GUIOCHON and VIDAL-Madjar [15] and SHCHERBAKOVA [16] at the present symposium. Therefore I will only point here to the great convenience of using the graphitized thermal carbon black in the separation of non-polar as well as polar homologues with a temperature programmed column, in separating isomers and in separating usual and deuterated hydrocarbons. It can be seen from-the chromatograms obtained by YASHIN, MIGUNOVA and SAVINOV [17, 18](Fig. 2) that temperature programming makes it possible to separate alcohols (group D molecules capable of strong specific interaction with second and third-type adsorbents and with each other) as well as alkanes (group A molecules incapable of specific interaction) on graphitized thermal carbon black, because the surface of graphitized thermal carbon black has no centres capable of forming a hydrogen bond with alcohols. It is evident from the chromatograms (Fig. 3) obtained by BORTNIKOV, EGOROCHKIN, MIGUNOVA, SHCHERBAKOVA and YASHIN that isomeric amyl naphthalenes, isomeric $C_{10}H_{16}$ hydrocarbons - adamantane and trimethylene norbornane, and also cis- and trans- isomers of metal-organic compounds with silicon and germanium: $(CH_3)_3SiCH{=}CHSi(CH_3)_3$, $(CH_3)_3GeCH{=}CHSi(CH_3)_3$ and $(CH_3)_3GeCH{=}CHGe(CH_3)$ can be separated readily on graphitized thermal carbon black. The sequence of isomer elution is determined by the maximum number of possible contacts of the heavier atoms of the molecule with the basal face of the surface of graphite crystallites [3, 4, 11, 13, 16]. Therefore, the first to leave the column among the amyl naphtalenes, is the isomer with branching closest to the nucleus, while the last is the one with an unbranched side chain. Spheroidal adamantane molecules leave the column much earlier cap than the more elongated molecules of its isomer, trimethylene norbornane. The cis-position secures a better contact of the heavy atoms of silicon and germanium (next to the double bond) with the flat surface. Therefore, with the general elution sequence of metal-organic compounds $(CH_3)_3SiCH{=}CHSi(CH_3)_3$, $(CH_3)_3GeCH{=}CHSi(CH_3)_3$ and $(CH_3)_3GeCH{=}CHGe(CH_3)_3$, in the order of increasing molecular weights, cis-isomers are retained more strongly than trans-isomers.

The homogeneity and non-specificity of the surface of graphitized thermal carbon black, and the similarity of the properties of the surface of the faces of its polyhedral particles to the properties of the semi-infinite lattice of the ideal graphite crystal leads to the semi-infinite lattice of the ideal graphite crystal leads to the retention data which is possible to treat by the methods of the molecular theory of adsorption [9]. It is important for chromatography that this theory enables one to calculate the relative and absolute values of the retention volumes (the HENRY's adsorption constant by unit surface) [19]. Since the force centres of a solid surface (as distinct from the liquid phase molecules in gas-liquid chromatography) are fixed, the molecular statistics of nonspecific adsorption on a homogeneous surface is simpler than dissolution statistics. Thus the rather rigorous statistical calculation of the relative and absolute retention volumes, V_s, for unit surface at low surface coverages (small sample sizes) and high enough temperatures, is possible on the basis of the potential functions of adsorbate-adsorbent interaction.

In the case of interaction of a complex molecule with a semi-infinite graphite lattice, the interaction potential of the molecule is:

$$\Phi \approx \sum_{c} \sum_{i} \varphi_{c \ldots i} \tag{1}$$

where $\varphi_{C...i}$ is a pair interaction potential of the carbon atom of graphite with the i-th atom (or i-th group of atoms) of the molecule [20, 21]. In POSHKUS' communication [22], Φ and V_s are calculated for the first five alkanes adsorbed on graphite. The value of V_s, for the first three alkanes treated as quasi-rigid molecules, in the classical approximation, are expressed through Φ with the aid of the corresponding configuration integral

$$V_s \sim \frac{1}{8\pi^2 A} \int \left[\exp(-\Phi/kT) - 1\right] \quad \sin\vartheta \ dxdydzd\vartheta d\varphi d\psi \qquad (2)$$

where x, y, z - Cartesian coordinates of the centre of mass of the molecule; ϑ, φ, ψ - EULERIAN angles of its orientation (the plane x, y, z = O passes through the centres of carbon atoms of the external basal planes of the semi-infinite graphite lattice). KISELEV, POSHKUS and AFREIMOVICH [22, 23] calculated V_s for n-butane and n-pentane, for which rotational isomerism exists. Therefore, allowances was made for the angles of internal rotation and the possible differences in conformation on the surface and in the gaseous phase when calculating the corresponding configurational integrals for these molecules. These calculations yielded results which agree with experimental data.

It is sufficient to know only two potential functions of atom-atom interaction, i.e. $\varphi_{C...C}$ and $\varphi_{C...H}$ to calculate $\varphi_{C...i}$ for any hydrocarbons. The parameters of these functions can be found from the theoretical expression for the dispersion attraction constants in KIRKWOOD and MULLER's formula for the dipole-dipole term of dispersion atom-atom (C...C and C...H) attraction and from similar formula for the dipole-quadrupole term of this attraction. The repulsion constant can be found from the equilibrium condition [20] at a distance $z_0 = d/2 + r_{oC}$, where d = 3.35 Å - interplanar spacing in graphite, and r_{oC}-VAN-DER-Waals radius of the corresponding atom in the adsorbate molecule. POSHKUS [22] shows that such a theoretical evaluation of the parameters of attraction forces in the expression für Φ yields values of V_s deviating from the measured ones by only 15%. The introduction of corresponding corrections to the parameters of the atom-atom potential functions $\varphi_{C...C}$ and $\varphi_{C...H}$ results in good agreement between the calculated and measured values for all five n-alkanes.

POSHKUS has also shown (Fig. 4) that the potential functions of atom-atom interaction $\varphi_{C...C}$ in the adsorption of n-alkanes on graphite thus obtained are very close to the corresponding potential functions determined from the interplanar properties of the graphite lattice [24], that of molecular crystals of hydrocarbons [25], and from the second virial coefficient of gaseous methane [26].

Thus, these potential functions describe carious manifestations of molecular interactions and can be used in gas chromatography, particularly for calculating variations in the values of V_s from molecule to molecule. Some examples of simplified evaluations of the sequence of V_s values for molecules of different geometrical structure are given by SHCHERBAKOVA [16] and in the papers [13, 27].

Also of great interest is the prediction of the sequence of V_s values for isotope-substituted molecules. The theory of adsorption of CH_4 and CD_4 on graphite was considered by YARIS and SAMS [28] and OLIVIER and ROSS [29], who established that adsorption of CD_4

is weaker than that of CH_4. A theoretical calculation of the values of retention volume V_s on graphite for the hydrocarbons CH_4, CD_4; C_2H_6, C_2D_6; C_3H_8, C_3D_8 and C_6H_6, C_6D_6 has been carried out by KISELEV and POSHKUS [30]. At high enough temperatures, the main role here is played by quantum-mechanical, and not quantum-statistical effects (the latter, however, are quite important in the adsorption of hydrogen isotopes and isomers at low temperatures). If we denote by V_s^H and V_s^D the values of the retention volumes of the H- and D-forms of the corresponding hydrocarbon, then

$$(V_s^H - V_s^D)/V_s^H = 1 - S_1^D/S_1^H, \tag{3}$$

where S_1^D and S_1^H - configurational integrals for quasi-rigid molecules in equation (2) for the retention volume. Thus, as a classical approximation, the values $(V_s^H - V_s^D)/V_s^H$ are determined by the difference of the potential functions Φ^H and Φ^D of the corresponding molecules, i. e. eventually by the difference in the interaction potentials $\varphi_{C...H}$ and $\varphi_{C...D}$.

The following assumptions were made with regard to the potential functions of isotopic molecules, as a classical approximation [30]: (1) The potential function is practically independent of the coordinates x and y of the centre of mass of the molecule and of the angle ψ of its rotation parallel to the graphite surface. (2) The oscillation of the centre of mass of the molecule perpendicular to the surface are taken to be harmonic. In this case

$$\Phi = \Phi_0(\vartheta,\psi) + \frac{1}{2}\Phi''_z (z-z_0)^2, \tag{4}$$

where $\Phi_0(\vartheta,\psi)$ and $\Phi''_z(\vartheta,\psi)$ - the values of Φ and its second derivative in z at an equilibrium distance $z_0(\vartheta,\psi)$ of the centre of mass of the molecule from the surface at fixed orientation angles ϑ and ψ. (3) It was also assumed that Φ''_z is independent of ϑ and ψ and, consequently, it is equal to Φ''_z at ϑ_0 and ψ_0. In the expression

$$\Phi_0(\vartheta,\psi) = \Phi_{00}(z_0, \vartheta_0, \psi_0) + \Delta\Phi_0(\vartheta,\psi), \tag{5}$$

where Φ_{00} - the value of Φ at the deepest minimum of the potential at the values of z_0, ϑ_0 and ψ_0, and $\Delta\Phi_0$-deviation of Φ_0 from Φ_{00} on deviations of ϑ and ψ from ϑ_0 and ψ_0. (4) In addition it was assumed that the functions $\Delta\Phi_0$ are practically the same for H- and D-forms of the molecule. Under these assumptions (1)-(4) the expression for the relative difference of the retention volumes is:

$$(V_s^H - V_s^D)/V_s^H = 1 - (\Phi_z''^H/\Phi_z''^D)^{1/2} \exp.(\Phi_{00}^H - \Phi_{00}^D)/kT \tag{6}$$

while the expression for the difference of the isosteric heats of adsorption is simply:

$$Q_{st}^H - Q_{st}^D = -N_A(\Phi_{00}^H - \Phi_{00}^D). \tag{7}$$

Consequently, the value $(V_s^H - V_s^D)/V_s^H$ increases exponentially with increasing difference in the potential energies of adsorption of the H- and D-forms of the molecule (equal to the difference of their heats of adsorption) and decreases exponentially with rising temperature. For molecules CH_4 and CD_4, the corresponding quantum corrections were taken into account [30].

The mean statistical polarizability α of the methane molecule on complete deuteration decreases by approximately 1.45 % [31], and that of benzene, by 1.2 % [32]. At the same time, the ratio $\alpha / \varkappa$ of polarizability α to magnetic susceptibility $\varkappa$ remains practically unchanged [30]. Therefore, the dispersion interaction constants of the H- and D-forms with a semi-infinite graphite latice change By a constant factor $\gamma_{H,D} = \alpha^H/\alpha^D$. One should also take into account the difference in the bond lengths: C-D is shorter than C-H by about 0.004 Å [31, 33], this affects the equilibrium distances which determine one of the parameters of the potential functions $\varphi_{C...H}$ and $\varphi_{C...D}$.
The calculated values of $(V_s^H - V_S^D)/V_s^H$ for the adsorption of CH_4, CD_4 and C_6H_6, C_6D_6 on the basal face of graphite [30] and the corresponding experimental values for CH_4, CD_4 on graphitized thermal carbon black [34] are shown in Fig. 5. The calculated values for CH_4 and CD_4 only slightly exceed the experimental values, while the calculated values of $Q_{CH_4} - Q_{CD_4} \approx 20-25$ cal/mole practically coincide with the experimental ones. The theoretically predicted sequence of the elution of CD_4 and CH_4 has been found experimentally for a Saran carbon column (see below). For C_6H_6, C_6D_6, there are no reliable experimental data for quantitative comparison so far, but it can be seen from the chromatogram of a C_6H_6, C_6D_6 mixture obtained by SAKODYNSKY, KHUDYAKOV and YASHIN [35] that deuterobenzene is the first to eluate from the capillary column with graphitized thermal carbon black in accordance with the prediction of the molecular theory.
To separate light hydrocarbons and inorganic gases, it is necessary to increase the specific surface area of the adsorbents, i. e. to use a porous carbonaceous adsorbent. The Saran carbon is fairly homogeneous having a low degree of oxidation. It is obtained by thermal decomposition of polyvinylchloride. Due to the increase in the dispersion force potential in fine pores, the adsorption of lower n-alkanes by Saran carbon are about twice their heat of adsorption on graphitized carbon black [36]. Therefore Saran carbon is suitable only for the separation of light molecules.

The low specificity of Saran carbon can be judged from Fig. 6, which, as Fig. 1, shows the dependence of the heats of adsorption of molecules belonging to groups A, B and D on their electron polarizability [36]. In this case too, points for most of the B and D molecules lie near the straight line for A molecules. However, deviations for H_2O, NH_3 and SO_2 are rather considerable here because of the specific interaction of these molecules with oxygen complexes which are inevitably present in active carbon. These complexes are formed at places of rupture of the lattice of small crystallites on contact of the carbon with the air.

In accordance with its low specifity, which follows from Fig, 6, Saran carbon is suitable for good separation of certain inorganic gases and lower hydrocarbons (Fig. 7). Olefins eluate from a Saran carbon column before the corresponding alkanes.

Thus, this carbonaceous adsorbent ist fairly similar qualitatively to graphitized carbon black as regards its non-specificity. Therefore, in accordance with the above considered theoretical calculations 30 , the CD_4 retention on Saran carbon should be less than that of CH_4, as in the case of graphitized carbon black [34]. Indeed, Sakodynsky and Sevastyanova [37] have shown (Fig. 8) that CD_4 eluates from the Saran carbon column before CH_4, a good separation taking place near to 0 °C.

The role of the dispertion interaction in gas-adsorption chromatography on charcoal was considered by SNYDER [38]. The adsorption and gas-chromatographic properties of similarly structured BN are obviously close to those of graphite [39] with regard to group A and B molecules.

In spite of the homogeneity and thermal stability of these adsorbents, their application for separating high-boiling substances involves difficulties because of the excessive retention due to the very high non-specific interaction potential. The reason for this is a high concentration of atoms on the basal plane. This potential can readily be decreased by shielding the lattice with a dense monolayer of chemically grafted alkyl or alkylsilyl groups or molecularly adsorbed molecules or macromolecules [1-4, 6, 14, 40-44].

The modification of these adsorbents by dense monolayers of molecules or macromolecules of saturated hydrocarbons or alkylsilanes has not been carried out jet although the conditions of the experiments in [12] are probably close to this case. Porous polymers from saturated hydrocarbon or alkylsilane macromolecules für gas chromatography have not been used either. Non-specific adsorbents with a low adsorption energy have been obtained so far only by chemical modification of the surface of dispersed oxides with the aid of reactions with halosilanes and alcohols. By treating aerosil with trimethyl-chlorosilane very dense modifying layers werde obtained which considerably decrease the adsorption energy and retention of all, but especially large molecules [1, 2, 41, 42]. The formation of these layers and the adsorption properties of their surfaces werde considered at the previous symposia [1, 2, 41] and in the reviews [3, 4].

3. Gas Chromatography on Second-Type Adsorbents-Specific, Carrying Locally Concentrated Positive Charges on their Surfaces

Among these adsorbents, those which are suitable for gas chromatography, carry acidic hydroxyl groups with partly protonized hydrogen (hydroxylated surfaces on some oxides) or exchange cations (cationized zeolites) on their surface. It is possible to abtain the surfaces of this type through modification of first and second type adsorbents by adsorbed layers of certain alcohols and amines or by chemically grafted groups -ROH, $-RNH_2$, $-R_1R_2NH$. Porous polymers with functional groups OH and NH on their surface also belong to this type of adsorbents. The surfaces of adsorbents with strong acceptor centres (e. g. silica gels containing alumina impurity, porous glases containing boric oxide impurity, aluminosilicagels and decationized zeolites) adsorb molecules with strong donor centres too firmly. Here, the chemisorption takes place by a donor-acceptor interaction with charge transfer mechanism. Therefore such adsorbents are unsuitable for chromatographing molecules possessing a strong donor ability, without preliminary modification of their surface.

Gas chromatography of comparatively light group A molecules on cationized zeolites has been successfully performed (see the bibliography in [3, 4]). However, owing to the high concentration of the dispersion force centres on the walls of zeolite cavities (these are mainly oxygen atoms) and the increase in the potential energy of these forces because of the small sizes of the channel openings and cavities, the use in gas chromatography of the internal surface of such porous crystals for n-alkanes with four or more carbon atoms requires high column temperatures. The chromatography of heavier group A molecules or even comparatively light group B and D molecules is often entirely impossible on zeolites. At the same time chromatography on the external surface of very fine-porous zeolites, such as NaA, affords no advantages as compared with chromatography on non-porous, dispersed salts.

Theoretical calculation of the retention volumes on zeolites is possible due to their crystalline structure. In contrast to the non-porous and non-specific surface of the basal face of graphite the adsorption field in the cavities of fine-porous zeolite crystals is much less homogeneous, especially in respect of group B and D molecules, which interact specifically with the cations and with the negatively charged oxygen atoms of the zeolite lattice. It is shown in BROYER and LOPATKIN's communication at this symposium [45] that it is possible to evaluate the distribution of the adsorption potential in zeolite cavities taking into account in addition to dispersion forces the intensity of the electrostatic field set up by the charged particles of its skeleton and by the exchange cations. One can carry out molecular-statistical calculations of retentin volumes by developing appropriate expressions for the potential function. The further development of the molecular theory in this field requires a more precise determination of the parameters of the potential functions on the basis of accurate experimental data for a small number of typical group A, B and D molecules.

The specifity of cationized zeolites can be regulated by leaving a small amount of adsorbed water on the exchange cations. Thus, for instance, by subjecting CaX zeolites to preliminary heat treatment at 400 °C (which leaves a little water on this zeolite). GVOZDOVICH and YASHIN were able to separate atmospheric argon and oxygen (Fig. 9).

Highly dispersed non-porous salts are used in gas chromatography both in the pure state and as fillers or coatings applied to various carrier adsorbents (usually with a moderate surface area). The relevant works are considered in reviews [3, 4]. Here, we will also mention the work of GEISS, VERSINO and SCHLITT [46] in which inorganic salts were successfully used for the separation of terphenyls, and the work of HARGROVE and SAWYER [47] in which salts Na_2SO_4 and Na_3PO_3 were deposited on alumina, and the cariation in their specifity as compared with graphitized carbon black was observed. In the work [48], chlorides of alkaline metals were deposited on macroporous silica gel. Filling of silica gel with salts LiCl and CsCl led to the narrowing of the peaks and some changes in the relative retention times (Fig. 10). The deposition of KOH on silica gel enabled the determination of the divinyl impurity in unsaturated C_4 hydrocarbons [49]. In Fig. 11, the relevant chromatogram is compared with the chromatogram of the same unsaturated hydrocarbons on graphitized carbon black [50].

Non-modified macroporous silica gel and porous glasses are suitable for the separation of saturated and aromatic hydrocarbons (see reviews [3, 4, 51]). For the separation of strong organic bases, however, silicas of higher purity are required. Aerosilogels are an example of high-purity macroporous and extremely thermostable silicas [52, 53]. These adsorbents are discussed in NIKITIN's communication to this symposium [54]. Therefore, here we will restric ourselves to one example of separation on aerosilogel. Fig. 12 shows the separation of a number of group B organic molecules, including some with high dipole moment, which elute in quite symmetric and narrow peaks.

The specific adsorption of group B and D molecules on the hydroxylated surface of pure silica is by hydrogen bond formation (see reviews [3, 4, 6, 9]). It is of interest to compare the strength of the hydrogen bond for interactions of organic bases with the hydroxylated surface of silica and between protonic acids and the same organic bases in solution. In the case of adsorption the hydrogen bond energy was determined as the difference of the heats of adsorption on the hydroxylated and dehydroxylated silica surface at half-coverage (see bibliography in [3, 4, 6, 9]). In the case of solution, this energy can be determined by the method of gas-liquid chromatography [55]. The hydrogen bond has also been investigated spectroscopically, as its formation causes a shift in the frequency of the stretching mode ν_{AH} of the acid groups AH (in particular, the surface groups OH of silica) in the infrared spectrum. For different acids and bases, a linear dependence between the enthalpy of the formation of the hydrogen bond und the corresponding increment in the root of molar intensity Γ of the adsorption band of the stretching mode ν_{AH} is observed throughout the known range of hydrogen bond energies [56]:

$$-\Delta H_{AH...B} = \text{const}\,(\Delta\Gamma)^{1/2} \qquad (8)$$

The value of $(\Delta\Gamma)^{1/2}$ is equal to the root of difference between the molar intensity Γ of the stretching mode ν_{AH} in infrared absorption of the H complexes formed and Γ_0 of the free acid AH. In the region of moderate hydrogen bond energies (up to $-\Delta H \approx$ 5 kcal/mole and $\Delta\nu \approx$ 400 cm^{-1}) the dependence of $-\Delta H$ on $\Delta\Gamma$ (or $\Delta\nu_{AH}$) is nearly linear [3, 4, 6, 57, 58].

IOGANSEN [59] established a simple parametric relation in the energies of formation of the hydrogen bond in solutions. The enthalpy of formation of the hydrogen bond in the complex $A_iH...B_j$ can be expressed as the product of independent factors characterizing A_iH and B_j:

$$\Delta H_{ij} = \Delta H_{11} P_i E_j$$

where P_i - electron-acceptor factor of the acid A_iH; E_j - electron-donor factor of the base B_j, and the coefficient ΔH_{11} - enthalpy of the standard H complex, for which, by convention, $P_i = E_j = 1$. (The adopted standard is the complex phenoldiethyl ether in a CCl_4; solution for this $-\Delta H_{11}$ = 5.4 kcal/mole). In the work of DAVYDOV, DZHIGIT, GALKIN, KISELEV, LYGIN and MUTTIK [3, 4, 6, 57, 58, 60], the values of the energy of formation of the hydrogen bond with the OH groups of the silica surface, and of the corresponding values of $\Delta\nu$ and, partly, $\Delta\Gamma$ werde determined. From the specific interaction energy of one of the organic bases studied in these works (e. g. diethyl ether) it is possible to evaluate from equation (9) the electron-acceptor factor P_{SiOH} = 0.95 (by increasing E_j for diethyl ether in the CCl_4 solution by 0.20 for the solvent CCl_4 relative to a vacuum). Knowing PSiOH and the values of E_j determined for H complexes in solutions (with a correction for the solvent relative to a vacuum), Iogansen used equation (9) to calculate further the values of $\Delta H_{SiOH...B}$ for complexes SiOH...B of OH surface groups of silica with other organic bases B_j. These values are in agreement with those determined from the difference in the heats of adsorption on hydroxylated and dehydroxylated surfaces. In Fig. 13, the corresponding results are compared for the interaction of a number of acids and organic bases in solution and for the adsorption of a number of organic bases on the hydroxylated surface of silica. The adsorption data are in line with the general rule, except for pyridine. In the latter case, as probably for other polyfunctional molecules, a substantial role is played by the impossibility of free orientation of the donor molecule with respect to the acceptor group of the surface (group SiOH). This is associated with the orientation of the molecules on the surface under the effect of the total interaction with the adsorbent, rather than just the specific interaction.

In order to obtain more symmetrical peaks for organic bases from aerosilogel, it is expedient to dehydroxylate its surface by heat treatment at high temperatures, since in this case the distance between the remaining hydroxyl groups is greater and they are energetically more homogeneous. On exhaustive dehydroxylation of aerosilogel, its specific surface area remains unchanged (when heat-treated in vacuo up to at least 1000 °C), but such dehydroxylation results in the appearance of active sites of chemisorption on the silica surface, the number of these sites increasing with increasing heat treatment temperature to 1000 °C. These active spots have been shown to chemisorb water and tetrahydrofuran [61]. Heat treatment in air does not lead to such prominent chemisorption effects. However heat -treated aerosilogel should be first tested for the absence of chemisorption of the components of the mixture being separated before usw.

On the other hand, such chemisorption or strong specific adsorption by the donor-acceptor mechanism can be used for preliminary modification of the adsorbent surface for the purpose of shielding strong acceptor centres. For instance, adsorption in a chromatographie column, of monoethanolamine and decyl alcohol from the carrier gas on macroporous silica gel containing alumina, creates strong acceptor centres on the surface, results in the formation of a modifying layer [62]. It can be seen from Fig. 14 that this modifying layer is highly thermostable: heating to 300 °C does not produce any noticeable change in the background current in the flame-ionization detector. At the same time, these very substances deposited in amounts used in gas-liquid chromatography begin to evaporate rapidly at much lower column temperatures. It should be noted that separation on such dense monolayers is based on separation on their surface, and not on dissoltution. This can be seen from the change in the elution sequence of methanol and tertiary butyl alcohol when passing from gas-liquid (40% glycerol on macroporous silica gel with s = 60 m 60 m^2/g) to gas adsorption chromatography (10% glycerol on the same silica gel, which corresponds to a dense monolayer) (Fig. 15). Separation on monolayers occurs at comparatively moderate temperatures.

4. Gas Chromatography on Third-Type Adsorbents - Specific. Carrying Locally Concentrated Negative Charges on their Surfaces

Adsorbents of this type can be obtained by depositing a monolayer of group B molecules or macromolecules (with π-bonds or lone electron pairs, but without protonized hydrogen) on the surface of first- or second-type adsorbents or by chemical grafting to the surface of second-type adsorbents in hydroxyl place of groups of compounds containing such groups B, for instance CN. Both these methods have been used by us [3, 4, 6, 43, 44, 62, 63]. The advantage of the modification of the surface by dense monolayer is that all surface-modifying molecules are strongly retained by the adsorption forces of the carrier adsorbent. Another method of obtaining third-type adsorbents was used by Vidal-Madjar and Guiochon, who deposited thick layers of solidifying organic substances, such as anthraquinone, on the surface of graphitized carbon black [64]. This method is similar to that used by Scott [65] in modifying aluminia by benzophenon, but the application of a non-porôus carrier adsorbent is more convenient, since it does not result in a substantial reduction of the surface area. Vidal-Madjar and Guiochon also deposited think layers of copper phthalocyanine on graphitized carbon black, The adsorbent so modified was successfully used for separating polynuclear aromatic hydrocarbons and phenols [66, 67]. It is important to find the minimum thickness of these layers which ensures satisfactory shielding of the carrier adsorbent and at the same time the minimum volatility. Our work indicates that it would suffice to deposit a dense monolayer on the surface [40, 43, 44].

The methods of modyfying the carrier adsorbent by deposition of adsorbed dense monolayers (for instance from solutions [43, 44]) are simpler than the methods of chemical modification of the surface. At the same time, these methods make it possible to regulate the specificity of the surface within a very wide range by selecting molecules or macromolecules with suitable functional groups or π-bonds. Molecules particularly suitable for modification by adsorbed monolayers are those with. The modification of surfaces by monolayers of optically active substances is also of great interest.

Another important method of obtaining third-type adsorbents is the synthesis of porous polymers with phenyl groups or with groups having lone electron pairs. These include polyphenyldisiloxane [68], copolymers of styrene with divinylbenzene [69] - the organic part of Chromosorb 102 [70], and similar polymers-Polysorbs [71], some Poropaks, as well as polyacrylonitrile [72]. The gas-chromatographic characteristics of some of these polymers are presented, in particular, in the communication of YASHIN, BEBRIS and GVOZDOVICH at this symposium [72], therefore we will only dwell here on the illustrations to the operation of columns with porous polymers and on certain comparisons.

Fig. 16 displays chromatograms of certain inorganic gases, a number of aldehydes and water, the alcohols C_1-C_4 and 1,4-dioxane, an Chromosorb 102 while Fig. 17 shows chromatograms of an ammonia impuritiy in water, and of water in ethanol and in benzene. Ammonia and water are particularly wedkly retained by Chromosorb 102, which points to a very weak specifity of adsorption on this porous polymer. The determination of the specificity type of this adsorbent is assisted by an investigation of the heats of adsorption at low surface coverages (as determined from the temperature dependence of the maximum of narrow symmetric peaks for small sample sizes) and of adsorption isotherms determined at large sample sizes.

Fig. 18 demonstrates the dependence of heat of adsorption at low surface coverages on Chromosorb 102 [70], on polyphenyl-disiloxane and on polyacrylonitrile [73] for group A molecules (n-alkanes), a number of group B molecules and for group D molecules (water and n-alcohols) [74]. Comparing this illustration with Fig. 1 for graphitized thermal carbon black and with Fig. 6 for Saran carbon, one can draw the following conclusions. First, the heats of adsorption of n-alkanes on the first two polymers with phenyl groups are quite high. This is probably due to the penetration of n-alkane molecules into the fine clearances between the macromolecules (in statical adsorption experiments, a noticeable swelling of these polymers in hydrocarbon vapours was observed). Second, the heats of adsorption of n-alkanes on polyacrylonitrile are not high (this polymer does not swell in hydrocarbons). Third, on all these polymers, the heats of adsorption of group D molecules (n-alcohols), in contrast to adsorption on graphitized carbon black (Fig. 1), greatly exceed the heats of adsorption of group A reference molecules (n-alkanes). At the same time, no increase in the heats of adsorption for a number of group B molecules is observed. The higher heats of adsorption of group D molecules are due to the formation of weak hydrogen bonds with phenyl groups (as well as with possible inclusions of oxygen groupings) of Chromosorb 102 and polyphenyldisiloxane and of considerably stronger hydrogen bonds with CN groups of polyacrylonitrile. It should be noted that CN groups also strongly and specifically adsorb group B molecules having very large dipole moments (and containing CN and NO_2 groups [3, 4, 6, 73].

Fig. 19 shows adsorption isotherms of water at different temperatures, and of water, diethyl ether and n-butanol at the same column temperature, on Chromosorb 102, obtained from chromatograms by GVOZDOVICH, KISELEV, MIGUNOVA and YASHIM [70]. The adsorption isotherms of water are convex toward the pressure axis, which indicates weak adsorbate-adsorbent interaction and a rather strong adsorbate-adsorbate interaction. A slight temperature increase, only to 100 $^{\circ}$C, makes the water isotherm near-linear over a fairly wide range of vapor pressures. All this points to a weak specificity of Chromosorb 102. Nevertheless, this specificity does exist and it becomes prominent when comparing the adsorption isotherms of diethyl ether and n-butanol, which have identical molecular weights. In both cases adsorption is much higher than for water because of an abrupt increase in the non-specific adsorbate-adsorbent interaction due to the hydrocarbon part of these molecules. However, the adsorption of n-butanol is much higher than that of ether, due to the presence of the hydroxyl group which form a weak hydrogen bond with π-electrons of the phenyl group of the polymer. Thus, Chromosorb 102 belongs to weak specific adsorbents of the third type and is quite suitable for the separation auf the B and D molecules and for determining impurities of water, ammonia and other similar molecules.

5. Preparative Gas-Solid Chromatography

The separation of mixtures and the purification of products on the basis of preparative chromatography requires homogeneous, selective, thermostable and readily regenerable adsorbents. We shall consider here only a few examples of preparative gas-solid chromatography on some of the abovedescribed adsorbents.

BRAZHNIKOV and SAKODYNSKY [75] successfully used graphitized carbon black for the purification of germanium tetraiodide at 240 °C. The investigation by VOLKOV, ZELVENSKY and SAKODYNSKY [76] showed that for preparative chromatography of some of the C_6 hydrocarbon isomers the optimum value of specific surface area s of macroporous silica gel [77] was about 30 to 50 m^2/g, while the optimum separation temperature is close to the mean boiling point of the mixture components. The removal of impurities of 3-methylpentene-2 from 3-methylpentane was carried out on macroporous silica gel with s = 44 m^2/g. The initial product contained about 5% olefin, this considerably impairing its optical properties. Over 200 ml of 3-methylpentane containing less than 0.005% 3-methylpentene-2 was obtained. The high purity of the product is also indicated by the practical absence of absorption up to wavelengths of 220 mμ . The high selectivity of such a specific adsorbent as silica gel with a hydroxylated surface was especially prominent. In the separation of this mixture.

A similar problem arose in separating an equimolar mixture of n-hexane and n-hexene-2. In this case, the difference in the boiling points of the components is 0.5 °C. This mixture completely fails to separate on a immobile liquid phase of medium polarity (dinonylphtalate). Macroporous silica gel exhibits satisfactory separation with a rather high yield (0.7 ml/min). Macroporous silica gel is particularly suitable for separation of high-boiling compounds. VOLKOV, ZELVENSKY and SAKODYNSKY [76] successfully used this silica gel for purification of n-hexadecane at 230 °C, when the use of most of the liquid phases for preparative purposes is already impossible. These authors later used macroporous silica gel for purification of labelled benzene, when it was necessary to avoid the contamination of the valuable product by the evaporation or decomposition products of liquid phase, and also to ensure minimum losses. The purity of the deuterobenzene obtained was at least 99.99 %.
Still better results were obtained in separation on another type of wide-pore silica - extremely chemically pure and geometrically homogeneous aerosilogel [52-54] (see Fig. 12). The satisfactory symmetry of the peaks of the equimolar mixture of cyclohexane and benzene made it possible to obtain cyclohexane containing less than 0.001 % benzene on aerosilogel with s = 90 m^2/g.

VOLKOV, ZELVENSKY and SAKODYNSKY [76] used Polysorb-1, which is a copolymer of styrene and divinylbenzene (40 %) with s ≈ 250 m^2/g as adsorbent for preparative chromatography. The preparative separation of five classes of compounds-n-alkanes, alcohols, ketones, fatty acids, and aromatic hydrocarbons - was carried out. A chromatogram of separation of water, acetic, and butyric acid mixture at 162 °C is shown in Fig. 20 as an example.

6. Open Gas-Solid Capillary Columns

In this brief survey we are unable to discuss open capillary adsorption columns. The relevant works are covered in [3, 4, 10, 78] and later work has been reviewed in the papers of HEINE and HALÁSZ and HOLDINGHAUSEN and HALÁSZ also submitted to this symposium [79].

7. Chromatography of Polymer Solutions on Macromolecular Sieves

Geometrically and chemically homogeneous adsorbents with controllable specificity will undoubtedly find wide application in the molecular chromatography of liquid solutions as well. The only unsuitable adsorbents will be those modified by adsorption monolayers or thick layers of organic substances soluble in the corresponding solvents. For liquid molecular chromatography, the adsorbent surface should be resistant to the solvent of the mixture being chromatographed.

The interrelationship between gas-solid and liquidsolid chromatography was considered by SNYDER and FETT [80]. The development of liquid molecular chromatography on adsorbents requires new experimental and theoretical work on adsorption from solutions (see review [81] and [82-85]) and on the theory of molecular solutions in general. The development of the latter is greatly assisted by gas-chromatographic investigations (see, for instance, [55, 86]. The development of a molecular-statistical theory of adsorption from solutions and of molecular chromatography of solutions on adsorbents is particularly urgent.

One of the examples of application of adsorbents already developed for gas chromatography is chromatographic fractionation of polymer solutions on macroporous silica gels - macromolecular sieves. The applications of such macromolecular sieves are discussed in the communication of ELTEKOV and BOGACHOVA [87]. We will here limit ourselves to one example Fig. 21 exhibits the curves of pore volume distribution of four silica gels according to the sizes of their effective diameters, Fig. 20b shows frontal chromatograms of a polystyrene solution in CCl_4, which were obtained by BOGACHOVA, GAVRILOVA, KISELEV, KHOKHLOVA and ELTEKOV [88]. A comparison with the results of the determination of the viscosity of the solutions leaving the column shows that the first portions of these solutions in the case of wider-pored, macroporous silica gels are enriched in macromolecules of lower molecular weight, and in the case of finer-pored gels, of higher molecular weight. For sufficiently wide-pore samples, equilibrium is attained rapidly, while the adsorption value per unit surface of the adsorbent is practically independent of the pore sizes and coincides with the corresponding value per unit surface of a non-porous silica gel-aerosil [89]. In the case of finer-porous silica gel, only small macromolecules are capable of penetrating into the pores, this process proceeding very slowly. These differences in the kinetics and in the equilibrium adsorption value of macromolecules of different sizes on silica gels with pores of different sizes determine the chromatographic fractionation of the polymer (Table 1).

Table 1. Frontal chromatography of polystyrene solution of mean molecular weight $M \approx 300.000$ on macroporous silica gels with mean pore diameter d. Amount of polymer m of molecular weight M in filtrate batches of 5 cm^3.

No. of filtrate batch	$\alpha \approx 550$ Å		d ≈ 4300 Å	
	m, mg	$M \cdot 10^{-3}$	m, mg	$M \cdot 10^{-3}$
1	8	1.200	3	100
2	15	1.000	10	150
3	15	850	24	170
4	18	600	40	210
5	20	500	42	230
6	25	470	45	270
7	32	410	48	280
8	40	370	50	290

The fractionation of polystyrene was also successfully carried out on silica gels 90 and porous glasses [91] by elution chromatography.

8. Conclusion

The foregoing shows that the method of gas-solid chromatography has firmly established itself as an important method of analysis and preparative separation of gaseous and some solid substances. Basic methods have been outlined for developing and improving various adsorbents for gas-solid chromatography ensuring their high geometrical and chemical homogeneity and thermal stability with controllable specificity. These adsorbents are also necessary for molecular chromatography of liquid molecule and macromolecule solutions.

References

1. Kiselev, A. V. in: Gas Chromatography, Trudy I. Vsesoyuznoy Konferenzii (in Russian, U.S.S.R. Acad. Sci. Press, Moscow, 1960, p. 45.
2. Kiselev, A. V., and Shcherbakova, K. D.; in R. E. Kaiser and H. G. Struppe: Gas Chromatographie 1959, Material zum 2. Symposium über Gas Chromatographie in Böhlen, p. 198, Akademie-Verlag, Berlin, 1959 and in M. Schroter and K. Metzner: Gas Chromatographie 1961, p. 207, Akademie-Verlag, Berlin 1962.
3. Kiselev, A.V., in J. C. Giddings and R. A. Keller: Adv. Chromatography, Dekker, N. Y., vol. 4, p. 11 3, 1967.
4. Kiselev, A. V., and Yashin, Ya. I.: Gas-Adsorption Chromatography (in Plenum Press, New York 1969.
5. Kiselev, A. V.: Rev. générale du Caoutchouc, 41, 377 (1964).

6. Kiselev, A. V., in A. Goldup: Gas Chromatography 1964, p. 238, Inst. Petroleum, London, 1965.
7. Everett, D. H.; in A. Goldup: Gas Chromatography 1964, p.219, Inst. Petroleum, London, 1965.
8. Barrer, R. M.: J. Coll. Interface Sci., 21, 415 (1966).
9. Kiselev, A. V.: Zh. Fizich. Khimii, 41, 2470 (1967).
10. Halasz, I., and Heine, E.; in J. C. Giddings, R. A. Keller: Adv. Chromatography, Dekker, N. Y., 1967, vol. 4, p. 207.
11. Kiselev, A. V., and Yashin, Ya. I. in H. G. Struppe: Gas Chromatographie 1965, p. C 103, Akademie-Verlag Berlin 1965.
12. Schneider, W., Bruderreck, H., and Halasz, I.: Anal. Chem., 36, 1533 (1964).
13. Petov, G. M., and Shcherbakova, K. D., in A. B. Littlewood: Gas Chromatography 1966, p. 50, Inst. Petroleum, London 1967; A. V. Kiselev, G. M. Petov. and K. D. Shcherbakova, Zh. Fizich. Khimii, 41, 1418 (1967).
14. Kiselev, A. V., in A. B. Littlewood: Gas Chromatography 1966, p. 422, Inst. Petroleum, London, 1967.
15. Guiochon, G., and Vidal-Madjar, C.: Appendix in this proceedings, respectively in the preprints „Gas Chromatographie 1968", p. 295.
16. Shcherbakova, K. D.: Appendix in this proceedings, respectively in the preprints „Gas Chromatographie 1968", p. 533.
17. Kiselev, A. V., Migunova, I. A., Savinov, P. M., and Yashin, Ya. I.: Neftekhimiya, in press.
18. Kiselev, A. V., Migunova, I. A., and Yashin, Ya. I. in: Gasovaja khromatographiya, NIITEKHIM, issue 6, p. 84, Moscow, 1967.
19. Poskus, D. P.: Zh. Fizich. Khimii, 39, 2962 (1965): Gasovaya khromatografia. Trudy III Vsesoyuznoy konferenzii po gasovoy khromatografiy (Gas Chromatography, Proe. 3d all-union conference on gas-chromatography), p. 145, Izd. Dzerzhinskogo filiala OKBA, Dzerzhinsk 1966.
20. Avugul, N. N., Kiselev. A. V., Lygina, I. A., and Poshkus, D. P.: Izv. Akad. nauk SSSR. Otd. Khim. nauk, 1959, 1196.
21. Kiselev, A. V., and Poshkus, D. P.: Trans. Faraday Soc., 59, 428 (1963).
22. Poshkus, D. P.: Appendix in this proceedings, respectively in the preprints „Gas Chromatographie 1968", p. 443.
23. Kiselev, A. V., Poshkus, D. P., and Afreimovich, A. Ya.: Trans. Faraday Soc., in press.
24. Crowell, A. D.: J. Chem. Phys., 29, 446 (1958).
25. Kitaigorodsky, A.: J. Chim. Phys., 1966, 9.
26. Borisova, N. P., and Volkenshtein, M. V.: Zhurn. structurn. khimii. 2, 346 (1961)
27. Kiselev, A. V., Kuznezov, A. V., and Shcherbakova, K. D.: Zh. Fizich. Khimii, in press.
28. Yaris, R., and Sams, J. R.: J. Chem. Phys., 37, 571 (1962).
29. Olivier, J. P., and Ross, S.: Proc. Roy, Soc., A 265, 447 (1962).
30. Kiselev, A. V., and Poshkus, D. P.: Zh. Fizich. Khimii, in press.
31. Bell, R. P.: Trans. Faraday Soc., 38, 422 (1942).

32. Dixon, J. A., and Schiessler, R. W.: J. Am. Chem. Soc., 76, 2197 (1954); Ingold, C. K., Raisin, C. G., and Wilson, C. L.: J. Chem. Soc., 1936, 915.
33. Laurie, V. W., and Herschbach, D. R.: J. Chem. Phys., 37, 1687 (1962).
34. Constabaris, G., Sams, J. R., and Halsey, G. D.: J. Phys. Chem., 65, 367 (1961).
35. Sakodynsky, K. I., Khudyakov, V. L., and Yashin, Ya. I.: Kernenergie, in press.
36. Gvozdovich, T. N., Kiselev, A. V., and Yashin, Ya. I.: Neftekhimiya, 8 (1968).
37. Sakodynsky, K. I., and Sevastyanova, A. S. in: Gazovaja khromatographiya, issue 9, NIITEKHIM, Moscow.
38. Snyder, L. R.: J. Chromatography, 1968
39. Curthoys, G., and Elkington, P. A.: J. Phys. Chem., 71 (1967) 1477; Pierotti, R. A., and Smallwood, R. E.: J. Colloid Interface Sci., 22, 469 (1966).
40. Kiselev, A. V.: Discuss. Faraday Soc., 40, 205 (1965)
41. Kiselev, A. A., and Shcherbakova, K. D.; in M. Schröter und K. Metzner: Gas Chromatographie 1961, p. 241, Akademie-Verlag, Berlin, 1962.
42. Babkin, I. Yu., Kiselev, A. V., and Korolev, A. Ya.: Dokl. Akad. Nauk SSSR, 136, 373 (1961); I. Yu. Babkin and A. V. Kiselev: Zh Fizich. Khimii, 36, 2448 (1962).
43. Belyakova, L. D., Kiselev, A. V., Kovaleva, N. V., Rozanova, L. N., and Khopina, V. V.: Zh. Fizich. Khimii, 42, 177 (1968).
44. Kiselev, A. V., Kryukova, O. G., Kovaleva, N. V., and Khopina, V. V.: Kolloidn. zh., 30 (1968).
45. Broyer, P., and Lopatkin, A. A.: Appendix in this proceedings, respectively in the preprints „Gas Chromatographie 1968", p. 129.
46. Geiss, G., Versino, B., and Schlitt, H.: Chromatographia, 1/2, 9 (1968).
47. Hargrove, G. L., and Saeyer, D. T.: Anal. Chem., 40, 409 (1968).
48. Kiselev, A. V., Krylov, B. K., Nikitin, Yu, S., Tomnikov, A. P., and Yashin, Ya. I. in: Gazovaya khromatographiya, issue 6, p. 78, NIITEKHIM, Moscow 1967.
49. Bruk, A. I., Vyakhirev, D. A., Kiselev, A. V., Nikitin, Yu. S., and Oliforenko, N. M.: Neftekhimiya, 7, 145 (1967).
50. Kiselev, A., Migunova, I. A., and Yashin, Ya. I.: Neftekhimiya, 7 807 (1967).
51. Yachin, Ya. I., Zhdanov, S. P., and Kiselev, A. V. in H. P. Angélé and H. G. Struppe: Gas Chromatographie 1963, p. 402, Akademie Verlag, Berlin, 1963.
52. Bebris, N. K., Kiselev, A. V., and Nikitin, Yu. S.: Kolloidn. zh., 29, 326 (1967).
53. Bebris, N. K., Zayzeva, G. E., Kiselev, A. V., Nikitin, Yu. S., and Yashin, Ya. I.: Neftekhimiya 8, 3 (1968).
54. Nikitin, Yu. S.: Appendix in this proceedings, respectively in the preprints „Gas Chromatographie 1968", p. 391.
55. Ioganson, A. V., in A. B. Littlewood: Gas Chromatography 1966, Inst. Petroleum, London, 1967, 35.
56. Ioganson, A. V., and Rassadin, B. V.: Zh. Prikl. Spektroskopii, 6, 492 (1967); 7, 801 (196).
57. Davydov, V. Ya., Kiselev, A. V., and Lygin, V. Y.: Dokl. Akad. Nauk SSSR, 147 (1962) 131; G. A. Galkin, A. V. Kiselev and V. I. Lygin, Trans. Faraday Soc., 60 (1964) 431; V. Ya. Davydov, A. V. Kiselev and B. V. Kuznezov, Zh. Fizich. Khimii, 39, 2058 (1965).

58. Davydov, V. Ya., Kiselev, A. V., and Kuznezov, B. V.: Zh. Fizich. Khimii, 42 (1968).

59. Ioganssen, A. V.: Dokl. Akad. Nauk SSSR, 164, 610 (1965); A. V. Ioganssen, G. K. Kurkchi and O. V. Levina, Zh. Fizich. Khmii, 42 (1968).

60. Dzhigit, O. M., Kiselev, A. V., and Muttik, G. G.: Kolloidn. zh., 23, 504 (1961)

61. Day, R. E., Kiselev, A. V., and Kuznezov, B. V.: Trans. Faraday Soc., in print

62. Kalmanovsky, V. I., Kiselev, A. V., Sheshenina, G. G., and Yashin, Ya. I.; in: Gazovaja Khromatographiya issue 6, p. 45, NIITEKHIM, Moscow, 1967.

63. Dzhavadov, S. P., Kiselev, A. V., and Nikitin, Yu. S.: Kinetika i kataliz, 8, 238 (1967).

64. Vidal-Madjar, C., and Guiochon, G.: Separation Sci., 2, 155 (1967).

65. Scott, C.; in M. van Swaay: „Gas Chromatography 1962", p. 36, Butterworth, London, 1962.

66. Vidal-Madjar, C., and Guiochon, G.: Nature, 215 (1967).

67. Vidal-Madjar, C., and Guiochon, G.: Bull. Soc. Chim. France, 1967, 26.

68. Vinigradov, G. V., Titkova, A. V., Akshinskaya, N. V.; Bebris, N. K., Kiselev, A. V., and Nikitin, Yu. S.: Zh. Fizich. Khimii, 40, 84 (1966).

69. Hollis, O. L.: Anal. Chem., 38, 309 (1966); Hollis, O. L., and Hayes, W. V., in A. B. Littlewood: Gas Chromatography 1966, p. 57, Inst. Petroleum, London, 1967.

70. Gvozdovich, T. N., Kiselev, A. V., Migunova, I. A., and Yashin, Ya. I.: Chromatographia, in press.

71. Moseva, L. I., Kovaleva, M. P., and Sakodynsky, K. I., in: Gasovaya Khromatographiya issue 7, p. 126, NIITEKHIM, Moscow, 1967.
Moseva, L. I., Grosheva, L. A., and Sakodynsky, K. I., in: Gasovaya Khromatographiya, issue 9, NIITEKHIM, Moscow, 1967.

72. Yashin, Ya. I., Bebris, N. K., and Gvozdovich, T. N.: Appendix in this proceedings respectively in the preprints „Gas Chromatographie 1968", p. 33

73. Bebris, N. K., Kiselev, A. V., and Nikitin, Yu, S.: Neftekhimiya, in press.

74. Kiselev, A. V.; J. Colloid and Interface Sci., in press.

75. Brazhnikov, V. V., and Sakodynsky, K. I. in: Gasovaya Kromatographiya, issue 8, NIITEKHIM, Moscow, 1968.

76. Volkov, S. A., Zelvensky, V. Yu., and Sakodynsky, K. I.: Neftekhimiya, 7, (1967) 153; in: Gasovaya Kromatographiya, issue 9, NIITEKHIM, Moscow, 1968.

77. Kiselev, A. V., Nikitin, Yu. S., Petrova, R. S., Shcherbakova, K. D., and Yashin, Ya. I.: Analyt. Chem., 36, 1526 (1964).

78. Ettre, L. F.: Open-tubular columns in gas chromatography, Plenum Press, N. Y., 1965.

79. Heine, E., and Halász, I. in this proceedings and also: Appendix in this proceedings, respectively in the preprints „Gas Chromatographie 1968", I. Halász and F. Holdinghansen, p. 311.

80. Snyder, L. R., and Fett, E. R.: J. Chromatography, 18, 461 (1965).

81. Kipling, J. J.: Adsorption from Solutions of Non-electrolytes, Academic Press, London, N. Y., 1965.

82. Kiselev, A. V., and Pavlova, L. F.: Izv. Akad. Nauk SSSR, Otd. Khim. Nauk, 1965, 18.
83. Everett, D. H.: Trans. Faraday Soc., 60, 1803; (1964); 61, 2478 (1965).
84. Eltekov, Yu. A., and Kiselev, A. V.; in: T. E. Symes: „Molecular Sieves", Staples, Ketering, Northants, London 1968.
85. Everett, D. H., Kiselev, A. V., and Khopina, V. V.: Trans. Faraday Soc., in print.
86. Cruickshank, A. J. B., Grainey, B. W., and Young, C. L.: Trans. Faraday Soc. 64, 337 (1968).
87. Eltekov, Yu. A., and Bogacheva, E. K.: Appendix in this proceedings; respectively in the preprints „Gas Chromatographie 1968", p. 163.
88. Bogacheva, E. K., Gavrilova, T. B., Kiselev, A. V., Khokhlova, T. D., and Eltekov, Yu. A. in: Vysokomolekulyarnye soedineniya, Moscow, in press.
89. Bogacheva, E. K., Kiseley, A. V., Nikitin, Yu, S., and Eltekov, Yu. A.: Zh. Fizich. Khimii, 39, 1777 (1965); in: Vysokomolekulyarnye soedineniya, Moscow, in press.
90. Unger, K., Vogel, K., and Kulschutter, H. W.: Z. Naturforschung, 226, 8 (1967).
91. Cantow, M. J. R., and Johnson, J. F.: J. Appl. Polymer. Sci., 11, 1851 (1967).

On Problems of Polarity

The Method of Linear Combination of Interaction Forces (LCIF)

E. sz. KOVÁTS

Institut de Chimie Physique, Ecole Polytechnique de l'Université de Lausanne

The still today predominating dualistic division of the intermolecular forces of interaction is based upon two works:
In 1921 DEBYE [1] showed that between electrically noncharged „polar" molecules, i. e. between molecules with permanently inhomogeneous electric field, forces can be observed; nine years later EISENSCHNITZ & LONDON [2] found the explanation for the fact that also „apolar" particles may attract each other by momentary fluctuation of their electric fields. Organic molecules are kept together chiefly by dispersion forces (London forces), while the DEBYE portion mostly only causes a relatively small additional interaction, whose energy remains of the order of the thermal energy kT. The action of the dispersion forces does not depend on a specific alignment of the molecules, and, consequently, free rotation is caused in liquids by the thermal energy. MARTIN [3] recognized that in such a system an additivity of the two part forces may be expected: The dispersion share and the additional "polar" share combine to the force acting between the molecules. If this additional force share is defined as polarity mass, this simple model will permit of measuring the „polarity" of a molecule, if the two shares can be separately measured with success by a suitable measuring arrangement.

However, in the following years it was seen that the London and Debye forces alone are insufficient to explain the intermolecular forces of attraction. The extremely week chemical bonds between molecules caused by specific properties of the molecules must also be counted among the interaction forces, if the bond energy remains of the order of kT. These additional potentialities of interaction are, for the purpose of this paper, advantageous divided into the following three groups [4]:

1. Interaction resulting from the shifting of a suitable nucleus of the one molecule (donator) against a region rich in electrons (mostly a lonely pair of electrons) of the other molecule (acceptor).*)

*) If this nucleus is a proton, we speak of hydrogen bridge proton donator and proton acceptor.

2. Transfer of charge: Interaction resulting from the shifting of „one" electron of the one molecule (electron donator) against the other, electron-affine molecule (electron acceptor).

3. LEWIS salts: Overlapping of an orbit occupied with two electrons of the one molecule (base, donator) with an empty orbit of the acceptor (acid).

These interaction terms are generally included into the polarity of the molecules. However, it seems of little value to again consider terms as belonging to polarity, which are responsible for the interaction besides the latter and act independent of it, even if they always represent addition properties of polar molecules, The model stated above for division of the interaction forces, however, may be logically enlarged: The molecules (electrically neutral) are attracted by dispersion forces plus „polar" forces; however, in addition to both the interaction forces enumerated under 1. through 3. may occur.

In this manner the measurement of polarity is reduced to the measurement of a specific interaction term, one single measurement being mostly able only to ascertain the sum of all terms. This sum must then be subdivided by suitable methods, the model used being of decisive importance for the subdivision. In the framework of our model a molecule is characterized by the following series of numbers, which will be called property vector in total:

$$\begin{pmatrix} & & H_a & E_a & L_a \\ D & P & & & \\ & & H_d & E_d & L_d \end{pmatrix} \equiv \varepsilon \qquad (1)$$

here the individual figures are to indicate the ability of the molecule to participate in a specific interaction (D: dispersion force; P: polarity; H_a and H_d: acceptor and donator ability for hydrogen bridges*); E_a and E_d: ability of electron transfer; and L_a and L_d: LEWIS salt formation ability). If sticking to the dualistic division desbribed above and if correcting the measured interaction sum only by the term D, a linear combination of all other terms is obtained as an „additional measure of interaction". The majority of the measures of polarity introduced today represent such sums. The indicated one number characterizes the additional interaction ability of the molecule against another specific molecule, since in the measurement of interaction forces both the measuring instrument and the substance to be characterized represent combinations of molecules. Obviously the number measured also depends on the specific interaction ability of the substance used as a measuring instrument, i. e. it can hardly be expected that this number permits of calculating also the interaction force against other molecules.

It is common practice, notably for the interaction terms H, E and L, to imagine that the place of the interaction is concentrated in a specific part of the molecule. This idea may be generalized by assuming that all interactions, with exception of D, are caused by „adhesion zones" [5, 6], which are distributed over the surface of the molecule like

---

*) Throughout this paper the term „hydrogen bridge" is to serve as a collective term for the interaction desbribed under 1.

islands. In this connection an adhesion zone may act by polarity alone, but it may also possess several adhesive properties simultaneously such as a lonely pair of electrons, which will display a force P by its permanent polarity, but at the same time serve as a proton acceptor (H_a) and LEWIS base (L_d).

From this initial discussion follows that there are two ways for calculating the interaction forces:

a) On the basis of the chemical structure the adhesion zones of the two interacting molecules are ascertained. Under the assumption that the interaction of each adhesion zone of the one Molecule with each adhesion zone of the other molecule results in an independent share, and assuming additivity, the total force may be readily calculated. The increments for the interaction of a larger number of adhesion zones with the ether adhesion zone of polyethylene glycol tabulated by WEHRLI [5] permits of a suprisingly good prediction of the additional interaction energy. In this connection the effect of steric inhibition could be taken into account additively as a negative contribution. The drawback of this method is that every chemical property of a molecule must be defined as an adhesion zone (e. g. the adhesion zones „six-membered ring", „benzene", „alcohol" etc.), whereby the calculations become somewhat circumstantial, and that an illustrative characterization of average interaction ability of the molecules is difficult. The advantage of the method consists in that the contributions from the chemical structure can be calculated by the investigator himself.

b) From the measurements the seven coefficients of the property vector as mentioned under 1. are found for each substance. They characterize the specific abilities of a substance to display interaction. If the simplest linear set-up for the sum of all interactions is wanted, first of all the fundamental difference between D (dispersion forces) and P (permanent polarity) on the one hand and H (hydrogen bridge), E (electron transfer) and L (LEWIS salts) on the other hand must be noted: The intensity of D and P may be characterized each by one figure, for stating H, E and L, on the other hand, two figures N are required, the one indicating the donator ability and the other the acceptor ability of the molecule. An acceptor can interact only with a donator. Of course, the same molecule may have both properties, just think of the proton donator-acceptor ability of an alcohol.

The linear set-up of zero approximation for the interaction force between the molecules 1 and 2 is, hence, given by

$$E_{WW} = D_1D_2 + P_1P_2 + (H_{a,1}H_{d,2} + H_{d,1}H_{a,2}) + (E_{a,1}E_{d,2} + E_{d,1}E_{a,2}) + (L_{a,1}L_{d,2} + L_{d,1}L_{a,2}) \quad (2)$$

mixed terms such as $D_1P_2 + D_2P_1$ *) having been neglected. The effect of steric inhibi-

*) Corresponds to the interaction mentioned under 1. β on page 71.

tion is included in the figures D, P, H, E and L. An equivalent, however, somewhat clearer form of writing the equation (2) by means of determinants is:

$$E_{WW} = \begin{vmatrix} D_1 & O \\ O & D_2 \end{vmatrix} + \begin{vmatrix} P_1 & O \\ P & P_2 \end{vmatrix} + \begin{vmatrix} H_{a,1} & -H_{a,2} \\ H_{d,1} & H_{d,2} \end{vmatrix} + \begin{vmatrix} E_{a,1} & -E_{a,2} \\ E_{d,1} & E_{d,2} \end{vmatrix} + \begin{vmatrix} L_{a,1} & -L_{a,2} \\ L_{d,1} & L_{d,2} \end{vmatrix} \qquad (3)$$

When accepting this specific kind of multiplication of the individual terms as a definition of the multiplication of the property vectors, symbolization of (3) is possible as follows:

$$E_{WW} = \varepsilon_1 \; \varepsilon_2 \qquad (4)$$

The evaporation enthalpy of the pure substance may then be expressed by

$$H_{V,1} = \varepsilon_1^2$$

The method of ROHRSCHNEIDER [8] for calculating the interaction forces discussed at the end of this paper is a version of the method LCIF summarized in the equation (4), with the difference that the five coefficients of ROHRSCHNEIDER are purely emperical values that are difficult to interpret physically.

Interaction forces are measured in binary mixtures, one of the substances of the latter being arbitrarily defined as the measuring instrument. From this point of view the measurement may be imagined as follows:

A point molecule is passed as a measuring instrument over the surface of the molecule to be characterized. The molecule employed as the measuring instrument is attracted by dispersion forces. However, there may exist zones, over which additional forces are acting. If the additional force does not exceed the order of magnitude of kT at the temperature T of the experiment, the rotary motion of the measuring instrument is not essentially disturbed, i. e. the instrument is not bound to an „adhesion zone", which would correspond to chemical binding. This free motion caused by thermal energy will be indicated by the measuring instrument on sensing all positions over the molecule as a resulting force of attraction consisting in good approximation additively of the dispersion forces and of the additional attraction by the adhesion zones such as A and B. In this connection it has been assumed that above the adhesion zones the share of the dispersion forces does not change. The form of the surface potential, however, is a function of the properties of the molecule used as a measuring instrument. The latter may, in turn, also possess several adhesion zones such as C and D. Obviously the roles of the measuring molecule and the molecule to be characterized may be changed. The resulting additional forces between the two molecules are then given by summing up the effects of the adhesion zones A-C, A-D, B-C and B-D [1].

From this discussion follows the following model of the interactions in binary mixtures [5].

a) The substance and the measuring substance from „regular solutions" with ideal solution entropy.

b) In an „apolar solvent" only dispersion forces act on the dissolved molecules, and a non-polar molecule is retained in every solvent only by such forces.

c) In a polar solvent additional small forces act on the dissolved molecules, which are classified as follows:

1. α · Polarity: attraction caused by the permanent inhomogeneity of the electric fields of the interacting molecules.
   β · Attraction of the polarizable zines, in which due to the induction through the permanently inhomogeneous field of the one molecule a shifting of the charge has been caused.
2. „Hydrogen bridge".
3. „Charge transfer".
4. „Lewis salts".

d) Steric effects influencing (reducing) the forces mentioned under b) and c 1) through c 4).

If the forces mentioned under c 1) through c 4) are of the order of kT, the condition stated under a) is fulfilled.

As already mentioned before, the interaction force measured in solutions must first by divided into the dispersion share and the residue. For this purpose a non-polar solvent with the property vector

$$\varepsilon_{ap} = \begin{pmatrix} & & 0 & 0 & 0 \\ D & 0 & & & \\ & & 0 & 0 & 0 \end{pmatrix}$$

would be required. In a solvent like this the dispersion share could be measured directly. Strictly speaking, such non-polar substances do not exist. It is, therefore, necessary to select the „most non-polar" substance and to define it as a non-polar standard. Among the organic substances the selection of the paraffins as the „most-polar" substances has given excellent results in several instances [5, 8, 9, 10], and it is for this reason that the following definition is used also in this paper:

Straight-chain or branched-chain paraffins of the elementary formula C_nH_{2n+2} (and only such) are non-polar substances. - Hence non-polar solvents are pure paraffin hydrocarbons or their mixtures. It is assumed that the interaction force between a paraffin and any other substance Y is expressed by the equation (6):

$$E_{WW} = \varepsilon_{paraffin}\,\varepsilon_Y = D_{paraffin}\,D_Y$$

Below it is attempted to qualitatively understand the today existing measures of polarity in the light of the method of LCIF. These measures of polarity may be divided into three groups: or kinetic, spectroscopic and gas-chromatographic origin. The measures of polarity of the first two groups are discussed on the basis of examples selected at random; for a complete survey we refer to the paper quoted under [11] .

1. Kinetic Measures of Polarity

Reactions, where, on passing the transition state, ion charges are developed, are accelerated, if the chemical potential of the activated complex is lessened by interactions with the surrounding medium. The ratio of the velocity constants k/k_{ap} of a selected standard reaction in the non-polar medium (k_{ap}) and in the medium to be characterized (or, better in k/k_{ap} at constant temperature) serves as the measure for the additional interactions of the activated complex with the surrounding medium and has been suggestedt for characterizing the polarity of the latter. The activated complex has in most cases a relatively great dipole moment, so that it should be expected first that the chief interaction is effected by the term $P^{\#}\ P_{medium}$. However, it should be considered that anions are strongly solvatized by hydrogen bridges, likewise they may also interact as bases with the corresponding adhesion zones of the medium. For characterizing the polarity a standard reaction of the first order is selected that is to satisfy the following requirements:

a) the starting products (A) are non-polar; their interactions with the medium are caused by dispersion forces only:

$$E_{WW} = D_A D_{medium} \tag{7}$$

b) The activated complex is composed of the starting products, however, it is distinguished by a separation of charges. The share of the dispersion forces in the interaction of the complex with the medium, however, remains equal to the sum of the interaction forces of the starting products:

$$D^{\#}\ D_{medium} = \Sigma D_A\ D_{medium} \tag{8}$$

If these drastically simplified assumptions are fulfilled, the energy used as a measure of polarity is expressed by

$$P_{kin} = P^{\#}\ P_{med} + \begin{vmatrix} H_a^{\#} - H_{a,\,med} \\ H_d^{\#} \quad H_{d,med} \end{vmatrix} + \begin{vmatrix} E_a^{\#} - E_{a,med} \\ E_d^{\#} \quad E_{d,med} \end{vmatrix} + \begin{vmatrix} L_A^{\#} - L_{a,med} \\ L_d^{\#} \quad L_{d,med} \end{vmatrix} \tag{9}$$

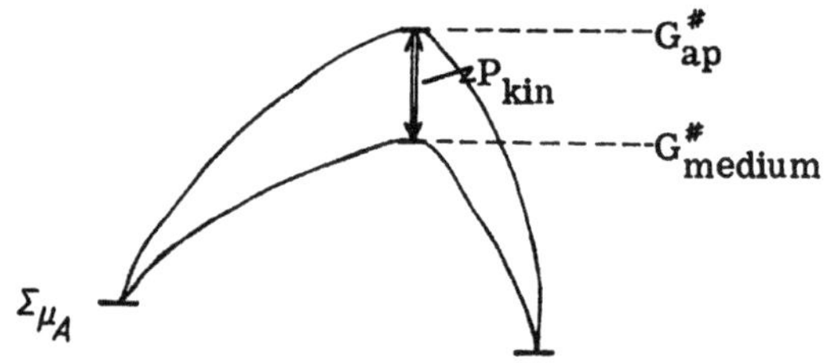

This value can be readily ascertained from experimental data, if assuming that the entropy of the reaction in the two mediums (non-polar medium and medium to be characterized) is about equal, so that

$$\Delta G^{\#}_{medium} - \Delta G^{\#}_{non-polar} \approx \Delta\Delta H \approx P_{kin} \qquad (10)$$

The following examples be mentioned in this connection:

The Y-values of GRÜNWALD and WINSTEIN [12]:

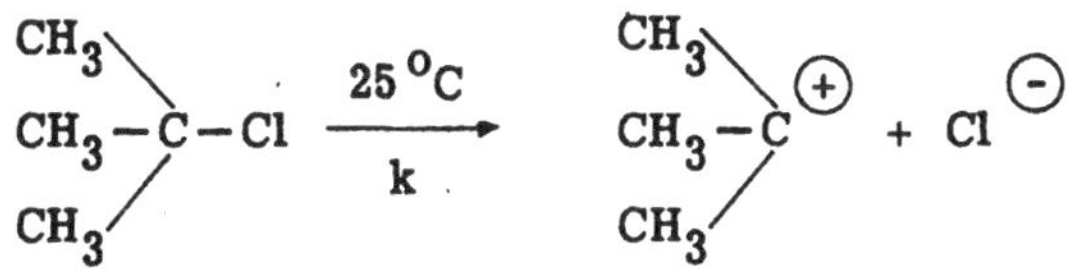

prod.

Standard solvent: ethanol-water (k_o)
Temperature: 25 °C

$$Y = \log K - \log k_o \qquad (11)$$

By the selection of the highly interactive standard medium the Y-values differ from the initially discussed interaction measure by a constant amount ($\Delta Y = \log k_{ap} - \log k_o$). This corresponds to a simple linear transformation.

The log k_{ion} values of SMITH, FAINBERG and WINSTEIN [13]:

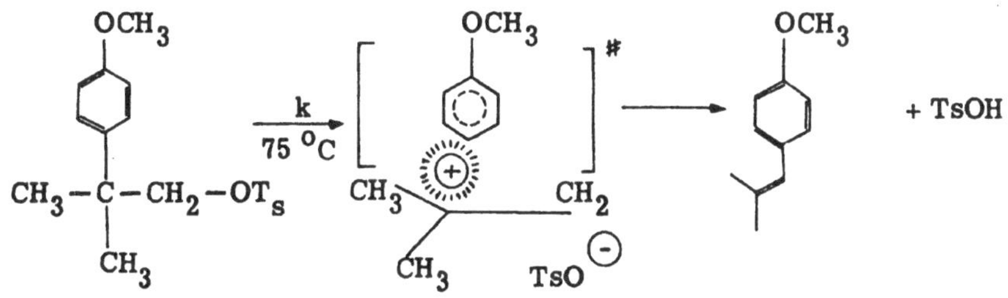

non-polar standard: not selected
temperature: 75 °C

The value of log k_{ion} as a measure of polarity is, if compared with P_{kin}, also shifted by a constant amount (log k_{ion} - log $k_{ion,ap}$), however, log $k_{ion,ap}$ being not known. Similar considerations hold for the X-values of GIELEN and NASIELSKI [14] (reaction of halogens with tetraalkyls of tin) and also for the Ω-values of BERON, HAMLET and MUELLER [15] (ration of endo and exo product in the DIELS-ALDER addition of cyclopentadiene on methyl acrylate).

2. Spectroscopic Measures of Polarity

Solvatrochromy will be displayed by such substances in a particularly high degree, where the basic state is distinguished by a high interaction ability, while the energized state is relatively non-polar. In an interacting solvent the chemical potention of such a substance in the basic state is lessened, while the chemical potention of the energized state is, in contrast, almost independent of the solvent chosen. Consequently the position of absorption associated with such a spectral transition is shifted towards the shorter wave-lengths in an interacting medium, if compared with the non-polar medium. This shifting is used as a measure of polarity. On the basis of similar considerations as for the kinetic measurements it can be easily shown that the spectroscopic polarity measured in this manner represents the additional interaction energy of the molecule in the basic state with the interacting medium. The corresponding energy may be

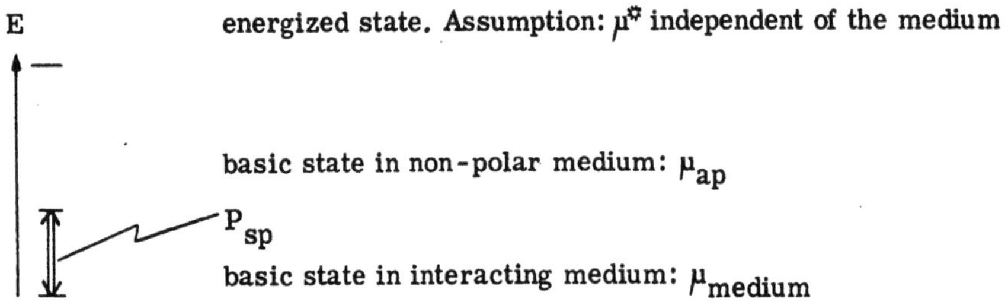

calculated from the shift of the position of absorption, its meaning within the framework of our model is:

$$P_{sp} = P^G P_{med} + \begin{vmatrix} H_a^G - H_{a,med} \\ \\ H_a^G \quad H_{d,med} \end{vmatrix} + \begin{vmatrix} E_a^G - E_{a,med} \\ \\ E_d^G \quad E_{d,med} \end{vmatrix} + \begin{vmatrix} L_a^G - L_{a,med} \\ \\ L_d^G \quad L_{d,med} \end{vmatrix} , \qquad (12)$$

the coeeficients $P^G \ldots L_d^G$ depending on the „polarity" of the basic state and represent constants for a given standard dyestuff. It is assumed that the dispersion forces between molecules in basic state/medium and molecules in energized state/medium are identical and that the energized state interacts with the medium by such forces only. The following examples be mentioned:

The Z-values of KOSOWER [15]:

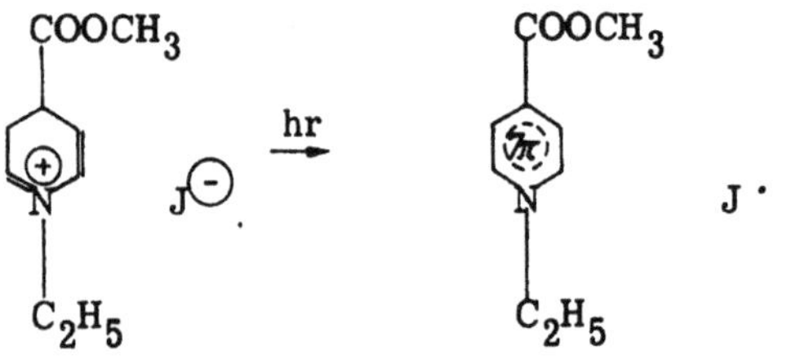

Z (iso-octane) = 60.1 may be used for conversion of the Z-values in order to obtain a measure corresponding to the interaction sum discussed above. Hence Z' (iso-octane) ≡ 0 and, e. g., Z' (CH_3CN) = 71.3-60.1 = 11.2 and Z' (H_2O) = 34.5.

The E_T-values of DIMROTH, REICHARDT, SIEPMANN and BOHLMANN [17]:

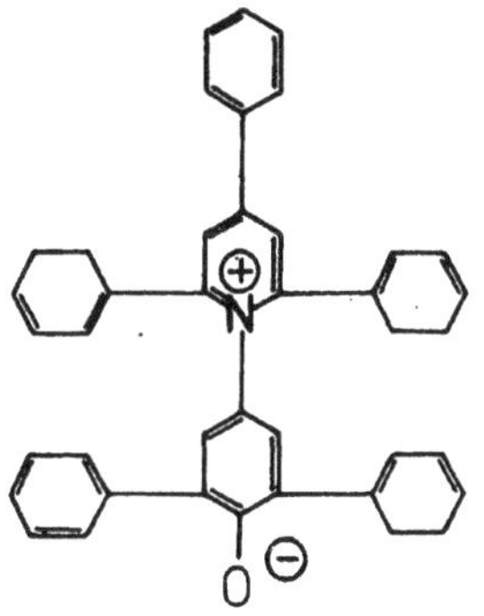

A standard solvent is not selected.
E_T (hexane) = 30.9; hence
E'_T (hexane) becomes = 0. As examples let us mention:
E'_T (pyridine) = 9.3, E'_T (CH_3CN) = 15.1 and E'_T (H_2O) = 3.2.2. kcal mol^{-1}

3. Gas-chromatographic Measure of Polarity

The polarity of CHOVIN and LEBBE [8]: If the logarithms of the retention volumes of the n-paraffins are plotted against the C number, this representation, especially for the higher members of the series shows on each stationary phase a remarkable linearity, i.e.

$$\ln \frac{V_N(P_{z+1})}{V_N(P_z)} = \text{const.} \tag{14}$$

From the relation (14) and the relation of the net retention volume V_N with the coefficient of HENRY for ideally dilute solutions h:

$$V_N = \frac{m_{SP} RT_K}{h} \tag{15}$$

it follows that

$$RT \ln \frac{h(P_z)}{h(P_{z+1})} = \Delta\mu(P_{z+1}) - \Delta\mu(P_z) = \Delta\mu(CH_2) \tag{16}$$

also represents a constant value. The free evaporation enthalpy of a „non-polar methylene group", thus calculated is the smaller that stronger are the mutual interactions between the molecules of the stationary phase. By these interactions between the molecules of a solvent an own structure is caused. This own structure is disturbed on dissolving a foreign molecule; it can be imagined that foreign molecule; it can be imagined that for this molecule a hole in the solvent must be prepared. The formation of this hole is accompanied by a loss of energy, which is the higher, the more specific the own structure of the solvent caused by interaction forces is. On the other hand energy is gained by the interaction between the foreign molecule and the stationary phase. If assuming, in turn that the „non-polar methylene group" in every stationary phase dispersion forces are in any case of the same order of magnitude, the reduction of $\Delta\mu(CH_2)$ is a measure for the destruction of the own structure of the solvent.

In order to express this „measure of destruction" in terms of numbers, squalane is used as the non-polar standard (the structure of squalane is effected by dispersion forces only, the same amount, hence, being recovered on on dissolving of the methylene group): polarity ≡ O. For establishing the scale ßß'-oxydipropionitrile was selected as the second solvent: polarity ≡ 1. Hence this polarity was calculated by simple interpolation:

$$P\ (\text{CHOVIN \& LEBBE}) = \frac{\Delta\mu\ (CH_2\ \text{in X}) - \Delta\mu\ (CH_2\ \text{in squalane})}{\Delta\mu\ (CH_2\ \text{in ODP} - \Delta\mu\ (CH_2\ \text{in squalane})} \tag{17}$$

In the framework of our model this polarity is, consequently, a measure for

$$P^2 + H_a H_d + E_a E_d + L_a E_d = \text{additional interaction} \tag{18}$$

of the solvent (stationary phase) with itself and declares: The stricter the internal structure of a solvent, the higher its polarity.

A similar measure of polarity is the „internal pressure" of a solvent introduced by HILDEBRAND and SCOTT [18, 19] and lately used by ROHRSCHNEIDER [20], which is defined as follows:

$$\delta^2 = \frac{\text{evaporation energy}}{\text{unit volume}}\ (\text{cal cm}^{-3}) \tag{19}$$

The latter might for volatile substances serve also as a „measure of destruction of the internal structure" and has a similar significance as the measure of polarity of CHOVIN and LEBBE; however, it is not corrected for the term D.

The polarity of ROHRSCHNEIDER I [9]: ROHRSCHNEIDER was the first to use as standard stationary phases the pair squalane (P = O)/ßß'-oxydipropionitrile (P = 100). The polarity of a stationary phase is found by determining the relative retention behaviour of the pair of substances butatiene/n-butadiene in the two standard phases an in the stationary phase that is to be characterized

$$RT \ln \frac{V_N\ (\text{butadiene})}{V_N\ (\text{butane})} = RT \ln \frac{h\ (\text{butane})}{h\ (\text{butadiene})} = \Delta\mu' \tag{19a}$$

Herefrom the „polarity" is again calculated by simple interpolation:

$$P\ (\text{ROHRSCHNEIDER I}) = \frac{\Delta\mu'\ (\text{in X}) - \Delta\mu'\ (\text{in squalane})}{\Delta\mu'\ (\text{in ODP}) - \Delta\mu'\ (\text{in squalane})} \cdot 100 \tag{20}$$

A correct analysis of this measure is difficult. If assuming that for butadiene the property vector takes the following form:

$$\varepsilon = \begin{pmatrix} & & H_a & 0 & 0 \\ D & P & & & \\ & & 0 & 0 & L_d \end{pmatrix}, \tag{21}$$

moreover, that butane with the stationary phase displays the same dispersion forces as butadiene, finally, that butane immediately distrubs the internal structure of the stationary phase as does butadiene, this polarity could have about the following meaning (butadiene = B):

$$P\ (\text{ROHRSCHNEIDER}) = P_B\, P_{\text{medium}} + H_{a,B} H_{d,\text{med}} + L_{D,B} L_{a,\text{med.}} \tag{22}$$

4. The measure of polarity $\Delta I^{\text{Cl, Br}}$ [21]:

The retention index of a compound Y on a stationary phase is identified as

$$I^X(Y) = 100\, \frac{\log V_N(Y) - \log V_N(P_z)}{\log V_N(_{z+1}) - \log V_N(P_z)} + 100\, z \tag{23}$$

Combining (23) with (14) and (15) gives

$$I^X = 100\, \frac{\Delta\mu(Y) - \Delta\mu(P_z)}{\Delta\mu(P_{z+1}) - \Delta\mu(P_z)} + 100\, z = 100\, \frac{\Delta\mu(Y) - \Delta\mu(P_z)}{\Delta\mu(\text{CH2})} + 100\, z, \tag{24}$$

i. e. the index scale is proportional to an energy scale: it measures the free evaporation enthalpy of the substance Y on the scale of the free evaporation enthalpy of the n-paraffins.

It was assumed that on a non-polar stationary phase only dispersion forces are active, so that:

$$I^{ap}(Y) = D_Y\, D_{ap} \tag{25}$$

On the basis of the arguments stated further above the „free evaporation enthalpy of a non-polar methylene group" is a measure for the dispersion share methylene group/stationary phase, reduced by the amount of energy that is required for destroying the own structure of the stationary phase. However, in the retention index this same value is used for calibrating the retention scale (divisor of the equation (24) or (23), respectively), and therefore the free evaporation enthalpy as measured by the index is corrected for this loss of energy. Because of the specific nature of the correction the share of the dispersion forces on every stationary phase is indicated by the index I^{ap}. Hence the share of the additional interactions may be readily calculated and is given by:

$$\Delta I^X(Y) = I^X(Y) - I^{ap}(Y)\,, \tag{26}$$

i. e. the ΔX-values are a measure for the additional interaction of the substance Y with the stationary phase X. In the framework of the method LCIF, therefore, the meaning of the ΔI-values is given by:

$$\Delta I^X(Y) = P_Y P_X + \begin{vmatrix} H_{a,X} - H_{a,Y} \\ H_{d,X} \quad H_{d,Y} \end{vmatrix} + \begin{vmatrix} E_{a,X} - E_{a,Y} \\ E_{d,X} \quad E_{d,Y} \end{vmatrix} + \begin{vmatrix} L_{a,X} - L_{a,Y} \\ L_{d,X} \quad L_{d,Y} \end{vmatrix} \tag{27}$$

If now a stationary phase could be found whose molecules can display addition interactions only by the inhomogeneity of their field, i.e. with the property vector

$$\varepsilon = \begin{pmatrix} & & 0 & 0 & 0 \\ D & P & & & \\ & & 0 & 0 & 0 \end{pmatrix},$$

The ΔI-value obtained on such a stationary phase would alone account for the polarity of the substance. Among the organic substances primary chlorides and bromides almost fulfil this condition and simultaneous have sufficient thermal stability, so that they may be used as stationary phase. ΔI-values on 1-chlorohexadecane have therefore been suggested as a measure for polarity:

$$\Delta I^{Cl} = I^{C_{16}H_{33}Cl}(Y) - I^{hexadecane} = P_{C_{16}H_{33}Cl}\, P_Y \qquad (28)$$

The correctness of the above assumption is confirmed by the fact that the ΔI-values measured on 1-bromohexadecane are in good agreement with the values for ΔI^{Cl}.

The polarity of the two phases would alone be caused by the dipolarity of the carbon-halogen bond. The dipole moments of the C-Cl and C-Br bonds are equal.

The polarity scale of ROHRSCHNEIDER II [10]:

As already shown, the ΔI-value is a measure for the sum of all additional interactions. ROHRSCHNEIDER has shown that these ΔI-values may be calculated in advance, if a stationary phase and a substance are characterized by five constants each. Then we have:

$$\Delta I = \frac{\Delta I(\text{benzene})}{100}\cdot b + \frac{\Delta I(\text{ethanol})}{100}\cdot \tilde{a} + \frac{\Delta I(\text{methyl ethyl ketone})}{100} + \frac{\Delta I(\text{nitromethane})}{100}\cdot n + \frac{\Delta I(\text{pyridine})}{100}\cdot p \qquad (29)$$

The substance coefficients b, $\tilde{a}$, m, n and p are calculated from a larger number of gas-chromatographic data by means of an equalization formula, while the coefficients of the stationary phase are ascertained as ΔI-values for five reference substances. For a stationary phase Z with still unknown properties, consequently, the retention indices of all substances (for which the substance coefficients are known) can be predicted, if the retention indices (i. e. the ΔI-values) of the five standard substances of this phase are known. Then is:

$$I^Z = I^{ap} + \Delta I^Z_{calc.} \qquad (30)$$

The surprising success of this method testifies to the correctness of the assumption of additivity of the intermolecular forces, but at the same time also to the functionability of the method of LCIF described in this paper. Obviously the ROHRSCHNEIDER coefficients $\frac{\Delta I}{100}$ as well as b ... p are, in turn, lucky linear combinations of the coefficients P, H, E and L. In the framework of our model, however, seven coefficients would be required for calculating the additional interactions; the ROHRSCHNEIDER method makes use of only five of them.

Table 1

Substance	P	H a	H d	E a	E d	L a	L d
Benzene	+	+	o	o	o	o	(+)
Ethanol	+	+	+	o	o	o	+
Methyl ethyl ketone	+	+	o	+	o	(+)	+
Nitromethane	+	+	(+)	+	o	(+)	+
Pyridine	+	+	o	+	o	o	+

Table 1 shows a qualitative comparison of the two series of coefficients, which is derived purely speculatively: For the interactions that are to be expected from the substance used as a standard a + - sign has been entered. Thus, e. g. it is expected for ethanol that it displays polar interactions (P), that it can form hydrogen bridges (H_a and H_d) both as a proton acceptor and donator, that it is a good Lewis base (L_d), however, that it will interact but negligibly through E_d, E_a and L_a. The table shows that in the ROHRSCHNEIDER system two of the LCIF factors are poorely occupied, namely E_d and L_a.

The figures of the ROHRSCHNEIDER system are at present the best characterization of the interaction ability of a substance, however, with the disadvantage that the physical meaning of the factors is difficult to interpret. But it may be easily imagined that by the selection of suitable stationary phases the individual LCIF factors are separable. A first attempt in this direction is the polarity measured by means of the ΔI^{Cl}-values.

In this place we express our gratitude to the Swiss National Fund for the Promotion of Scientific Research (Proj. No. 4986.2) for the financial assistance.

References

1. Debye, P.: Physik. Z. 22, 302 (1921).
2. Eisenschnitz, R., and London, R.: Z. phys. Chem. B11, 222 (1930); Trans. Faraday Soc. 33, 8 (1937).
3. Martin, A. J. P.: Biochemical Soc. Symposia vol. 3, page 4, Cambridge, 1949.
4. cf.: Hirschfelder, J. O., Curties, Ch. P., and Bird, R. B.: „Molecular theory of Gases and Liquids", J. Wiley & Sons, Inc., New York, 1954.
5. Wehrli, A.: Diss. Eidg. Techn. Hochschule, Zürich, 1959; Wehrli, A., and Kováts, E. sz., Helv. 42, 2709 (1959).
6. Schauer, H. K., and Bulrisch, R.: Z. Naturforsch. 10b, 683 (1955), 13b, 328 (1958); Naturwissenschaften 43, 34 (1956).
7. Pierotti, G. J., Deal, C. H., Derr, E. L., and Portar, P. E.: J. Am. Chem. Soc. 78, 2989 (1956).
8. Chavin, P., and Lebbe, J.: Avant-Edition des Conférences aux Journées d'Etude de Séparation Immédiate, p. 30, Paris, 1961.

9. Rohrschneider, L.: Z. Anal. Chemie 170, 256 (1959).
10. Rohrschneider, L.: J. Chromatog. 22, 6 (1966).
11. Reichardt, Chr.: Angew. Chem. 77, 30 (1965).
12. Grünwald, E., and Winstein, S.: J. Am. Chem. Soc. 70, 846 (1948).
13. Smith, S. G., Fainberg, A. H., and Winstein, S.: J. Am. Chem. Soc. 83, 618 (1961).
14. Gielen, M., and Nasielski, J.: Recueil Trav. Chim. Pays-Bas 82, 228 (1963).
15. Berson, A., Hamlet, Z., and Müller, W. A.: J. Am. Chem. Soc. 84, 297 (1962).
16. Kosower, E. M.: J. Am. Chem. Soc. 80, 3253 (1958).
17. Dimroth, K., Reichardt, C., Siepmann, T., and Bohlmann, F.: Liebigs Ann. Chem. 661, 1 (1963).
18. Hildebrand, J. H., Scoty, R. L.: „Solubility of Nonelectrylytes" 3 rd ed. Reinhold, New York, 1950.
19. Hildebrand, J. H., and Scott, R. L.: „Regular Solutions", Prentice-Hall, New Jersey, 1962.
20. Rohrschneider, L.: J. Gas Chromatogr. 6, 5 (1958).
21. Kováts, E. sz., and Weiß, P. B.: Ber. Bunsenges. 69, 813 (1965).

Specific Detectors

A. B. LITTLEWOOD

School of Chemistry, The University, Newcastle upon Tyne

1. Introduction

Specific detectors are those which respond sensitively only to certain specified groups of compounds, or only to compounds containing specified elements. There are three main classes of such detectors, (a) ionisation, (b) servomechanisms voulometers, (c) electro-chemical, and emphasis in this review is on groups (b) and (c), which have hitherto not received so much attention in the literature as (a), but which have great potentialities for future development and use.

2. Detector Response

In almost every practical gas-chromatographic detector, the „Response" consists of the variation of an electrical coutput as a result of the presence of a vapor in the carrier gas stream. If the detector is such that the total response for a peak is proportional to the total mass of vapor, and the constant of proportionality varies little with the chemical nature of the vapor, then the detector is said to be „General". If, instead, the constant of proportionality is such that it varies considerably with the chemical nature of the vapor, then the detector is said to be „Specific". In the latter case, it may be that, though the proportionality constant between response and mass varies with the vapor, the proportionality constant between response and some other property is invariant. For example, in the hydrogenation coulometer described below, the proportionality constant between response and number of double bonds is invariant, but that between response and mass depends on the degree of unsaturation.

In any discussion of detectors, „Response" needs more detailed definition. In some detectors, both general and specific, the electrical response, i. e., the voltage or current produced as a result of presence of the vapor, is proportional to the concentration of vapor in the detector. In others, the response is proportional to the rate of flow of vapor into the detector. The choice between these two general classes has been discussed in detail by HALÁSZ [1].

A result of the discussion is that, if detector sensitivity is defined in terms of the smallest detectable sample,concentration-dependent detectors have advantages at slow rates, while mass-flow-sensitive detectors have advantages at fast flow rates. In general, detectors are concentration-dependent if they measure some physical property of the column effluent within a cavity inside the detector, since the physical property is likely itself to depend on concentration. The mass-flow sensitive detectors include all those in which detection involves the consumption of the vapor, and in addition, any case in which the actual process of detection is concentration-dependent, but in which also the effluent entering the detector mixes with a scavenging stream of gas before detection; thus, various forms of ionisation detector employing non-homogeneous electric fields and scavenging gases are also approximately mass-flow-dependent. The mass-sensitivity of concentration-sensitive detectors can be increased by reducing their volume, but this cannot be done with mass-flow-sensitive detectors.

For most present practical purposes, the distinction made between concentration - sensitivity and mass-sensitivity is adequate, but in some cases the distinction can become blurred. For example, an electrochemical cell acting as a detector is fed by diffusion; if diffusion is fast compared to the flow of gas, then all the reactive vapor in the gas stream reacts, with the result that the cell is mass-sensitive. If, however, the gas flow is fast compared to the diffusion, the response depends on the rate at which molecules can diffuse to the electrodes, which depends on the concentration at the cell surface.

The distinctions between selective and general, and between mass-sensitive and concentration-sensitive, are exemplified for various detectors in Table 1. including generally familiar detectors and others described in this essay.

Table 1. Classification of Detectors

	Concentration Dependent	Mass-flow Dependent
	Catharometer	Micro-argon Detector
General	Cross-Section Detector	Combustion Voulometer
	Direct weighing mass detector	Flame Ionisation Detector
	Ultra-sonic detector	
	Large Argon Detector	
		Hydrogenation Coulometer
Specific	Electron Capture Detector	All liquid coulometric systems
	Electron Mobility Detector	Electrochemical cells in condition of complete reaction
	Most practical electrochemical cells	

For analytical purposes, systems in which a general detector is combined with a specific detector are particularly revealing, for with this combination, the general detector provides a quantitative analysis, while the specific detector provides some measure of qualitative identification. Specific detectors can be divided into three main classes:

a) Those working by ionisation,
b) Those working by some form of servomechanism,
c) Those which depend on some kind of specific electrochemical cell.

Of these three classes, the first is the most familiar, and includes, principally, the electron capture detector, the electron mobility detector, and modifications of the flame ionisation detector, especially that in which sodium is introduced into the flame. These forms of specific detector have been extensively reviewed in the recent literature, e. g. by SCHOMBURG in the last symposium of this series [2], and in articles by JANÁK et al. [3] by CREMER [4] and by GUDZINOWICZ [5]. In this article, therefore, the emphasis is on the other two classes, except in the summary, where all are discussed.

3. Servomechanisms using chemical reactions

One can broadly classify measuring instruments into two classes: a) open loop, and (b) closed loop. In the first type, the property to be measured in converted into a measurement in a series of stages; in the first stage, the property is converted by a transducer to a different property in one-relation to it, and in subsequent stages, further one-one transductions are performed, till the final transduction gives a number which is thus characteristic of the original property. For example, for a catharometer whose output is recorded on a strip-chart recorder, the set of properties in one-one relation is:

a) Concentration of vapor,
b) Thermal conductivity of gas containing vapor,
c) Temperature of hot element,
d) Resistance of hot element,
e) Out-of-balance EMF across bridge,
f) Movement of meter,
g) Numbered scale on meter or chart paper.

For a flame ionisation detector, the series is:

a) Mass flow of vapor,
b) Stages not fully understood, but involving at length, electron concentration in the flame,
c) Conductivity of flame,
d) Current in load resistor,
e) Input EMF to electrometer tube,
f) Numerous stages in amplifier,
g) Movement of meter,
h) Numbered scale.

In these, and all other open loop systems, the final reading is only proportional to the value of the property measured if every separate transduction is linear. In practice, it is found expensive, or sometimes indeed impossible to secure good enough linearity. For this reason, the universal tendency in recent years has been to substitute where possible closed loop measurement systems, for in these, non-linearity of any transduction inside the loop does not affect the linearity of the instrument.

In a closed loop measuring system, the property to be measured is compared in a transducer with an equivalent property which can so react with the property that at perfect equivalence ther is no output. With conditions other than perfect balance, ther is an output from the transducer, called the „error signal", which is applied via further transducers to the transducer producing the equivalent property, in such a way as to cause the error signal to converge to zero. One can see in recent years many instances where open loop systems have given way to closed loop systems. For example, the potentiometric recorder has displaced the recording moving coil meter for the majority of applications and most IR, UV, and visible spectrometers use a simple servo-method for recording, displacing methods in which spectra were directly amplified. It seems likely, therefore, that closed loop methods will become more and more used in GC.

Closed loops are used ubiquitously for control, and their theory is well worked out, mainly in the context of electrical engineering („Control Theory"). In this context they are usually called „Servomechanisms". There exist already several „Servo-detectors" for GC, and most of these are to some extent specific.

In present GC Servodetectors, the property of the vapor by which it is detected is usually a chemical property, and the equivalent property used to balance it is a chemical reagent which will react stoichiometrically with it. The general scheme, therefore, of the servo-system is as shown in Figure 1. Vapor from the column enters a reactor, and either in the reactor or beyond it, thereis a sensor which is sensitive to the vapor. An error signal in the sensor promotes via an amplifier production of reagent in a generator which is fed to the reactor till the sensor no longer gives an output.

The earliest detector ever used for GLG was a servodetector designed to titrate acid vapors automatically [6, 7, 8]. The apparatus was a straightforward automatic titrator, in which vapor from the column entered a titration cell; change in color of the indicator was noted by a photocell, which thus formed the sensor. The error signal operated a servo-amplifier which operated a motor-driven syringe buret filled with alkali. In this scheme the only transduction occurring outside the loop is that by which the volume of alkali delivered is converted to a linear measurement, and this, clearly, is not subject to any reasonable doubt. This apparatus has been occasionally used since [9, 10, 11, 12, 13] and a commercial automatic titrator has also been applied successfully to GC effluents [14].

More suitable for the small samples produced by GC are systems in which the reagent required to react with the vapors is produced coulometrically. Coulometric methods were first used extensively by LIBERTI et al [15, 16, 17, 18, 19] and by JURANEK [20] for GC effluents, and subsequently by many workers. Coulometric methods can be applied to any type of compound for which a suitable coulometric reaction exists. The standard method for liquid coulometry of GC effluents can be illustrated by that for the coulometric determination of halogens, using coulometrically generated silver [21, 22, 23, 24]. The titration cell (Figure 2) contains four electrodes; two for the generation of reagent, silver ions, and two for sensing whether or not there is still excess halide. The sensor electrodes consist of a silver/silver chloride electrode, together with a calomel electrode for reference. The generator electrodes consist of a silver cathode and a silver-plated platinum anode. The circuit is as shown in Figure 2 [24]. The output of the sensor cell is amplified, and applied to a servo motor which works a potentiometer which supplies current to the generator electrodes. When the potential of the sensor cell falls to a value set by the backing off circuit A, the generation of current ceases. The current passing is monitored on a strip chart recorder, which thus produces peaks.

In the majority of applications of solution coulometry to GC effluents, the effluent passes through a reactor where the vapors are converted into compounds which the coulometer can handle. Thus, compounds containing only C, H, and O are converted to carbon dioxide and it is this which is coulometrically estimated. The method is at its best for halogens and for sulfur compounds, which are estimated as halogen, free halogens, or sulfur dioxide. For analysis of halogenates, a silver cell is used [22, 23, 24]. Sulfur dioxide from sulfar compounds is determined by coulometrically generated iodine [25, 26], or via silver [28] or bromine [27]. For these substances, redox electrodes can be used as the sensors, as described above. If carbon dioxide is titrated directly as a weak acid, it appears that the small pH changes resulting from solution of carbon dioxide provide insufficient error signal to produce a viable servo-system. Carbon dioxide an, however, be titrated satisfactorily against coulometrically generated barium hydroxide, the barium carbonate being precipitated from solution. Coulometric titration of volatile strong acids or strong bases, e. g., amines, or even ammonia, is perfectly possible using a potentiometric sensor [29, 30].

The servor-coulometer has been made the basis of a commercial instrument specially for GC purposes.

The author and his colleagues have recently developed a servo-detector working entirely in the gas phase [31] callert the Reaction Coulometer; two forms of the system are shown in Figures 3 and 4. In the first system, carrier gas from the column joins an auxilliary gas stream which picks up a steady concentration of oxygen produced in the electrolytic generator by passage of a steady current. The two streams enter the reactor, which is a heated platinum catalyst, and thence pass to a cell which is sensitive to oxygen. The output of this cell is fed to an amplifier which supplies the current to the oxygen generator. In the absence of a vapor, a stable control loop is set up and the constant oxygen generation provides a steady base-line. When a vapor appears, however, it bums in the oxygen, causing a fall in the cell output, and thus an increase in the oxygen generation. The vapor acts, in fact, as a perturbation to the control, and the system, if properly designed, acts as a perfect control in completely removing the perturbation. The extra oxygen required to do this is equivalent exactly to the perturbing vapor, and by recording this, a chromatographic peak is obtained.

The above scheme is used as a general purpose detector, since any organic compound is combustible. The detector becomes selective, however, by replacing the generator by a unit which will generate other reagents, by replacing the cell by a unit which is sensitive to other reagents, and by replacing the reactor by a unit which will secure other reactions. The simplest is to reverse the generator, so that it produces hydrogen, to use a hydrogen electrode in the cell, and a hydrogenation catalyst in the reactor [32]. In this way, the same apparatus becomes a sensitive quantitative detector for hydrogenatable material. An example of the analysis of a mixture first by the general combustion coulometer, and second by the specific hydrogenation coulometer, is shown in Figure 5. The alkenes solely are measured by the hydrogenation system.

Coulometric methods have several advantages for practical GC:

1. Sensitivity. Coulometric methods are at present quite sensitive, and potentially can be made more so. At least 1 Faraday of electricity (96.540 coulombs) is transferred per mole of vapor, and quantities of electricity of the order of microcoulombs are easily measurable. Thus, quantities of the order of 10^{-11} mole may be measured. For an average vapor of molecular weight of the order of 100, this gives 10^{-9} gm. Thus, coulometric detectors are at present almost, but not quite as sensitive as the Argon or Flame Ionisation Detectors. The performance of an electrical detector can also be expressed as an „Ionisation Efficiency", i.e., the number of electron transfers obtained per molecule in the sample being detected. In coulometric detectors, the ionisation efficiency is a small integer, since every molecule reacts. In an ionisation detector, however, only a small proportion of the molecules react, so that ionisation efficiencies are small numbers, e.g., 10^{-4}. At present, the noise levels of ionisation detectors are smaller than those for coulometric detectors by about 1 power of ten more than the ratio of the ionisation efficiencies, thus making them slightly more sensitive overall. There seems no theoretical reason, however, Why the noise level of coulometric detectors should not be reduced below its present level of the order of 10^{-6} amp.

2. Absolute response. Since the response per molecule is an integral number of electrons, the molar response is exactly calculable, given the stoichiometry of the reaction. Since the number of coulombs in a peak is measurable absolutely, coulometric detectors are absolute, and not merely relative.

3. Linearity. The only stages which necessarily must be linear are the stoichiometric reaction, the production of reagent, and the read-out. Of these, errors in read-out linearity are easily avoided. Though there are regimens in which the stoichiometry of the reaction can err, e. g., formation of SO_3 rather than SO_2 during coulometric determination of sulfur [39], such cases are few and easily recognised. It is important to realise in a servo-system, the linearity of the sensor is completely irrelevant. In use, the response of the sensor is held nearly constant by the operation of the servo-system, and the only variations at the sensor are the small variations giving the error signal. So long as the sensor is a monotonic transducer in the immediate neighbourhood of its

range of operation, so that the sign of the error signal correctly represents the direction of departure from the steady state, the linearity is unimportant.

An important general advantage of the use of a servomechanism is its ability to correct for imperfections in components of the loop. For example, when appearance of a vapor begins to perturb the system, the amplifier can be designed to respond to the first or any higher partial with respect to time of the error signal, thus not having to wait for an equilibrium response from the sensor. The design of amplifiers in servo-loops thus to accelerate the control is well-established, and enables servodetectors to respond quantitatively faster than the individual components in the loop. As an example, Figure 6 shows some rapid peaks obtained with a gas-phase combustion coulometer occupying only a couple of seconds, using a cell the time constant of which is considerably longer than the total time occupied by the peak.

4. Electrochemical detectors

A general rule is that, if one can succeed in converting a quantity of material into an equivalent quantity of electricity, one will have a very sensitive measure of one's material, since electricity is so much more easily and sensitively measured than mass. Electrochemical cells which convert matter into electricity ought, therefore, to have great application in GC. Traditionally, electro-chemical cells have been studied with liquid electrolytes and reagents, and solid electrodes. Recently, however, there has been growing interest in cells using gaseous reagents. The prime example of such a cell is the Fuel Cell.

4.1. Hydrogen and Oxygen Cells

Cells for the measurement of hydrogen or oxygen in gas streams have been familiar for many years. The most familiar cell for oxygen is that of HERSCH [33, 34]. Basically the cell is:

O_2/Ag/KOH/Ni

in which there are electrodes of silver and of nickel in contact with an electrolyte of potassium hydroxide. The oxygen reacts at the silver electrode, and in practice, porous silver electrode, and in practice, porous silver is used in order to secure a large area of contact with the gas. In order to make the cell „dry", the electrolyte may be contained in some porous medium, e. g., porous polyvinyl chloride, or any other porous material not attacked by oxygen or alkali. In the absence of oxygen, this cell gives no current, but in the presence of oxygen a current is passed. This cell has been used directly as a GC detector for oxygen [35], and can be used as the sensing cell in the coulometric system described above.

Any cell which contains a hydrogen electrode in any form is, of course, sensitive to hydrogen, and hydrogen electrodes are thoroughly familiar in classical electrochemistry. In most hydrogen cells, the sensitive electrode is of platinum blacked platinum, which can work in either an acid or an alkaline medium, but preferably acid. For the Hydrogenation Coulometer, the author and his colleagues used an electrode of platinum blacked platinum operating in 2 N sulfuric acid. Such a cell gives zero current in absence of hydrogen, but passes a current in its presence.

Cells such as the Hersch cell or those with hydrogen electrodes are normally used as current transducers; in such a case, the current obtained may, in favorable circumstances, be a linear funktion of the concentration of gas to which they are sensitive, though they are better used in servomechanisms, where their dubious linearity does no harm.

The Hersch cell suffers in some applications from the fact that it employs an aqueous electrolyte; thus it cannot be used at elevated temperatures. Cells sensitive to oxygen or hydrogen, which operate at high temperatures, can be made using solid electrolytes consisting of defect lattices [36, 37, 38, 39,40]. For example, ceramics can be made in which an oxide of a metal of large valency, e. g. zirconium, is doped with a substantial proportion of the oxide of a metal of lower valency e. g. magnesium. In favorable circumstances, the lattice remains that pertaining to the higher valent oxide, and the lattice thus contains oxygen ion vacancies. At high temperatures, these vacancies become mobile. Thus, a cell may be made in which there are two electrodes one each in contact with the electrolyte. The electrodes may be of similar material, e. g., platinum. At temperatures of 600 °C or more, oxygen can equilibrate with the defect lattice, by the reaction:

$$O_2 \leftrightharpoons 2O^{--} = 4\,e$$

with the result that an electrode in simultaneous contact with the lattice and with the oxygen acquires a negative potential relative to the other electrode. This potential can be measured by a high-impedance meter. Alternatively, of the two electrodes are connected though a low impedance meter, an appreciable current flows, with resultant diffusion through the electrolyte of the oxygen ions via the lattice vacancies. Thus, oxygen diffuses from one electrode to the other through the lattice; since it diffuses in ionic form, there is necessarily a flow of electrons through the external current in order to preserve electrical neutrality. The diffusion of oxygen is therefore accompanied by an accurately stoichiometric flow of current.

4.2. Cells Specific to other Substances

Another type of cell also using a non-stoichiometric solid electrolyte, which responds specifically to halogens, halogen acids, or sulfur dioxide, has been described by BECHTOLD [41] and by CREMER [4].This uses the cell:

$$Pt/Ag/AgI/Ag_2S/Pt$$

In this, metallic silver with platinum connection forms one electrode, silver iodide is the electrolyte, and silver sulfide forms the other electrode, with platinum connection. The cell is operated at a temperature from about 200° to 450°, and at these temperatures, silver iodide is an ionic conductor of low resistance. The silver sulfide electrode is in contact with the gas to be measured, and its operation depends on the fact that silver sulfide forms a continuous phase in the range $Ag_{2.000}S$ to $Ag_{2.0025}S$ [42, 43]. At the silverich end of the range, the silver sulfide is in equilibrium with silver, and therefore the cell shows no potential, but at the sulfur-rich end of the range, the sulfide is in equilibrium with sulfur, so that the cell shows a potential corresponding to the free energy of formation of silver sulfide from its elements. Furthermore, silver can reduce sulfor dioxide to silver sulfide; this is the reaction by which silver tarnishes in air. Thus, if the cell is initially in the silver-rich form, the presence of sulfur dioxide induces a potential, which can be measured. In the same way, a halogen removes silver, forming silver halide, and this also increases the sulfur/silver ratio in the silver sulfide. As with the oxygen defect lattice described above, this can be measured either as a potential, or as a current.

In both of the cells described above, the potential was caused by the interaction of the vapor of interest either with the electrolyte or with the electrode, thus changing the concentration of an ionic component. In both cases, interaction of the vapor with the solid occurred by an ionic reaction, and, any current given by the cell measured the extent to which the reaction occurred. It is possible also to contrive cells which similar mechanisms work within aqueous electrolytes. Such cells are commonplace in classical electrochemistry, and they have been studied extensively for use in gas chromatography by BERTON [44, 45, 46, 47].

A suitable cell for GC use as described by Berton has two platinum electrodes, between which is the aqueous electrolyte. Cne electrode is composed of gauze, or other such finely divided form, and the gas to be studied is bubbled over it. In the steady state, the cell shows no potential, but as soon as the gas bubbling over the electrode contains some species which can react with the electrolyte, the cell acquires a potential. By using different electrolytes, different reactions can be induced, so that different classes of vapor can be caused to react. Table 2 lists chemical types for which specific electrolytes are available. Though many systems use aqueous electrolytes, there is no reason why other ionising solvent should not be used, and, indeed, Berton describes some non-aqueous electrolytes, as seen from the Table.

Table 2. Specific Electrolytes [45]

Vapor to be Detected	Electrolyte
Acids, Anhydrides, or Bases	Any saline solution, e.g., Sodium or Ammonium Nitrate
Weak acids, e. g., carbon dioxide	Sodium hypochloride; specific response to CO_2 blocked by adding sodium bicarbonate.
Oxidising vapors	Potassium iodide + Ferric Chloride
Nitrobenzene	Sodium sulfate
Hydrogen sulfide, formaldehyde	Sodium hydroxid
Hydrazine	Formamide
Hydrogen chloride	Concentrated nitric acid
Hydrogen chloride or hydrogen fluoride	Copper sulfate (Current reverse for HF)

Vapor to be Detected	Electrolyte
Oxidisable non-electrolytes, e.g., EtOH, CH_3CHO, esters, Et_2O, etc.	Nitrochromic mixture
Aromatics and unsaturates	Iodine pentoxide in sulfuric acid
Anilines and phenols	Potassium perchlorate in sulfuric acid.

4.3. General Considerations

In electrochemical cells used for GC, the vapor causes a change in chemical potential of an electrochemically reactive species at one electrode. This change in chemical potential by measuring the change in potential of the cell. In this case, the cell is connected to a high impedance measurer, e. g., a potentiometer, which can conveniently be a servo-potentiometer, or by a high-impedance voltmeter, Alternatively, the current of the cell can be monitored by a low-impedance current monitor between the electrodes. These two methods are fundamentally different.

Current monitoring is necessary in the few cases where a cell is used completely to remove all of the species to be measured by means of a stoichiometric reaction, as for example, in the Hersch cell used quantitatively. In this case, the total quantity dielectricity passed measures the amount of sample. In the majority of cases, however, either the electrode reaction is slow, or the rate of diffusion to the reaction zone is too slow for this to be practical. In this case, the concentration of reagent at the reaction zone is virtually zero, and the rate of supply of reagent to the reaction zone is determined solely by diffusion, the rate of which depends on the concentration of reagent in the bulk of the gas. In these circumstances, the current of the cell is directly proportional to concentration. This regime is probably the easiest to attain. A case where the cell response is proportional to concentration is given by the author and his colleagues [31].

If the potential is monitored, so substantial quantity of reagent is consumed at the electrodes, so that the response is not determined by diffusion. During a measurement of a potential, there is a set of equilibria between the potential, and every species involved in the electrochemical reaction, which set is defined by the relation that their chemical potentials are all equal and equal to the electrical potential.

The chemical potential of the vapor is related to its concentration (strictly, activity) by the equation:

$$\mu = \mu_0 + RT \ln c$$

with the result that, in potential measurements, the response is proportional to the logarithm of the concentration. This has the drawback that potential measurements are essentially non-linear. It has the advantage, however that cells can be made to register extremely small concentrations; for example, the change in potential occurring in a cell for a change in concentration of tenfold between 1 part per thousand of electrochemically reactive vapor in the gas and 1 part per hundred is exactly the same as the change in potential occurring for a change between one part per million and ten parts per million.

The change in potential occurring as a change in concentration depends on the ratio of the two concentrations, and not on the magnitude of either. Reduction in the concentrations affects only the internal impedance of the cell which rises with decreasing concentration. The use of the oxygen cell with a defect lattice thus to measure minute concentrations of oxygen has been described [36].

5. Summary

5.1. Qualitative Performance

There are two general ways in which the various selective detectors perform their selection; first, there are those which respond selectively to perticular elements in the molecule, and second there are those which respond to particular molecular groupings, or even to particular molecules, on the whole, the selective ionisation detectors, such as the electron capture detector, the electron mobility detector, or the sodium-modified flame ionisation detector, belong to the first class, and respond selectively to elements. The coulometric detectors respond to particular elements of their use is preceded by some sort of destructive reactor, e.g., a combustion tube, but otherwise, they respond to particular chemical classes. Similarly, the various types of electrochemical cells respond either to the simple products of a degradative reaction, or to particular chemical types.

In addition to the three general classes of selective detectors which were developed primarily for gas chromatography, there has been great application to GC of analytical instruments developed for qualitative analysis and qualitative identification, such as the various classes of spectrometer, or the polarograph. These cannot truly be classed as „Chromatographic Detectors", but their use in connection with GC is common and well known [48, 49].

Table 3. Summary of selective Detectors.

Detector name	Selective for:	Ref.
a) Ionisation		
Electron Capture Detector	Aromatics, Halogen compounds	[50]
Electron Mobility Detector	Permanent gases (But has universal response also)	[51]
Sodium-Modified Flame Ionisation Detector	Halogens, phosphorus	[52]
Flame-Photometric Detector	Halogens, phosphorus, sulfur, metals	[53]
Beilstein Detector	Halogens	[54]
b) Coulometric		
Liquid Coulometer	Especially for sulfur, phosphorus; also any other suitably reactive grouping	

Hydrogenation Coulometer	Any hydrogenatable material	
c) Electrochemical		
Hersch Cell	Oxygen	
Hydrogen cell	Hydrogen	
Silver Sulphide Cell	Sulfur, halogens	
Cells with reactive electrolyte	(See Table 2)	
d) Analytical Instruments		
Mass-Spectrometer	Any compound	[48]
I. R. Spectrometer	Any compound	[49]
U. V. Spectrometer	Any compound	[55]
Polarograph		[56]
e) Other		
Radiation Counters	Radioactive substances	[57]
Surface Potential Detector	Polar and, apparently, all substances that smell	[58]
Semiconductive Film Detector	Electron donor molecules	[59]

In Table 3 is given a list of selective detectors or analytical instruments together with the elements or substances to which they respond. In the case of those not discussed in this essay, one selected reference is given; further references may be obtained from it, or from other reviews [1, 2, 3, 4, 5, 48, 49].

5.2. Sensitivity and Linearity

In the search during the last fifteen years for devices to use as GC detectors, workers have rarely found great difficulty in producing devices which can detect small quantities or small concentrations of vapor. Thus, all the devices listed in classes a), b) or c) of Table 3 can detect quantities of the order of 10^{-9} to 10^{-10} mole of vapor. Of the spectrometers, the massspectrometer is the most sensitive, and if used non-dispersively, is as sensitive as an ionisation detector. The infrared spectrometer, however, is considerably less sensitive, and normally quantities of the order of 10^{-6} mole are required.

Whereas sensitivity is easily achieved, linearity is much harder to attain. Among the ionisation detectors of Table 3, none is in principle linear, and some are in principle definitely non-linear; for example, the relation between concentration and current (which is monotonically negative) for the electron capture detector is:

$$I = I_0 e^{-kC}$$

which I is the current for a concentration C, and I_0 pertains to C = O. As with any other essentially non-linear device, approximate linearity may be achieved over a range small anough not to show curvature, and if this range happens to be that which includes the origin, then the detector may be described as „linear". In a useful instrument, the dynamic range which is „linear" in this sense may include concentrations spanning several power of ten between the noise and the concentration at which nonlinearity becomes experimentally manifest. With the electron capture detector, this dynamic range is small, e.g., 10^3, though with the flame detector may well be larger. The result of this basic nonlinearity of many of the specific detectors is that they should only be used for quantitative analysis with caution. In most applications, any quantitative analysis is performed by the accompanying general detector; usually, the common general detectors have a greater range of effective linearity.

The above restriction on linearity do not, of course, apply to the coulometric servo-detectors. Of the spectrophotometers, the mass-spectrometer has a large range of linearity, but the response of absorption spectrometers is, of course, related with concentration non-linearly by BEER's law.

References

1. Halász, I.: Anal. Chem. 36, 1428 (1964).
2. Schomburg, G.: „Gas Chromotographie 1968", Proceedings of the Fifth Symposium, Akademie-Verlag Berlin, 1966. Ed. Struppe, H. G., p. 281.
3. Svojanovsky, V., Krejci, M., Tesarik, K., and Janák, J.: Chromatographic Reviews 8, 90 (1966).
4. Cremer, E.: J. of Gas Chromatog. 5, 329 (1967).
5. Gudzinowicz, B. J.: „The Practice of Gas Chromatography", Ed. Ziatkis, A. and Ettre, L. S., Interscience, 1967, p. 239.
6. Insull, W., and James, A. T.: „Advances in Gas Chromatography", A. C. S. Symposium, New York, 1957, p. D 111.
7. James, A. T., and Martin, A. J. P.: Analyst, 77, 915 (1952).
8. James, A. T., and Martin, A. J. P.: Biochem. J. 50, 679 (1952).
9. Heines, V., JuHasz, R., O'Leary, M. A., and Schramm, G.: Trans. Kentucky Acad. Sci. 18, 1 (1957).
10. Hughes, R. B.: J. Sci. Food and Agr. 11, 47 (1969).
11. Kamer, J. H., Gerritsma, K. W., and Wansink, E. J.: Biochem. J. 61, 174 (1955).
12. McInnes, A. G.: „Vapour Phase Chromatography", Proceedings of the First International Symposium of the GC Discussion Group, London, 1956, Ed. Desty, D. H., Butterworths 1957, p. 304.
13. Tilley, J. M. A., Canaway, R. J., and Terry, R. A.: Analst 89, 363 (1964).
14. Aunstrup, K., and Djiertoft, R.: Brugmesteren 17, 137 (1960).
15. Liberti: Anal. Chim. Acta 17, 247 (1957).
16. Liberti, A., and Cartoni, G. P.: Atti accad. natz. Lincei. Rend. Classe sci. fis. mat. e. nat. 20, 787 (1956).
17. Liberti, A., and Cartoni, G. P.: „Gas Cnromatography 1958". Proceedings of the Second International Symposium of the GC Discussion Group, Amsterdam, 1958, Ed. Desty, D. H., Butterworths, London 1958, p. 321

18. Liberti, A., Cartoni, G. P., and Palotta, U.: Il Latte 30, 581 (1956).
19. Liberti, A., and Cartoni, G. P.: Chim. e Ind. (Milano), 39, 821 (1957).
20. Juranek, J.: Tech. Dig. 11, 14 (1960).
21. Cipriano, C., and Hayes, W. J.: J. Agr. Food Chem. 19, 366 (1962).
22. Coulson, D. M., Cavanagh, L. A., DeVries, J. E., and Wlater, B.: J. Agr. Food Chem. 6, 399 (1960).
23. Hayes, W. J., Dale, W. E., and LeBreton, R.: Nature 199, 1189 (1963).
24. Coulson, D. M., and Cavanagh, L. A.: Anal. Chem. 32, 1245 (1960).
25. Martin, R. L., and Grant, J. A.: Anal. Chem. 37, 644 (1965).
26. Bosin, W. A.: Anal. Chem. 35, 833 (1963).
27. Klaas, P. J.: Anal. Chem. 33, 1851 (1961).
28. Fredericks, E. M., and Harlow, G. A.: Anal. Chem. 36, 263 (1964).
29. Martin, R. L.: Anal. Chem. 38, 1209 (1966).
30. Thielemann, H., Behrens, V., and Leibnitz, E.: Chem. Tech. (Berlin) 14, 162 (1962).
31. Burton, G., Littlewood, A. B., and Wiseman, W. A.: „Gas Chromatography 1966", Proceedings of the Sixth International Symposium of the GC Discussion Group, Littlewood, A. B., Ed., Institute of Petroleum, London 1967, p. 193.
32. Littlewood, A. B., and Wiseman, W. A.: J. Gas Chromatog. 5, 334 (1967).
33. Hersch, P.: Brit. Pat. 880, 965 (196ˆ), 707, 323 (1952), 750, 254 (1953), 913, 473 (1960).
34. Hersch, P.: Anal. Chem. 32, 1030 (1960).
35. Phillips, T. R., Johnson, E. G., and Woodward, H.: Anal. Chem. 36, 450 (1964).
36. Bates, R. E., and Littlewood, R.: Acta Imeko, 1964.
37. Kiukkola, K., and Wagner, C. J.: J. Electrochem. Soc., 104, 379 (1957).
38. Peters, H., and Möbius, H. H.: Z. Phys. Chem. (Leipzig) 209, 298 (1958).
39. Weissbart, J., and Ruka, R.: Rev. Sci. Inst., 32, 593 (1961).
40. Schmalzneid, H.: Z. Elektrochem. 66, 572 (1962).
41. Bechtold, E.; Z. Anal. Chem. 221, 262 (1966).
42. Wagner, C.: J. Chem. Phys. 21, 1910 (1953).
43. Kobayashi, H., and Wagner, C.: J. Chem. Phys. 26, 1609 (1957).
44. Berton, A.: Ind. Aliment. Agr. 78, 521 (1961).
45. Berton, A.: Chim. Anal. (Paris) 45, 585 (1963).
46. Berton, A.: Chim. Anal. (Paris), 41, 351 (1959).
47. Berton, A.: Bull. Soc. Chim. France, 1959, 536.
48. Littlewood, A. B.: Chromatographia 1, 37 (1968).
49. Littlewood, A. B.: Chromatographia 1, 133 (1968).
50. Landowne, R. A., and Lipsky, S. R.: Anal. Chem. 34, 726 (1962).
51. Lovelock, J. E.: Anal. Chem. 3, 162 (1961).
52. Karmen, A.: Anal. Chem. 36, 1416 (1964).
53. Juvet, R. S., and Durbin, R. P.: J. Gas Chromatog. 1 (12), 14 (1963).
54. Gunther, F. A., Blinn, R. C., and Ott, D. E.: Anal. Chem. 34, 302 (1962).
55. Merritt, J., Comendant, F., Abrams, S. T., and Smith, V. N.: Anal. Chem. 35, 1461 (1963).

56. Mairanovskii, V. G., and Yanatovskii, M. T.: Russ. J. Phys. Chem., 37, 370 (1963).

57. Nelson, D. C., Ressler, P. C., Jr., and Hawes, R. C.: Anal. Chem. 35, 1575 (1963).

58. Griffiths, J., James, D. H., and Phillips, C. S. G.: Analyst. 77, 897 (1952).

59. Seiyama, T., Katu, A., Fujishi, K., and Nagatani, M.: Anal. Chem. 34, 1502 (1962).

The Importance and Prospects of Gas Chromatography in Catalytic Researches

S. Z. ROGINSKY, M. I. YANOVSKY and A. D. BERMAN

Institute for Chemical Physics of the Academy of Sciences, Moscow

The outstanding scientific and practical importance of catalysis is generally know. It suffices to mention that in many industrially developed countries approx. 20 per cent of the total industrial production are manufactured with the help of catalysts and that the vast number of the fine and completely biochemical processes taking place in living matter are of a catalytic nature.

Only to maintain the high level of the already existing catalytic plants and to complete them, a great deal of research work is required. To the latter, the highly topical, but also very difficult, researches required for finding catalysts for unknown processes are added. Here, great obstacles have to be overcome, as the innermost mechanism of catalysis has not been elucidated as yet. But this is connected with particular specific difficulties of the catalytic researches, where the decisive role is played by labile forms that are difficult to identify and are little known, which warrant the (in the ideal case - infinite) regeneration of the catalyst during the chemical process.

In this respect, the catalyst behaves like phoenix, the fabulous bird of Egypt, that burns itself an infite number of times and always rises rejuvenated from the ashes. Only that the periods of combustion and regeneration of the active centers of the catalyst are much shorter and are measured in small fractions of a second.

The second essential difficulty is connected with the unusually high susceptibility of the catalytic process to the microchemical stability of both catalyst and substrate. In actual catalysis, the complicating influences of various macrokinetic factors, of diffusion and heat transfer, are additionally felt.

Because of these difficulties the theory of catalysis is not yet able to meet the requirements of the field, notably for complex organic reactions. This situation calls for a general intensification of catalytic studies and the mobilization of the new methods of investigation. Within the latterly developed bulk of methods for the investigation of catalysis, gas chromatography deserves a place honor.

It is for these reasons that an analysis of the successes of chromatography in catalysis and the prospects of its development seems to be of interest.

In this paper, which I have prepared together with my collaborators M. JANOVSKY and A. BERMAN, brief reference is made to the main directions of the application of gas chromatography in catalysis; the applications of common and also of isotopic chromatography in the studies of kinetics and the mechanism of catalysis are being treated in greater detail.

1. Traditional Applications of Gas Chromatography in Catalysis

1.1. The problem of purity

Of extraordinarily great importance in catalysis are the problems of the high degree of purity of the catalysed substrate and of the introduction in doses of desirable micro admixtures into the catalyst and the substrate [1].

I take the liberty to mention here two examples:

a) First, it is the HABER-BOSCH-MITTASCH synthesis of ammonia on iron catalysts, where the high degree of purity of the gases is so important that some American companies are purifying the hydrogen by diffusion through palladium alloys.

b) Only the introduction of microscopic admictures of organic halogen compound into an ethylene-oxygen mixture ensures a stable and high selectivity of the formation of ethylene oxide on silver catalysts.

Therefore, the employment of preparative chromatography has gained great importance in catalytic research. In fact, the conditions at which the major part of the catalysts were examined and the catalytic reactions were carried through, have not been sufficiently pure until now, and all of them must be studied anew on a higher level of purity.

1.2. Relative vacuum

One of the most difficult tasks in the studies of catalysts is the production of the latter in studies of catalysts is the production of the latter in a totally reproducible chemical starting condition. The surface of the catalyst contains as a rule various foreign admixtures that are difficult to control (introduced, adsorbed substances, etc.). Partly these admixtures will form surface films covering the surface with a coat. Latterly the chemists succeeded in attaining for metals and a few high-melting binary compounds a high degree of purification in respect of casual admixtures as well as a good reproduction of the stock by application of heat together with an ultra-high vacuum of approx. 10^{-10} Torr. In this connection, new interesting phenomena could be observed, and a number of incorrect views have been revised. The impulse method with the employment of pure carrier gases is opening up interesting possibilities [2]. Thus, for example, when using very pure helium as a carrier gas, quasi vacuum conditions in respect of all other gases can be maintained in the stream of the indifferent gas entraining the desorbing substances. Here the degassing of the surfaces takes place under the conditions of a very good heat dissipation. This is probably also the reason why in our studies [3, 4] and also in the studies of the institute of physical chemistry of the Ukrainian Academy of Sciences [5] a systematic increase of the velocity constant of catalytic processes by 1-3 orders of manitude were observed, if these processes were observed under impulse conditions [1].

1.3. Utilisation of the Analytical Advantages of Gas Chromatography

In the study of every catalytic process, the measurements of the chemical consistency of the reacting substances and of the admixtures leaving the contact apparatus serve as the starting data for the determination of kinetic values and of the activity and selectivity of the catalyst.

The application of analytical gas chromatography for this purpose is widely practiced already both in scientific studies and in the field. As concerns the completeness of analysis chromatography is second only to mass spectrometry (but only in principle), with the latter being incomparably more complicated, not only as regards the apparatus required but also in respect of the interpretation of the results obtained. For complex mixtures, where chromatography is particularly effective, the application of mass spectrometry has to overcome great difficulties. When using good analytical chromatographs for the investigation of organic catalytic reactions, information is, as a rule, not only obtained more quickly, but also its scope is substantially widened [6].

As an example, let us mention a group of papers [7] on the nature of difficultly evaporating condensation films forming in consequence of the action of hydrogen on alkyl-aromatic compounds; on the basis of these papers a new type of processes has been discovered, the so-called „inaccurate depolymerization". Mention may be made, in this connection, also of the study of the reactions of the soft contact oxidation, where the chromatographic methods have revealed quite a number of attendant processes that interlink the various types of olefin oxidation with each other, namely the formation of unsaturated aldehydes, the dehydration by oxidation, the coupled formation of acrylonitriles and saturated nitriles and also a complex of side reactions: isomerization, deep oxidation, etc. Finally the intermediate formation of cyclenes and cyclodienes on the dehydrogenation of cyclohexanes and the hydrogenation of aromatic compounds were observed, what was formerly long disputed for theoretic reasons, etc.

But greatest is the importance of analytical chromatography in the search of new catalysts. This question will be discussed further below in the section dealing with a closer coupling of chromatography with catalysis.

1.4. Investigation of the Surface of Catalysts and Their Adsorption Characteristics

Somewhat more specific, though not only suited to catalysis alone, are the chromatographic methods of investigating the structure and size of the surface of the catalysts and the adsorption charcteristics of catalysts. A substantial part of these methods is based on the chromatographic definition of adsorption isothermals. There exist various versions of this procedure, the one being based on the well known equation of DE VEAU

[8-10], the others on the frontal method of Schilov-Tiselius-Klesson. In velocity and versatility the chromatographic methods of determining the adsorption characteristics are superior to all other methods. For the time being they are inapplicable to such catalysts whose pores are of molecular dimensions such as the most microporous zeolites. On the basis of the adsorption isotherms and isosteres of the individual gases, as is generally known, the differential adsorption heats of these gases can be determinde [11, 12].

The isotherms furnish starting data for determination of the size of the specific surface. By means of specifically adsorbed molecules it is possible to determine the size of the surface of the active component in complex contacts and carrier contacts [12a]. The enormous efficiency of the gas-chromatographic methods in the measurement of specific surfaces, in combination with the vacuum method of determining the nature of the pores, enabled Karnaukhov to carry through, in a comparatively short time, a comparative study of a very great number of operative catalysts; in the course of this study the existence of a small number of characteristic structural types (the globular type, the spongy type, etc.) was identified and unusual forms of distribution of the active substance Pt and Po on the carriers were proved [12, 13].

Of great fundamental importance for catalysis is the radical simplification of the measurements of adsorption equilibriums of gas mixtures. Because of the great deal of work required for such measurements, our knowledge on this question is very limited. But during catalysis, binary, ternary or still more complex mixtures are always present on the surface. However, as the measurements of the leaving effort of the electrons and also of the optical spectrums of adsorbed molecules have shown, the reaction components are forming various addition complexes on the surface of the catalyst. For studying the adsorption of mixtures a method for the direct determination of true and quasi equilibrium adsorption isotherms of the components of mixtures has been elaborated in our laboratory [14]. This method (see Fig. 1) is based on the principle of the so-called „vacant" chromatography (which was suggested by A. ZHUKHOVITSKI for analytical chromatography) [15].

2. On the Application of Gas Chromatography to Studing the Kinetics and the Mechanism of Catalytic Reactions

The specific gas chromatographic methods of investigating catalytic activity possess a number of peculiar features. First, these methods are not stationary. In the most widespread impulse method (often called the „microcatalytic" method) the concentration in a randum section through the reactor varies with time. As a rule, also the catalyst itself does not remain constant, as it is repeatedly regenerated in the intervals between the introduction of the impulses into the reactor. In the limiting case the data obtained by the chromatographic method characterize the surface of the catalyst that has not yet been exposed to the action of the reagent.

The question for the ratio of the results obtained by means of the impulse chromatography and the results obtained by flow methods is, consequently, at the bottom, the question for the ratio of the stationary and the non-stationary methods of studying catalytic reactions. There it is difficult, for the general case, to arrive at a complete coincidence of the results obtained by means of chromatographic methods on the one hand and with the help of the conventional stationary methods on the other hand.

Frequently the chromatographic methods offer a number of positive advantages over other methods and furnish information that cannot at all be ascertained by means of the other methods.

Particularly useful are the chromatographic methods for the solution of the most important task - the extensive study of new systems for the search of catalysts. With them it is quite easily possible to ascertain information on the adsorption properties and catalytic properties of the active solid body that has not yet been affected by the action of the reagent, and also on the subsequent changes of this body. This fact is very important for an understanding of the nature and the formation of the active surface. The second advantage offered by chromatography in the search of catalysts is the rapid and sensitive identification of products contained in very low concentrations. In this manner it is possible to identify first traces of the desired product as well as of other, unexpected products, and by suitable variation of the conditions, the desirable processes can be intensified.

Experimental data have shown that the acticity series of the catalysts obtained by means of the chromatographic methods are usually quite close to those obtained with the more conventional methods. Today we cannot imagine any more or less extensive investigation in search of catalysts for new reactions without employing the chromatographic method of determining the catalytic nature and activity of the catalyst.

Thanks to the very small amounts of the reacting substance introduced with the impulse into the stream of the carrier gas it is easy to carry out the catalytic process under practically entirely isothermal conditions. It is for this reason that the chromatographic method is of particular importance for the study of the catalytic reactions involving high dissipation of heat. The application of chromatography to studies of the kinetics and the mechanism of catalysis becomes specially effective, if chromatography is combined with the isotopic (notably in the form of radio-chromatography) [16-18]. Our first model was demonstrated at the International Exposition 1958 in Brussels. Fig. 2 shows an improved model which we have developed in collaboration with the design office for the construction of analytical apparatus. Frequently, successful work is possible with simpler and smaller arrangements.

For more than 12 years the radio-chromatographic method has been systematically employed in our laboratory for identifying mechanismus of the catalytic processes. With the help of these works we were able to determine the main routes of a few catalytic reactions, to identify stable intermediate compounds and to find out a number of genetic relationships between different substances. Moreover, it has been shown that the evaluation of the chromatographic data by means of computers, with consideration of information obtained by radio-chromatography, permits of increasing the certainty of the results and making the latter more unequivocal [19]. The recently issued paper by HEIT-

AUER, et al. [20] is another example of the successful application of the radio-chromatographic method to catalytic studies.

In the study of the chromatographic conditions of catalytic reactions, the topic of my paper read at the symposium in Jena, we have, for the first time, used chromatograms taken immediately behind the reactor for investigation into catalytic processes. Latterly it has been possible to show that on suitable mathematical processing such „catalytic chromatograms" can, on principle, furnish valuable information on the role of the adsorption and desorption kinetics of the starting materials and of the reaction products and also on the role played by the diffusion factors in catalytic conversions.

Besides the data on the degree of conversion obtained by the analysis of the catalysate, the analysis of the form of the catalytic starting curves is still of great importance. This is a new and, apparently, promising direction in the application of chromatography to catalysis.

Let us now consider a few concrete results that have been obtained in our laboratory and which illustrate the promising prospects of the application of gas chromatography to catalysis.

3. Study of the Kinetics of the Catalytic Oxidative Dehydrogenation of Butylene in Divinyl on a Bi-Mo Oxide Catalyst

In order to describe the reaction kinetics of the oxidative dehydrogenation of n-butene with sufficient completeness, it is necessary to determine the values of six isomerization constants and of three oxidative dehydrogenation constants (see scheme of Fig. 3).

The calculation was based on the following assumptions resting on the results of provisional experiments and on information from literature:

1. The oxidative dehydrogenation is not inhibited by the divinyl.
2. It proceeds according to the first order as butylene and according to zero order in respect to oxygen.
3. No inverse reaction of the conversion of divinyl into butylene takes place.

On the basis of the assumptions mentioned above the following system of differential equations was written:

$$\left.\begin{aligned}
\frac{d[A]}{dt} &= -(K_1 + K_3 + K_7)[A] + K_2[B] + K_4[C] \\
\frac{d[B]}{dt} &= -(K_2 + K_5 + K_8)[B] + K_2[A] + K_6[C] \\
\frac{d[C]}{dt} &= -(K_4 + K_6 + K_9)[C] + K_3[A] + K_5[C] \\
\frac{d[D]}{dt} &= K_7[A] + K_8[B] + K_9[C]
\end{aligned}\right\} \qquad (1)$$

where [A], [B], [C], [D] are the respective concentrations of butene - 1, trans-butene - 2, cis-butene - 2, and divinyl. K_i (i = 1, 2 ... 9) are the velocity constants of the individual reactions (see schema 1, Fig. 3).

For determination of the velocity constants experiments with mixtures of different starting composition were carried through. The approximated values of the velocity constants were obtained according to statistical methods of levelling the eyperiments [22]. Further, these values were made more pricise by means of the method of non-linear programming from the condition for the minimum mean square error. Typical calculated kinetic curves and experimental points are shown in Fig. 4.

The values of the velocity constant are contained in Table 1. From the data of Table 1 it follows that butene-1 is substantially more readily oxidized to divinyl than cis-butene-2 and trans butene-2. This has been confirmed by radio-chromatographic experiments, of which some results are shown in Fig. 5 and Fig. 6. In the one series of experiments a mixture of non-labelled butene-1 and labelled cis-butene-2 was introduced into the reactor, while in the other experimental series this mixture consisted of non-labelled butene-1 and labelled trans-butene-2. In all cases no activityl could be found in the resulting divinyl (within the error limits of the experiments), as it was formed largely from non-labelled butene-1.

Table 1. Velocity constants (K) and activation energies for oxidative dehydrogenation of butenes

K $[sec^{-1}]$	310 °C	396 °C	436 °C	496 °C	[E kcal/mol]
K_1	4,0	9,0	15,0	18,0	8,0
K_2	3,0	6,0	7,0	10,0	6,0
K_3	6,0	11,0	11,0	17,0	5,0
K_4	6,0	7,0	8,0	9,0	2,0
K_5	10,0	13,0	14,0	16,0	2,0
K_7	4,0	10,0	33,0	69,0	13,0
K_8	1,0	1,0	3,0	17,0	28,0
K_9	1,0	3,0	7,0	23,0	21,0

To the same conclusions lead the kinetic experiments with binary mixtures of radioactive butene-1 and non-radioactive cis-butene-2.

For evaluation of the velocity constant the material balance equations of the reacting substances in respect to radioactive or non-radioactive substances were simultaneously solved.

The velocity constants have been selected on the basis of the condition that the calculated and experimental values of the concentrations or the specific radioactivities should lie as closely together as possible. Therefore, it was possible to make the calculations somewhat more precise. Table 2 contains the experimental and calculated values of the specific radioactivities of the products of oxidative dehydrogenation of the mixture consisting of butene-1 and labelled cis-butene-2 at 396 °C. A good agreement of the calculated or experimentally obtained values can be seen. The use of radio-chromatographic data is very expedient, as it makes the determination of the constants more certain and helps to avoid errors in the selection of the conversion scheme.

Tabel 2. Comparison of the experimental and calculated specific radioactivities (imp/min) of the different hydrocarbons in oxidative dehydrogenation of butene-1 and cis-butene-2 and $C_4^{14}H_8$ mixtures at 396 °C

Contact time	butene-1		trans-butene-2		cis-butene-2		divinyl	
in seconds	exp.	calc.	exp.	calc.	exp.	calc.	exp.	calc.
0,0065	9,4	9,4	92,0	92,0	155,0	155,0	20,0	20,0
0,0090	11,0	12,7	93,0	93,6	151,0	152,4	30,0	32,8
0,0118	15,0	16,3	94,0	95,2	149,0	149,0	39,0	42,8
0,0144	17,5	19,5	94,5	97,0	145,0	146,2	48,0	50,0
0,0190	23,0	25,0	96,0	98,0	140,0	141,0	58,5	59,6

4. Differential Chromatographic Method of Determination of Velocity Constants of Heterogeneous Catalytic Reactions

Until recently the evaluation of the kinetic data obtained by the impulse-chromatographic method was very difficult. Quantitative calculations, indeed, could be made rather simply for reactions of the first order only, for which the degree of conversion does not depend on the distribution of the catalyst layer and on the type of the introduced impulse. A few single cases of reactions of other orders [23] were formerly dealt with in the papers of our laboratory. It is possible to show that at low degrees of conversion under the conditions of ideal linear chromatography, there arises the possibility to evaluate kinetic data for reactions of any desired constant order and for any type of introduced impulse [24, 25]. This purpose serves the mean velocity of the heterogeneous catalytic reaction $\overline{W}$ (see equation 2)

$$\overline{W} = \frac{\Delta m F}{V_{Imp} V_{Kat}} \quad (2)$$

corresponding to the average concentration $\overline{C}$ in the introduced impulse.

Δm - is here the quantity of a certain product that has formed; F - the volume velocity of the carrier gas; V_{Imp} and V_{Kat} - are the volumes of the specimen and of the catalyst, respectively, while $\overline{W}$ - is the mean reaction velocity.

Moreover, from the equation

$$\overline{W} = \overline{K} \Gamma^n \overline{C}^n \quad (3)$$

the product of the effective velocity constant E can be calculated with the coefficient of Henry Γ power n and also the reaction order, where n is the reaction order. The relation between the effective velocity constant, which has been determined by the method stated above, and the true velocity constant K is determined by the following equation:

$$K = \bar{K}\left[\frac{1}{T}\int_0^T \psi(t)\,dt\right]^n \Big/ \frac{1}{T}\int_0^T \psi^n(t)\,dt \tag{4}$$

where $\psi(t)$ is the form of the impulse introduced into the reactor, while T determines the duration of the impulse, according to the chromatogram, introduced into the reactor.

From (4) it follows that for the zero and first order the ratio of the integrals in the right part is unity and does not depend on the form of the impulse. It is possible also to show that this ratio equals unity also for square wave form impulses introduced into the reactor, for any desired reaction order.

From this follows a simple method for determining the velocity constants of the reaction of n-th order. Before the entrance into the reactor an arrangement must be procided ensuring the introduction of the square wave impulse into the reaction zone. If, for any reason whatever, it is difficult to produce a square wave form impulse, a detector must be arranged before the entrance into the reactor, by means of which the form of the entering impulse is identified. After the analysis of the reaction mixture the dependence of the mean reaction velocity on the mean concentration C in the specimen is graphically constructed. On the basis of this graphic representation the order of reaction is determined and also the value of the effective velocity constant. In order to find the true constant, a correction must be introduced in respect to the form of the introduced impulse according to the formula stated above.

Fig. 7 shows, how the reaction velocity of the oxidative dehydrogenation of isoamylenes on a complex oxide catalyst depends upon the mean concentration in the impulse. The data of Fig. 7 correspond to four series of experiments. Series I belongs to the experiments, where the impulse introduced into the reactor has square wave form. In series II the specimen of random form was introduced immediately before the reactor. In the series III and IV the volume of the specimen and catalyst varied, while the form of the impulse was random.

From Fig. 7 it may be seen that the experimental points obtained for different catalyst quantities, different volumes of the specimen, different amounts of the oxygen-amylene mixture and different forms of the impulse coincide well with a straight line within the system of coordinates: reaction velocity $\bar{W}$ vs. mean concentration $\bar{C}$. These data permit of drawing the inference that under our experimental conditions the reaction of the mild oxidation of amylenes takes place according to the first order. The first order is confirmed also by the fact that the velocity constant does not depend on the form of the impulse introduced.

By this method the reaction order for oxygen was also determined. It came out that in these experiments the reaction velocity does not depend on the concentration of oxygen, so that, consequently, a zero order is observed in respect of oxygen.

The chromatographic method described is practical for studying the processes of working-down of the catalyst, of its being poisoned with special poisons or with reaction products.

5. Chromatographic Determination of the Velocity Constants of Adsorption and Desorption During Catalytic Conversion

We have already mentioned that information concerning the character of the kinetics of adsorption and desorption may be obtained during the catalytic process by studying the form of the catalytic chromatograms of the starting materials and of the products. Since the equations describing such phenomena are complicated, we shall restrict ourselves to considering the cases of the linear adsorption isotherms and the reactions of the first order. The most promising method for the analysis of linear outputs is, in our opinion, the so-called „method of moments", which has been suggested for chromatography by TUNITSKY [26, 27] in the U.S.S.R. and by GRUBNER [28] in Czechoslovakia.

Though this method does not yield precise solutions, it nevertheless permits the writing ob an expression for the moments of the leaving curve (which is considered as the distribution density of the time the molecules are leaving the analytical chromatographic column).

Put briefly, the scheme of application of the „method of moments" for the analysis of catalytic processes under chromatographic conditions is the following:

The starting system of differential equations with partial derivatives is transformed according to Laplace. Here, a system of conventional differential equations is obtained for representation of the wanted functions. If the wanted function is called y (t), and its representation $\tilde{y}$ (s), the following formula applies:

$$m_n = \int_0^{\sim} y(t)\, t^n\, dt \qquad (5)$$

or

$$m_n = (-1)^n \left. \frac{\partial \tilde{y}(s)}{\partial s} \right|_{s = 0} \qquad (6)$$

The central moments are further found from the equation:

$$\mu_n = \frac{1}{m_o} \int_o^{\sim} \left(t - \frac{m_1}{m_o}\right)^n y(t)\, dt \qquad (7)$$

$$= \sum_{i=0}^{n} (-1)^{n-i}\, C_n^i\, m_i\, m_1^{\,n-1}$$

As is well known, the first moment can be considered as the leaving instant of the center of gravity of the chromatographic peak, while the second central moment (the dispersion) characterizes the washed-out condition of the band. Hence, the expression (5, 7) may be used for experimental, ord the expression (6) for analytical determination of the moments. Let us first consider the most simple case, where the equilibrium of adsorption is reached instantaneously and where no complications of diffusion exist.

In this case the processes taking place under chromatographic conditions when carrying through the reaction may be described by means of the following system of differential equations with partial derivatives:

$$\left.\begin{aligned}
&U\frac{\delta C_A}{\delta\varkappa} + \varkappa\frac{\delta C_A}{\partial t} + (1-\varkappa)\frac{\delta a_A}{\partial t} + (1-\varkappa)\quad \delta a_A = O\\
&a_A = \Gamma_A c_A\\
&U\frac{\partial c_B}{\partial\varkappa} + \varkappa\frac{\partial c_B}{\partial t} + (1-\varkappa)\frac{\partial a_B}{\partial t} - (1-\varkappa)\,Ka_A = O\\
&a_B = \Gamma_B c_B\\
&c_A(\varkappa, O) = O\\
&c_A(O, t) = \Psi(t)
\end{aligned}\right\} \qquad (8)$$

Here the indices A and B refer to the starting material or to the reaction product, respectively, c - is the concentration of the gaseous phase, a - is the concentration on the surface of the catalyst, $\varkappa$ - is the external porosity, Γ- is Henry's coefficient. The band of the starting material A moves along the catalyst layer with a velocity that is determined by HENRY's coefficient; the height of the band is decreasing exponentially with the time of stay. The reaction does not affect the velocity of motion and will not wash out the band of substance A. The band of the products B is, of course, not symmetric. The shape of this band is determined by the velocity constant of the reaction and by the ratio of the coefficients Γ_A and Γ_B. Theory permits it to calculate the leaving moment of the centre of gravity of the band for the products t^B:

$$t^B = t^A + \frac{\Gamma_A - \Gamma_B}{K\Gamma_A} = \frac{t^A}{\alpha}\left[\frac{\Gamma_A - \Gamma_B}{\Gamma_A}\right] \qquad (9)$$

where α is the degree of conversion.

Prior to the conversion of A into B each future molecule of the product B will move for some time with the velocity of the reacting substance A; but after the conversion it will move for some time at its own velocity, which is determined by Γ_B. The molecules B that are near the band A obviously existed over the major part of the time as A and are moving chiefly with a velocity that has been determined by Γ_A. The molecules contained in the other part of the band have moved, for the major part, with a velocity determined by Γ_B. Let is now consider two limiting cases:

Case „a" Fast reactions $\widetilde{\widetilde{K}} \rightarrow \infty$ and $\alpha \longrightarrow 1$. From (9) it follows:

$$t_B = t_A \frac{\Gamma_B}{\Gamma_A} \tag{9a}$$

Hence, band B moves as though it had been introduced simultaneously with substance A and there would not be a reaction.

Case „n". Slow reactions $K \longrightarrow O$ and $\alpha \approx kt_A$. In this case $t_B \longrightarrow t_A$, i.e., in the case of a slow reaction the product will move with the velocity of the starting material. Let is now consider the effect of the reaction on the elution of the band. As already pointed out above, in this ideal case, where no adsorption kinetics exist, the reaction does not affect the elution of the starting material A.

The elution of the product B is determined by the following formulas:

$$\sigma_B^2 = \left(\frac{\Gamma_A - \Gamma_B}{\Gamma_A}\right)^2 \left[\frac{1}{K^2} - \frac{t_A^2}{\alpha}\left(\frac{1}{\alpha} - 1\right)\right] \tag{10}$$

It will easily be seen that the fast reactions at a narrow entering impulse obey the equation $\sigma_B = O$ (10a). For slow reactions the washed-out·condition of the peak B increases. For small K the washed-out condition is determined by the following equation

$$\sigma_B^2 = \left(\frac{\Gamma_B - \Gamma_A}{\Gamma_A}\right)^2 \frac{t_A}{K} \tag{10b}$$

It is seen that the washed-out condition is directly proportional to the square of the difference of the relative adsorption coefficient and of time t_A (the retention of A), while it is inversely proportional to the velocity constant of the reaction K.

This simple example shows that the catalytic chromatogram furnishes essential and illustrative information on the nature of the catalytic process. Thus, for example, the ratio of the coefficients Γ_A and Γ_B during the catalytic conversion process can be immediately recognized from the relative position of the bands.

Let us now consider a more complicated case, where the equilibrium of adsorption of the product and of the starting material is not immediately reached. The results for this case were previously arrived at in the works of ROGINSKY and ROSENTAL [29] and in the thesis of ROSENTAL. In this case the starting system of equations is the following:

$$\left.\begin{aligned}
& u \frac{\partial c_A}{\partial x} + \varkappa \frac{\partial c_A}{\partial t} + (1-\varkappa) \frac{\partial a_A}{\partial t} + (1-\varkappa)\, K a_A = O \\
& \frac{\partial a_A}{\partial t} = K_a^A c_A - a_A (K^A + K_g^A) \\
& u \frac{\partial c_B}{\partial \varkappa} + \varkappa \frac{\partial c_B}{\partial t} + (1-\varkappa) \frac{\partial a_B}{\partial t} - (1-\varkappa)\, K a_A = O \\
& \frac{\partial a_B}{\partial t} = K_a^B c_B - K_g^B a_B + K a_A
\end{aligned}\right\} \tag{11}$$

$$\left.\begin{aligned} c_B(\varkappa, t) &= a_B(\varkappa, t) = 0 \\ a_A(\varkappa, t) &= 0 \end{aligned}\right\} \text{ if } t = 0$$

$$c_A(\varkappa, t) = \varphi(t) \qquad \text{if } t = 0$$

K_a^A, K_g^A, K_a^B, K_g^B are the velocity constants of the adsorption or desorption, respectively, of the starting material or the resulting reaction product. Now let us, under these presuppositions, write the expressions for the degree of conversion α, the retention time of the starting material t_A and the dispersion σ_A^2 of the band:

$$\alpha = \exp - \frac{\varkappa}{u}\left[\frac{(1-\varkappa)\, K_a^A\, K}{K + K_g^A}\right] \tag{12}$$

$$t_A = \frac{\varkappa}{u}\left[\varkappa + \frac{(1-\varkappa)\, K_a^A\, K_g^A}{(K + K_g^A)^2}\right] \tag{13}$$

$$\sigma_A^2 = \frac{\varkappa}{u} \; \frac{2\,(1-\varkappa)\, K_a^A\, K_g^A}{(K + K_g^A)^3} \tag{14}$$

In principle, if t_A, α, σ_A are known, K, K_a^A and K_g^A may be calculated directly. The expressions (6 - 8) show that the effect of the reaction on the form of the band is determined by the ratio of the velocity constants of reaction or desorption, respectively. If, for example, the surface reaction is the limiting state, $K \ll K_G^A$, then follows $\alpha = \exp \frac{\varkappa}{u}\ (1-\varkappa)\,\Gamma_A K$, what coincides with an analogous relation for the case of equilibrium. The velocity of motion of the starting material is determined by the effective value of the term Γ_A^{eff}, which, for $K \ll K_g^A$ degenerates into HENRY's coefficient.

However, if the velocity constants of the reaction or desorption, respectively, are comparable, the peak of the starting material moves more rapidly than in gas adsorption chromatography. The reaction also reduced the elution of the band of the starting material. Let us now consider the expression for the leaving time of the product. It takes the following form:

$$t_B = t_A - \frac{K_a^B\,(K + K_g^A - K_a^A\,(K + K_g^A)}{K\, K_g^B\, K_a^A} + \frac{\varkappa}{2}\; \frac{(1-\varkappa)}{u}\left[\Gamma_B\; \frac{K_a^A\, K_g^A}{(K + K_g)^2}\right] \tag{15}$$

For $\Gamma_B \approx \Gamma_A = \dfrac{K_a^A \cdot K_g^A}{(K + K_g^A)^2}$ follows

$$t_B - t_A = \frac{1}{t_g^B} + \frac{1}{K + K_g^A} \tag{16}$$

Thus it is seen that in this case, in spite of the equal adsorption coefficients, the peak of the resulting product is trailing behind the band of the starting material for a time that is required for the adsorption of the starting material, the surface reaction and the desorption of the product.

By employing the „method of moments" described above, expressions have been obtained for the leaving moment of the center of gravity of catalytic chromatograms for the dispersion and the degree of conversion under the conditions of adsorption with due consideration of longitudinal diffusion [30] and inner diffusion [31].

My paper is approaching its end. It comprises several sections of a very different nature. In the first section the importance of preparative chromatography for purification and of analytical chromatography for investigating the composition of the reaction products in common catalytic research have been considered quite in general (with various examples).

The second is devoted to the chromatographic studies of the surface of catalysts and their different adsorption characteristics, among them also an the adsorption of mixtures during the reaction process. In the third section the essential applications of gas chromatography for investigation of the kinetics and the reaction mechanism are considered, with employment of the microanalytical impulse method, without or with addition of radiochromatography. Also the importance of these methods for speeding up the researches for finding new catalysts is discussed.

Owing to these three fields of application, gas chromatography has gained great importance for catalysis and has become indispensable for a major part of catalytic researches.

The fourth and fifth sections are devoted to new and, as it seems to us, promising directions in the application of chromatography, viz.:

a) the detailed study of quantitative kinetic characteristics of heterogeneous catalytic reactions;
b) the chromatographic determination of the velocities of the essential auxiliary processes - adsorption, desorption and diffusion during the catalytic process. In these sections the results of our latest works are described: a) the theory of the differential chromatographic methods of determining the velocities of heterogeneous catalytic reactions has been developed, which permitted of determining the essential kinetic data and coefficients for processes of different reaction order;
c) the methods for determination of the kinetic characteristics by the form of the leaving impulses on the basis of the „method of moments" have been developed.

Their are no fundamental obstacles to the employment of this method for any complex systems whatever, but in order to ensure the „express" properties that are so typical and valuable for analytical gas chromatography a further improvement of the experimental methods, of the apparatus and of the method of computer calculation is still required. Once these problems will have been solved, the field of application of gas chromatography to catalysis will become still larger, and its importance may some day surpass even that of the conventional analytical methods.

References

1. Roginsky, S. Z., Altshuler, O. V., Vinogradova, O. M., Yanovsky, M. I., and Krivoruchko, O. P.: Izv. Akad. Nauk SSSR, Ser. khim., No. 2, p. 214, 1965.
2. Yanovsky, M. I., and Gaziev, G. A.: Sb. Problemy Kinetiki i Kataliza, Vol. 12, Publishers „Nauka", Moskow, p. 277, 1968.
3. Gaziev, G. A., Krylov, O. V., Roginsky, S. Z., Samsonov, G. V., Fokina, E. A., and Yanovsky, M. I.: Dokl. Akad. Nauk SSSR, p. 140, 863 (1961).
4. Roginsky, S. Z., and Yanovsky, M. I.: Tezisy Dokladov III Wsesoyuznoi Konferentsii po Gazovoi Khromatografii, Moscow, Publishers „Nauka", p. 5, 1964.
5. Gershingorina, A. V., Belousov, V. M., and Rubanik, N. Y.: Kataliz i Katalizatory, Publishers „Nauka dumka", Kiev, p. 41, 1966.
6. Lyu-Chzhun-Khuei, Roginsky, S. Z., Samsonov, G. V., and Yanovsky, M. I.: Neftekhimiya, p. 3, 845 (1963).
7. Rudenko, A.P.: Teoriya Samorazvitiya Otkrytykh Kataliticheskikh Sistem, Publishers M.G.U., Moscow, 1968.
8. Gregg, S. I., and Stock, R.: 2nd Symposium of Gas Chromatography 1958, Amsterdam, Butterworth London, p. 80
9. Roginsky, S. Z., Yanovsky, M. I., Lu-Pei-Chzhan, Gaziev, G. A., Zhabrova, G. M., Kadenatsi, B. M., and Brazhnikov, V. V.: Dokl. Akad. Nauk SSSR, p. 133, 878 (1958).
10. Yanovsky, M. I.: Zhur. Prikl. Khim., p. 24, 660 (1951).
11. Kiselev, A. V., and Yashin, Y. I.: Gazoadsorptsionnaya Khromatografiya, Publishers „Nauka", Moscow, 1967.
12. Cremer, E., and Huber, H.: Angew. Ch., Vol. 73, p. 461, 1961; E. Cremer, Ber. Bunsengesellschaft f. Phys. Ch., p. 69, 802 (1965); H. L. Gruber and A. Hansen, Zs. für Polymere p. 214, 66 (1966).
13. Buyanova, N. E., Gudkova, G. B., and Karnaukhov, A. P.: Sbornik „Methody Issledovaniya Katalizatorov i Kataliticheskikh Réaktsii, vol. 2, Publishers SO Akad. Nauk SSSR, p. 55, 1965; Kinetika i kataliz, 6, p. 1085, 1965.
14. Krivoruchko, O. P., Roginsky, S. Z., and Yanovski, M. I.: Trudy III Vsesoyuznou Konferentsii po Gazovoi Khromatografii, p. 18, 1964.
15. Zhukhovitsky, A. A., Turkeltaub, N. M., Mayasova, N. A., Semkina, M. S., Lankin, L. I., and Somov, A. P.: Tezisy III Vsesoyoznoi Konferentsii po Gazovoi Khromatografii, Moscow, p. 1, 1964.
16. Kokes, R. I., Tobin, H., and Emmett, P. H.: J. Am. Chem. Soc., p. 77, 5860 (1955).
17. Yanovsky, M. I., Kapustin, D. S., and Nogotkov-Ryutin, V. A.: Problemy Kinetik i Kataliza, p. 9, 331 (1967).
18. Roginsky, S. Z., Yanovsky, M. I., Zhabrova, G. M., Kadenatski, B. M., and Vinogradova, O. M.: Trudy I Vsesoyuznoi Konferentsii po Gazovoi Khromatografii, Moscow, 1962.
19. Zimin, R. A., Gagarin, S. G., Berman, A. D., and Yanovsky, M. I.: Kinetika i Kataliz, IX, p. 117 (1968).
20. Hightower, I.W., Gerberich, H. R., and Keit Hall, W.: J. of Catalysis, vol. 7, No. 1, 1967.

21. Gagarin, S. G., Zimin, R. A., Berman, A. D., and Yanovsky, M. I.: Zavodskaya Laboratoriya, vol. XXXIV, No. 1, p. 65 (1968).

22. Nalimov, V. V., and Chernova, N. A.: „Statisticheskie Metody Planirovaniya Ekstremalnykh Eksperimentov", Publishers „Nauka" Moscow, 1965.

23. Gaziev, C. A., Filinovsky, V. Y., and Yanovsky, M. I.: Kinetika i Kataliz, 4, 688 (1963).

24. Roginsky, S. Z., Aliev, R. A., Berman, A. D., Lokteva, N. K., Semenenko, E. I., and Yanovsky, M. I.: Dokl. Akad. Nauk SSSR, 176, 1114 (1967).

25. Jermain, I. E.: C. R.-Sc. Paris, Serie C, 27, 1966

26. Tunitsky, T. N., and Chernova, E. P.: Zhur. Fiz. Khim., 24, 1350 (1950).

27. Tunitsky, T. N., and Shenderovich, I. M.: Zhur. Fiz. Khim., 27, 1425 (1952).

28. Kucera, E., and Grubner, O.: Collection Czech. Chem. Commun., 29, 1782 (1964).

29. Roginsky, S. Z., and Rosental, A. L.: Dokl. Akad. Nauk SSSR, 146, 152 (1952).

30. Kocirik, M.: J. of Chromatog., 30, 459 (1967).

31. Denisova, T. A., and Rosental, A. L.: Kinetika i Kataliz, VIII, p. 441 (1967).

The Separating Properties of Stationary Liquids

L. ROHRSCHNEIDER

Research Department of Chemische Werke Hüls AG, Marl

1. Introduction

More than a hundred different separating liquids have been successfully used for gas chromatographic separation. The analytical chemist, faced with the problem of selecting the stationary liquid best suited to his separating task, is embarrassed by the immense number of separating liquids offered in the market. He remains with his three to five standard columns and increases, in case of poor resolution, the separating power in order to obtain the desired separation. This method, which has been employed in practice time and again, has given excellent results. However, an optimum solution, which frequently calls for a faster gas chromatographic analysis, is not always arrived at by this method.

When elaborating a method of gas chromatographic analysis we must, wherever possible, make use of the best of the available separating liquids. To this end it is necessary to survey the existing separating liquids.

At the present state of our knowledge we are able to describe the striking, essential features of separating liquids that enable retention ratios to be predicted with a mean error of 6 %. However, a stationary liquid is not only characterized by its retention behavior, but also by its influence on the separating power. It is of no use to attempt a difficult separation with a separating liquid of a good separating power but of little selectivity, if another separating liquid with high selectivity will do this separation practically „by itself".

The objective of every gas chromatographic separation is the attainment of a sufficient resolution between the components to be separated. The resolution RS is a function of the separating power and of the retention ratio $r = t'_2/t'_1$ of the substances to be separated. The resolution [1] is calculated as the relation of the peak distance $t'_2 - t'_1$ to the mean peak width $\bar{w}$ ($\bar{w} = (w_1 + w_2)/2$):

$$RS = \frac{t'_2 - t'_1}{w_1 + w_2} \cdot 2 \quad (1)$$

This equation can be rewritten [2] as

$$RS = \sqrt{\frac{\overline{N}}{16}} \cdot \ln r \quad (2)$$

$\overline{N}$ being the mean number of the effective bottoms [3] for both peaks, for $N = 16 \left(\frac{t'}{w}\right)^2$.

(t' = retention time measured from air peak; w = peak width at the zero line). The value $\sqrt{N/16}$ is a measure for the separating power of the column, ln r is a measure for the separating efficiency, and we may assume:

retention = separating power · separating efficiency.

What influence has the stationary liquid on the separating power and the separating efficiency?

2. Influence of the Stationary Liquid on the Separating Power

The influence of the separating liquids on the separating power can be determined by comparing separating columns which are filled with different stationary liquids, but whose other properties are identical. We have prepared 12 such columns which contained 20 per cent by weight liquid and were operated at 100 °C with hydrogen as the carrier gas. The separating power was ascertained with decane at various rates of flow. As a measure for the separating power the value $\sqrt{N/16} = t'/w = t'/1 \cdot 7 \cdot b_H$ decisive for the resolution was entered right from the beginning. (See Fig. 1). The highest separating power has a column filled with Squalan, while the poorest separation is observed with ethylene glycol bis-propionitrile ether as the separating liquid. This graph would suggest the erroneous assumption that non-polar separating liquids have a good separating power, while polar liquids have a poor one, if we do not take into consideration that decane has quite different distribution coefficients between the gaseous phase and the liquid phase in the separating column. The coefficient of distribution K between the liquid and the gaseous phase may be calculated from the specific retention volume.

$$K = V_g \cdot \frac{w}{V_e} \cdot \frac{T_s}{273} \quad (3)$$

We have calculated the coefficient of distribution for some separating liquids from the specific retention volume V_g, of the density at 100° and the column temperature T_s and entered them into our graph. It is clearly seen that with rising distribution coefficient the separating power increases, and with the latter, of course, the time of analysis. The stationary liquid has a substantial influence on the separating power and the required time via the distribution coefficient K of the dissolved substance, which may differ by more then one power of ten.

According to the simple VAN DEEMTER equations [4] the separating power is inversely proportional to the height of a theoretical bottom h ($h = 16\ (^{t}/w)^2$) consisting of three parts, the eddy diffusion A, the extension of diffusion B/u and the resistance to the mass transport Cu. For a high separating power all three terms, for quick analyses especially the factor C, should be as small as possible.

$$h = A + B/u + Cu \tag{4}$$

The type of stationary liquid used has no influence on A and B, factor C, however, is largely dependent on the separating liquid.

If the separating power ist not expressed by the height of the theoretical bottom but by the height of the effective one H ($H = 16\ (t'/w)^2$), a modified VAN DEEMTER's equation results

$$H = A' + B'/u + C'u \tag{5}$$

where

$$C' = \frac{2}{3}\ \frac{1}{k}\ \frac{d_f^{\,2}}{D_1} \tag{6}$$

holds for C'.
Here $k = K \cdot (V_{fl}/V_{gas})$ (V_{fl} and V_{gas} being the liquid or gas volume, respectively, in the column), d_f the thickness of the layer and D_1 the coefficient of diffusion of the dissolved phase in the separating liquid.

In case of homogeneous charging of the stationary phase with separating liquid, C' is inversely proportional to the coefficient of distribution and the coefficient of diffusion. As a k-value of 2-5 ensures particularly favorable separating-power-per-unit-time yields, the separating liquids that have a good dissolving power and, hence, a high coefficient of distribution, will give stationary phases with lower charge and smaller layer thickness and, consequently, also better separating power than separating liquids with poor dissolving power. We should bear in mind:

Good solubility gives good separation.

The coefficient of diffusion of the dissolved phase in the separating liquid is connected with the viscosity of the separating liquid at the respective temperature. A study by HAWKES and CARPENTER [5] of last year gives the following relationship between the coefficient of diffusion D_1 and the viscosity:

$$D_1 = c\,\frac{T}{\eta^x} \tag{7}$$

where T is the absolute temperature, c an empirical constant between 0.1 for tricresyl phosphate and 0.7 for silicone oils and the exponent x also an empirical value between 0,1 for silicone rubber and 1 for tricresyl phosphate. Hence the measured viscosity is no unequivocal measure for the coefficient of diffusion to be expected. The general rule remains that separating liquids of low viscosity give excellent separating effects. HAWKES and CARPENTER arrived at the conclusion that a viscosity below 50 cP does not affect the separating power.

In order to compare some separating liquids in the range of possibly similar k-values or similar coefficients of distribution, respectively, with each other, we have measured the separating power of columns with the same phase ratio (V_L/V_{gas}) in respect of methyl isobutyl ketone, for which very similar coefficients of distribution were observed on polar and non-polar separating liquids. In the following table 1 the optimum separating power $\sqrt{N/16}$ that is decisive for the resolution, the capacity factor $k = K (V_L/V_{gas})$ and the viscosity at 100 °C are given for some of the separating liquids studied. The rule of HAWKES and CARPENTER is confirmed; only Reoplex with a viscosity of more than 50 cP has a reduced separating power, while all other separating liquids, polar and non-polar ones, are giving similar separating powers in the range of similar coefficients of distribution. We may further note: With every separating liquid of suitable viscosity similar separating powers may be obtained for substances with similar coefficients of distribution.

Let me note that the discussion of VAN DEEMTER's equation above, has been simplified. But this does not change the fact that the coefficient of distribution K and the coefficient of diffusion D_l are the only value through which a separating liquid has influence on the separating power. In how much also the layer thickness d_t is dependent on the kind of separating liquid used has not yet been elucidated.

Table 1. Comparison of the separating power of packed columns with the same phase relation in respect of methyl isobutyl ketone

Separating liquid	Maximum separating power $\sqrt{N/16} = {}^{W}/2$	k	Viscosity $\eta_{100°}$ cP
Squalan	10.3	9.3	3.1
Ethylene glycol bis-propionitrile ether	10.0	9.4	2.4
Carbowax 1540	7.5	6.4	29.2
Reoplex 400	5.6	6.0	253
Silcine oil DC 200	8.7	4.6	96.2
Di (ethylhexyl)sebacate	11.2	12.5	2.9

3. Influence of the Stationary Liquid on the Separating Efficiency

While only ten years ago it seemed as if the retention behavior of the separating liquids were of a multifariousness that cannot be overlocked, today, after all, empirical and also theoretical connections have been found which permit the forming of an illustrative and practicale idea of the separating efficiency of the stationary liquids.

The starting point for this model has been the phenomenon of the „polarity of the stationary liquids" [6], which may be described by the following equation:

$$\log\left(\frac{t'_2}{t'_1}\right)_p = \log\left(\frac{t'_2}{t'_1}\right)_n + (a_2 - a_1) \cdot x \qquad (8)$$

The retention ratio of two substances in a separating liquid p is composed of the retention ratio of these substances on a non-polar phase and the product of the polarity times the separating liquid, which may be described by a logarithmic retention ratio, e.g.:

$$x = \log \frac{t'_{olefin}}{t'_{paraffin}} \quad \text{or } x' = \log \frac{t'_{paraffin\ (n-1)}}{t'_{paraffin\ (n)}}$$

and the difference $(a_2 - a_1)$ in the polarities of the substances to be separated. This simple model signifies that two substances of different polarity will change their retention ratio, if changing from a non-polar to a polar separating liquid.

Graphically seen, equation (8) is the equation of a straight line whose gradient corresponds to the difference in polarity of the substances to be separated.

This model shows a very coarse picture of the conditions which is not true for all separating liquids for a precalculation already of the simplest retention ratios of paraffins and olefins. Thus, for example, the silicone oils are a class be themselves that won't fit into this model. Fig. 2 shows the dependence of the retention ratio of cyclohexane and n-octane on the retention ratios of n-nonane and n-octane for various separating liquids. Though a linear dependence may be read from the graph, the approximation of equation (8) is nevertheless inadequate for a fairly serviceable calculation of retention ratios.

The simplest theory of solutions, the theory of regular solutions, already furnishes an equation for the retention ratio

$$RT \ln \frac{t'_2}{t'_1} = (V_1 - V_2)\, x + (V_1 \delta_1 - V_2 \delta_2)\, y, \qquad (9)$$

where the stationary liquid is described by two values x and y that are mutually interdependent [2]. $(V_1 - V_2)$ is the difference between the molal volumes of the substances to be separated, while δ_1 or δ_2, respectively, are the solubility parameters of the substances to be separated, which are calculated from the evaporation energy and the molal volume. The first term shows that the retention ratio depends on the difference between the molal volumes of the substances to be separated or on the difference of size - comparable with the „own volume" in the equation of state. The second term, corresponding to the „internal pressure" accounts for the force of attraction between the molecules or for the difference between the non-polar interaction energies between the substances to be separated. As both terms are of the same order of magnitude, the molecule size, which has been regarded little so far, is getting a substantial share in the retention volume and the separability of two substances.

The retention ratio $r = t'_2/t'_1$ shows, according to CREMER [7], the following relationship to the free solution enthalpies ΔG_2 and ΔG_1 of the substances to be separated in the stationary liquid:

$$RT \ln \frac{t'_2}{t'_1} = \Delta G_2 - \Delta G_1 = \Delta(\Delta G)_{2,1} \tag{10}$$

The difference of the free solution enthalpies of two substances in a separating liquid is the free separating liquid is the free separating enthalpy $\Delta(\Delta G)$, which determines the retention ratio r.

Now the equation 9 may be rewritten as follows:

$$RT \ln \frac{t'_2}{t'_1} = \Delta(\Delta G)_{21} = \Delta(\Delta G^R)_{21} + \Delta(\Delta G^E)_{21} , \tag{11}$$

in which $\Delta(\Delta G^R)_{21}$ represents the share in the separating enthalpy, which is caused by the difference of size or the space required by the components to be separated, and $\Delta(\Delta G^E)_{21}$ represents the share of energy caused by dispersion forces. Equation [9] may be formulated quite generally as the simplest basic equation for the retention ratio:

$$RT \ln \frac{t'_2}{t'_1} = (a_2 - a_1) \cdot x + (b_2 - b_1) \cdot y , \tag{12}$$

where x and y are values characteristic of the separating liquid, while a_i and b_2 are values characteristic of the dissolved substance.

From the equations (11) and (12) already results the basis of the model for the retention behaviour or the separating efficiency of the separating liquids. The free separating enthalpy is composed of several portions,

$$RT \ln \frac{t'_2}{t'_1} = \Delta(\Delta)_{21} = \sum_{1}^{n} (\Delta G_i)_{21} , \tag{13}$$

where each separating enthalpy portion $\Delta(\Delta G_i)_{21}$ is the product of a factor characteristic of the dissolved phase $(a_2 - a_1)$ and a factor characteristic of the separating liquid x_i [8]

$$\Delta(\Delta G_i)_{21} = (a_2 - a_1) \cdot x_i = a_2 x_i - a_1 x_i . \tag{13a}$$

By means of equation (12) and when using empirical figures, the retention behavior of simple substances (paraffins, cycloparaffins, olefins, aromatic hydrocarbons) can be described, which are unable to form hydrogen bonds, dipoledipole or other associates. From equation (12) a formula for the retention index system may be derived according to which

$$I_P^{RX} = I_u^{RX} + a\, \Delta I_{pu}^{B} \tag{14}$$

The retention index of the substance RX on the separating liquid p is composed of the retention index of this substance on the non-polar phase u and the product of the polarity of the substance a and the polarity of the column, observed by the index difference of benzene between phase p and the separating liquid u [9]. Despite the formal similarity of the equations (8) and (14) there is nevertheless a fundamental difference between them. While in equation (8) the separating liquid is incompletely defined by only one value, the index system already takes into account a property of the separating liquid by reference to the retention of the n-paraffins.

The second factor is contained in the index difference ΔI^B, so that in equation (14) two factors serve to characterize the separating liquid. From Fig. 3 it is seen, how well equation (14) describes the retention of the non-polar substances but also that of the polar ones. For a correct prediction of the retention ratios of all substances the equations (12) or (14) are not sufficient. In either case additional separating enthalpies must be taken into consideration in accordance with equation (13) for the ability of the separating liquid to display dipoledipole interactions with the dissolved phase, to form hydrogen bonds as both a donator and acceptor of protons and to form electron acceptor-donator complexes.

If characterizing a separating liquid by 5 characteristic index differences, namely those of benzene, ethanol. methyl ethyl ketone, nitromethane and pyridine, 660 retention indices of 30 substances on 22 separating liquids may be described with a mean error of 6 index units. Catalogues for stationary liquids are being published already, in which these characteristic data for the various stationary liquids are indicated.

If characterizing a separating liquid by 6 characteristic retention ratios, namely those of dodecane, benzene, ethanol, methyl ethyl ketone, nitromethane and pyridine relative to n-octane, 1500 retention ratios of 72 substances on 22 separating liquids may be calculated with a mean error of 6%;that is an error of 1/2 minute at a retention difference of 9 minutes. Also the coefficient of distribution K and, hence, the specific retention volume V_g are functions of the free solution enthalpy. Just like the free separating enthalpy the free solution enthalpy may be represented by a sum of products. In this connection also, the separating liquids may be characterized by a number of retention ratios.

Finally I should like to deal with the special selectivities for the different types of interactions, which go beyond the normal polarity response. PHILLIPS and MACEK [10] have proved with a graphic representation consistent with equation (8) an additional effect, which is exercised by trinitrophenetol on benzene.

In the following illustration (Fig. 4) it can be seen from the polarity graph in the index system for heptanone-4 and butanol that fluorosilicone oil QF 1 and the fluoroester Zonyl E 7 are selectively retaining ketones but, as has been shown, also esters, aldehydes, nitriles and nitro compounds, while the separating liquids quadrol, sorbitol, hyprose, diglycerin and docosanol are retaining all alcohols more firmly than corresponds to their polarity relative to benzene.

Let me summarize; separating liquids influence the separating power via the coefficient of distribution K and the coefficient of diffusion D_1. The separating efficiency of stationary liquids can be described by stating 5 index differences or 6 retention ratios. It would be welcomed, if these data were stated when publishing new separating liquids in order that new separating liquids could be compared with those already existing.

References

1. Scott, R. P. W.: Applied Gas Chromatography p. 429, Butterworths, London 1960.
2. Rohrschneider, L.: Chromatographia 1, 108 (1968).
3. Purnell, I. H.: Nature (London) 184, 2009 (1959)
4. van Deemter, I. I., Zuiderweg, F. I., and Klinkenberg, A.: Chem. Engng. Sci. 5, 271 (1956).
5. Hawkes, S. I., and Carpenter, D. I.: Anal. Chem. 39, 393 (1967).
6. Rohrschneider, L.: Z. anal. Chemie 170, 256 (1959).
7. Cremer, E.: Discussion remark. Z. anal. Chem. 181, 364 (1961).
8. Rohrschneider, L.: J. Chrom. 22, 6 (1966).
9. Rohrschneider, L.: Z. anal. Chem. in print.
10. Macek, A. O S., and Phillips, C. S. G.: J. Chrom. 29, 7 (1967).
11. Purnell, J. H., and Quinn, C. P. in Scott, R. P. W.: Gas Chromatography 1960, p. 184, London: Butterworths Sci. Publ. 1960.
12. Guiochon, G.: Anal. Chem. in print.

On Column Support Materials

H. ROTZSCHE

Institut für Silikon- und Fluorkarbonchemie, Radebeul

1. Introduction

Although solid supports have to be completely inert and should not contribute to retention it is known that, without preparatory treatment, solid support material based on Kieselguhr shows some adsorption effects and chemical reactivity with reference to certain dissolved substances. Problems of this type are briefly dealt with in the paper; then follows a concise survey of the various methods of deactivation, and a detailed discussion on silylation as well as on the use of trimethyl chlorosilane, hexamethyl disilazane, dimenthyl dichlorosilane, trimethylsilyl diethylamine and N,O-Bis(trimethylsilyl)-acetamide. Moreover, the studies comprise the influence of solvents on the silylation with hexamethyl disilazane and a new method concerning the chemical modification of column support material. Finally, the paper reveals the various advantages obtained with tetra-fluorethylene in the investigation of reactive compounds.

2. Support Materials

Solid supports serve to retain the liquid phase and to obtain a large number of theoretical plates with large interfaces between the mobile and the stationary phase without solute interactions. From a large number of investigated materials, only a few products based on silica (about 90 per cent of the entire column support material), PTFE and, to a lesser degree, glass or quartz beads proved successful. Some of the frequently used materials of the first mentioned type are compiled in Table 1.

It should be pointed out, however, that the above materials are not as inert as required for the investigation of polar substances at low liquid loadings which, for certain reasons, proved of advantage. Considering the chemical composition of these solid supports this is by no means surprising (cp. Table 2).

Table 1. Solid Supports Based on Silica

Kieselguhr
Celite 545
Embacel
Porolith
Chromosorb
Anachrom
Rysorb
Sil-O-Cel C 22 brick
Sterchamol

Table 2. Average Chemical Composition of Solid Kieselguhr Supports [1]

SiO_2	90.0%	MgO	0.5%
Al_2O_3	4.5%	$Na_2O + K_2O$	2.5%
Fe_2O_3	1.5%	moisture content	0.5%
CaO	0.5%		

(Porolith, Rysorb, and Sterchamol deviate from the given values [1] as their Al- and Fe-content is somewhat higher).

The non-inert behavior of solid supports is reflected in their chemical activity and adsorption effects.

3. Reactivity of Solid Supports

Various solid supports showed isomerization of terpenes, allenes, and dienes; cis-trans-isomerization of fatty acid methyl esters; chain isomerization of alkyl benzenes; dehydration of alcohols; conversion of benzoin into desoxy benzoin and benzil; as well as decomposition of amino derivates, alkaloids, acid halides, chlorosilanes and aminosilanes. But even liquid phase decomposed at elevated temperatures when applied to non-inert solid supports; especially through dehydration of hydroxy group containing phases as well as catalytic autoxydation of hydrocarbons, phthalates, and sebacates, caused by the solid support. This can already be observed at a temperature of 150 °C provided the carrier gas contains traces of oxygen.

The causes of such chemical interactions virtually depend on the surface area, the presence of active, catalytic sites (Al, Fe, SiOH), and the pH value of the solid support.

4. Adsorption Effects - Tailing

These effects primarily depend on size and nature of the surface area, but also on liquid phase, sample volume, type of substance, molar weight of the sample, liquid loading, and column temperature as well. Such phenomena are rather detrimental to column performance and efficiency, quantitative analysis calculation of physical data, qualitative evaluation (as the retention and the elution order markedly depend on the liquid loading and the sample volume), and especially to trace analysis. For instance, retention time changes with the sample volume and, with the latter being constant, the peak height increases after deactivation as the major part is eluated immediately rather than after desorption.

Adsorption occurs at the surface centers of the solid support. With loadings of less than 2 per cent of a stationary phase the retention of polar samples is predominantly determined by the coated support. Although this influence diminishes with rising polarity of the liquid phase, a certain influence of the solid support has been noticed on the retention of methanol, for example, at a 22 per cent loading of tricresyl phosphate.

Naturally, tailing can have other causes too, for instance, through diffusion processes of the solute in the liquid (in case of filled, narrow and deep pores - i. e., with a long diffusion distance - the equilibrium of the sample is reached only very slowly) or by dimerization in the liquid (with fatty acids).

Active centers are preferably formed by mineral contaminations present in the solid support (Al, Mg, Fe) which, as acceptors of free electron pairs opposite to donors of free electron pairs ($-NH_2$, $>NH$, $>N$, $>C=O$, $-CN$, $-NO_2$, unsaturated hydrocarbons) may release coordination forces. Metal atoms (Al, Fe) with an OH group may even cause chelate formation, provided the solute contains two or more OH or NH_2 groups spaced 2 to 3 C atoms apart (glycols, amino alcohols, etc.).

The second type of active centers comprise silanol groups and water molecules retained by coordination. Here hydrogen bonds with H-donors and H-acceptors are predominating.

Substances especially susceptible to tailing are (in falling order) polyamines, glycols, fatty acids, primary, secondary and tertiary amines, primary, secondary and tertiary alcohols, esters, ketones, and ethers. Tailing increases with falling molar weight. Hence, as compared to C_2H_5OH, increased tailing is observed with CH_3OH.

5. Polytetrafluoroethylene Supports

Difficulties arising from the described activity are best circumvented by using solid supports based on PTFE (cp. Table 3).

Initially, Teflon-1 of Du Pont was widely used which, owing to its small specific support surface, proved insufficient and contributed to the opinion that such a material is not suitable for this task, However, later works of LANDAULT and GUIOCHON [2], HORNING, MOSCATELLI and SWEELEY [3], STASZEWSKI and JANÁK [4], KIRKLAND [5], as well as studies in the laboratory of the author revealed that theoretical plates were possible down to 1.7 mm height equivalent. From Fig. 1 to 4 it will be seen how favorable solid supports of this type are, especially with reference to the investigation of reactive compounds. Their superiority is most clearly revealed in the trace analysis of the mentioned reactive substances. Moreover, compounds with pronounced hydrogen bonds were symmetrically eluated (cp. Fig. 5).

Table 3. Solid Supports Based on PTFE

Trade Name	Specific Surface (m^2/gm)	H_{min} (mm)
Teflon-1	0.23	10.0-50.0
Teflon-6	10.90	1.7- 3.4
Chromosorb T	7.80	
Halopak F	7.80	
Columpak T	4.80	
Fluoropak 80	1.30	
Heydeflon		2.1 (pyridine)
Fluoroplast 4		1.7 (benzene)
		3.8 (pyridine)
Firebrick/PTFE		3.1

Laboratory technicians are by no means too enthusiastic about charging columns, or loading and screening of such a solid support, as PTFE powder is rather soft and liable to electrostatic charging. Cooling to 0 °C, and avoiding glass vessels, facilitates the handling of PTFE powder to quite an extent. As the surface energy of this material is very low, high column performance prohibits the use of liquid phases with too high a surface energy (e. g., diglycerin) because they are unable to wet the solid support. Methyl silicone oils hydrocarbons, and polyethylene glycol, however, proved to be successful for this task.

6. Porous Polymers

Investigation on strongly polar substances with a lower boiling point is best accomplished by using porous polymers obtained through polymerization of styrene and cross linking with divinyl benzene (Porapak, Chromosorb 101, Chromosorb 102, Polypak). Polymers of this type show slight tailing only with amines. As only interactions due to dispersions prevail there is no H-association. Thus, water with its low molar weight is eluated at a rather early state. Owing to the fact, however, that the material not only serves as solid support but also participates in the entire process, it is not possible to further deal with this material within the scope of this paper.

7. Deactivation Methods for Kieselguhr Supports

Because a Kieselguhr support apparently yields the highest column performance this material was treated in different ways [1] so as to eliminate the undesirable activity, i. e., : -

Washing with acids	(removes part of the Al and Fe content from the support surface)
Loading with silver	(reduces adsorption)
Loading with tide detergent	(eliminates troublesome effects only in certain cases)
Loading with amines and amino alcohols	(triethanolamine, Armeen SD, THEED; very good deactivation, but low thermal resistance)
Loading with epoxides [4]	(resistant only up to about 100 °C)
Loading with small amounts of a polar liquid phase	(weakens the activity but - possibly with the exception of formamide suggested by JANÁK [6] - cancels activity completely only in a few cases)
Increased liquid loading	(not always possible and unreliable)
Saturation of the carrier gas flow with a polar substance	(H_2O, NH_3; may cause other reactions)
Frequent sample injection until all active sites are saturated	(this, however, is a pseudo-stable condition as the adsorbed substances are rinsed off as soon as the injection ceases and have to be replenished through a fresh injection)
Covering the support surfaces with polytetrafluoroethylene	(with Teflon on firebrick it was possible to obtain a height equivalent of a theoretical plate of H_{min} = 3.2 mm [7] - cp. Table 3 - and to eliminate tailing; KIRKLAND [5] found a similar value (3.4 mm) for this material but still observed some tailing at the H_2O peak. According to his findings, and in compliance with the author's investigations on PTFE-covered Sterchamol, it is unlikely that the entire support surface can be covered completely. Material of this type is superior to pure PTFE only insofar that it permits better handling)

Silylation is of great significance and dealt with in detail in the following section.

8. Silylation

Deactivation of active centers due to the presence of silanol groups is best accomplished through silylation where the hydrogen of the free SiOH group is replaced by organosilyl remnants. The residual $Si-O-SiR_3$ formed during this process remains thermally stable even at a temperature of more than 300 °C.

Silylation with trimethyl-chlorosilane [8, 9] (TMS) gives

$$-\overset{|}{\underset{|}{Si}}-OH + Cl-\overset{CH_3}{\overset{|}{\underset{\underset{CH_3}{|}}{Si}}}-CH_3 \longrightarrow -\overset{|}{\underset{|}{Si}}-O-\overset{CH_3}{\overset{|}{\underset{\underset{CH_3}{|}}{Si}}}-CH_3 + HCl \qquad (I)$$

and, same as the silylation with hexamethyl-disilazane [10] (HMDS) and its reaction

$$2-\overset{|}{\underset{|}{Si}}-OH + HN\left[Si(CH_3)_3\right]_2 \longrightarrow 2-\overset{|}{\underset{|}{Si}}-O-\overset{CH_3}{\overset{|}{\underset{\underset{CH_3}{|}}{Si}}}-CH_3 + NH_3 \qquad (II)$$

will not cover all SiOH groups owing to the voluminous $(CH_3)_3$ Si group [8, 11]. Hexamethyl-disilazane proved to be still more reactive than trimethyl-chlorosilane. If both substances are used together to increase the overall reactivity we obtain:

$$3-\overset{|}{\underset{|}{Si}}OH + (CH_3)_3SiCl + (CH_3)_3SiNHSi(CH_3)_3$$

$$\longrightarrow 3-\overset{|}{\underset{|}{Si}}-O-\overset{CH_3}{\overset{|}{\underset{\underset{CH_3}{|}}{Si}}}-CH_3 + NH_4Cl \qquad (III)$$

However, as a follow-up process, the silanized support must be heated to more than 335 °C in order to remove the NH_4Cl content.

Our findings concerning the silylation of OH containing substances with hexamethyl-disilazane alone revealed that the solvent has a pronounced influence on the rate of reaction II [12] (cp. Table 4).

Table 4. Influence of the Solvent on the Reaction of OH Groups with Hexamethyl-Disilazane (Measured by Means of NH_3 Generation)

dioxane	increasing reaction velocity ↓
dibutyl ether	
xylene	
butyl acetate	
chlorobenzene	
without solvent	
quinoline	
methylethylketone	
pyridine	rate of reaction approximately constant
m-nitrotoluene	
benzyl cyanide	
dimethyl sulphoxide	
dimethyl formamide	

Utilization of petroleum ether as well as working without solvents yields inferior results as compared to conversion in a polar solvent. This has been verified by the silylation of Porolith with HMDS (cp. Fig. 6 and 7). Using dimethyl formamide instead of petroleum ether yields a material which is less active and the CH_3OH_c content is already eluated quantitatively and more symmetrically during the first sample injection. This gains in significance as previous acid washing has been dispensed with.

Silylation with dimethyl dichlorosilane (DDS) [13, 14, 15] results in a very good inactivity. Perhaps this is due to an enhanced reactivity as compared to TMS, and to the better penetration of the pores as well as the deactivation of the vicinal SiOH groups. Thus, we obtain:

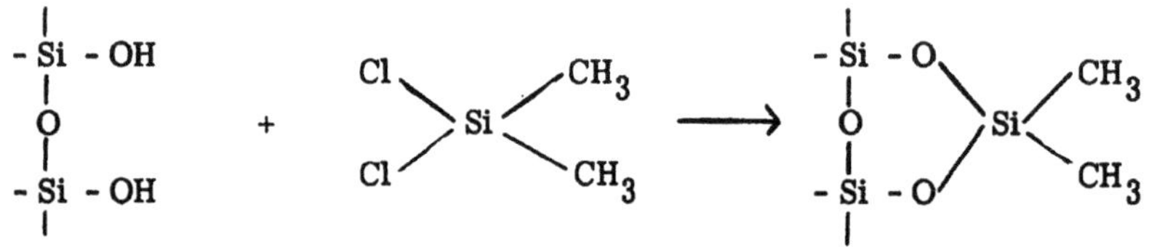

(IV)

Frequently it will be impossible for both Cl atoms to react and we find:

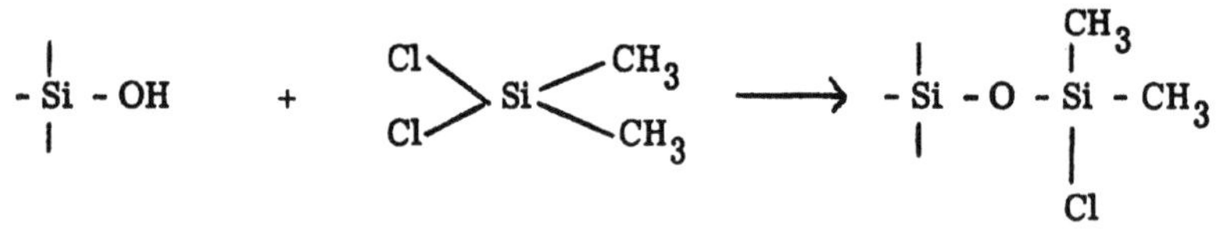

(V)

Here a reactive Si-Cl group remains which, through subsequent conversion with CH_3OH, has to be replaced by a $Si-OCH_3$ group. This yields:

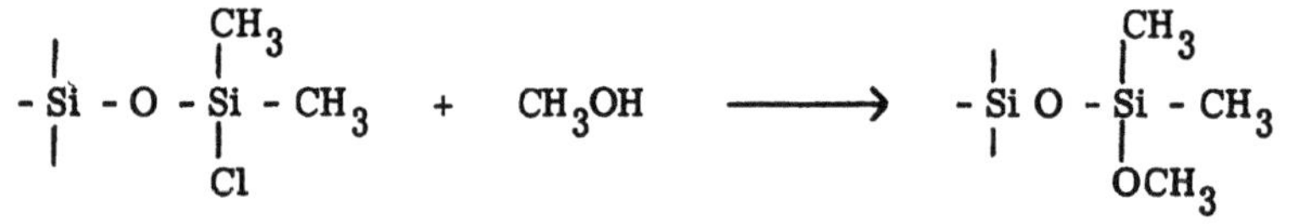

(VI)

Such a CH_3OH treatment serves to remove the generated hydrochloric acid at the same time. Unfortunately, this process has the disadvantage that the $Si-OCH_3$ group is subject to hydrolysis and - owing to the moisture content of the ambient - may be converted into Si-OH again.

Recently, other silylation reagents have been employed in situ for the rejuvenation of columns which have become active, such as trimethyl silyl diethylamine for the trimethylsilylation of amino acids [16]. This gives

$$-\overset{|}{\underset{|}{Si}}-OH + (CH_3)_3SiN(C_2H_5)_2 \longrightarrow -\overset{|}{\underset{|}{Si}}-O-Si(CH_3)_3 + HN(C_2H_5)_2 \quad (VII)$$

Moreover, the Pierce Chemical Corp. offers N, O-Bis (trimethylsilyl) acetamide for the same purpose which shows the following reaction:

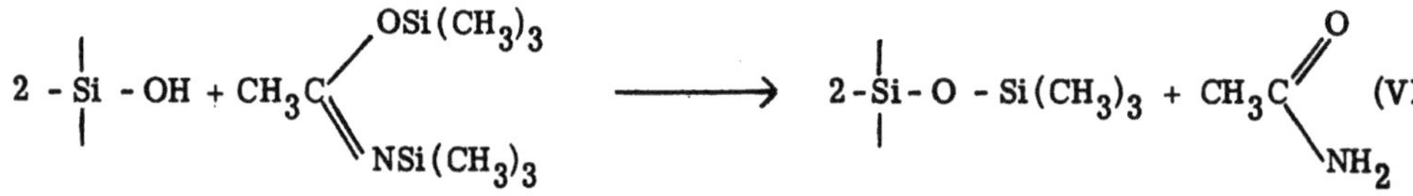

(VIII)

These products react with the active sites of the column support material, with the column walls, the sample injector, and with contaminations present in the column. Naturally they cannot be employed wherever they are liable to react with the liquid phase.

After previous acid washing and subsequent loading with 0.1 to 0.5 per cent of a $-\overset{|}{\underset{|}{C}}-OH$ or $-\overset{|}{\underset{|}{C}}-NH_2$ group containing liquid phase, the already described silylation methods yield rather inactive supports. On the other hand, loading of this type has a certain disadvantage especially with regard to thermal considerations. Hence, in our laboratories we tried to carry out the deactivation merely with silicon-organic groups.

Of course it was obvious to replace the commericially available solid supports with their mineral contaminations by other materials of a rather high chemical homogeneity. So we used technically pure methyl silicone acid of various granulation and - after silylation of the free Si-OH groups - investigated the products

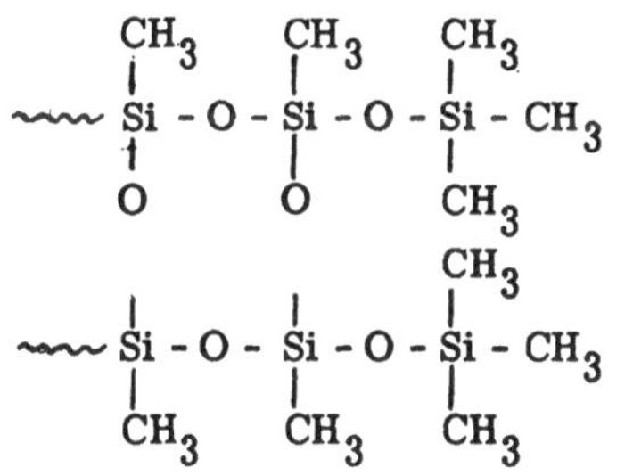
(IX)

with reference to their gas-chromatographic suitability. As, after trimethyl silylation, methyl silicone acid has been found free from OH groups, and without traces of Fe or Al, it is possible to obtain symmetric curves even with substances which are difficult to handle (cp. Fig. 8). However, the structure of the pores is more unfavorable, and the specific support surface smaller than with Kieselguhr. As a consequence it proved impossible to obtain comparable H_{min}-values. Thus, we tried to coat a commercial solid support of high H_{min}-value chemically with a similar substance as specified under (IX) [17].

At elevated temperature, Porolith was converted with methyl trichlorosilane (MTS) which is more reactive than TMS, HMDS or DDS. In cases where three SiOH groups are in such a vicinal arrangement that all three Cl atoms of the MTS react (which, however, is an exception) the process (X) runs as follows:

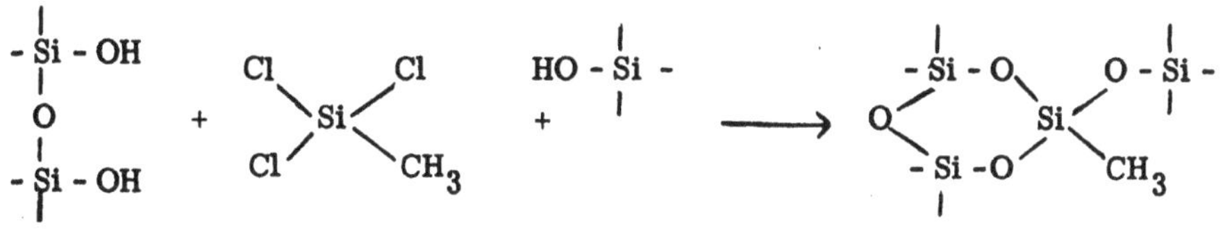
(X)

In general, one or two Si-Cl groups are maintained, viz.:

(XI)

With water, these groups are then converted into $-\overset{|}{\underset{CH_3}{Si}}-OH$ groups and then trimethyl-silylated by means of hexamethyl di-silazane:

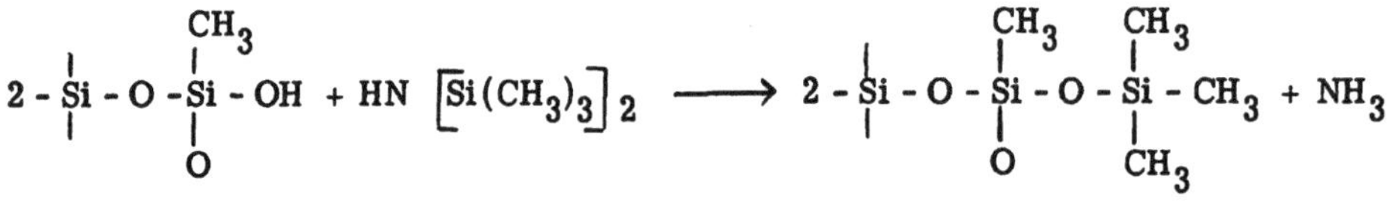

(XII)

In this way, MTS causes rapid conversion of the Si-OH groups located on the surface of the solid support. Moreover, the exchange of non-converted SiCl grops with $Si-OCH_3$ groups is eliminated which, in the presence of moisture, hydrolize anew anyway. Based on the reaction (XII), this gives a thermally stable solid support which is practically free from Si-OH groups and superior to those obtained with DDS, HMDS, or TMS treatment. Owing to its 'methylized' surface such a solid support is easily wetted by hydrocarbon or silicone phases. Hence, H_{min}-values of 1 mm or less can be obtained without difficulty. From Fig. 9 and 10 the excellent inactivity becomes clearly evident. In this context it should be mentioned that the Fe and Al containing Porolith [1] has been treated without previous acid washing. Moreover, it is known that silanized supports cannot be wetted thoroughly by a polar liquid phase, especially if the latter is subject to hydrogen bridge linkage. Replacing CH_3SiCl_3 in (X) and (XI) with either $NC-CH_2-CH_2-SiCl_3$, $NC-(CH_2)_3SiCl_3$ or $CF_3-CH_2-CH_2-SiCl_3$ serves to overcome this drawback and may improve coating even with liquid phases of this type.

Finally, it should be pointed out that inactive solumn support material alone won't suffice. Troublesome effects may arise from adsorption phenomena at the walls of columns, sample injectors, detectors and connecting lines, or contaminations contained therein. Such detrimental effects have been noticed especially with eutectics used as stationary phase as, in this case, bleeding of the liquid phase - which generally results in the saturation of the most active sites - does not occur.

References

1. Rotzsche, H.: „Gefüllte Säulen mit flüssiger Phase" in „Handbuch der Gas-Chromatographie", ed.by Leibnitz, E., and Struppe, H. G., Akademische Verlagsgesellschaft Geest & Portig KG, Leipzig, 1966, pp. 158-199.
2. Landault, C., and Guiochon, G.: J. Chromatog. (Amsterdam) 9, 133 (1962).
3. Horning, E. C., Moscatelli, E. A., and Sweeley, C. C.: Chem. and Ind. 1959, 751.
4. Staszewski, R., and Janák, J.: Collect. Czechoslov. Chem. Commun. 27, 352 (1962).
5. Kirkland: Anal. Chem. 35, 2003 (1963).
6. Novák, J., and Janák, J.: in Advances in Gas Chromatography 1965, ed. by Zlatkis, A., and Ettre, L. S., Preston, Ill., 1966, p. 64
7. Onaka, T., and Okamoto, T.: Chem. & Pharm. Bull. (Japan) 10, 757 (1962).
8. Kohlschütter, H. W., Best, P., and Wirzing, G.: Z. anorg. Chem. 285, 236 (1956).

9. Babkin, I. J., Vasileva, B. S., Drogaleva, I. V., Kiselev, A. V., Korolev, A. J., and Schtscherbakowa, K. D.: Dokl. Akad. Nauk USSR 129, 131 (1959).
10. Bohemen, J., Langer, S. H., Perrett, R. H., and Purnell, J. H.: J. Chem. Soc. (London) 1960, 2444.
11. Kiselev, A. V., and Schtscherbakowa, K. D.: Abhandlung der Deutsch. Akad. Wiss., Klasse Chem., Geol. u. Biolog. 1962, 1, pp. 207-240.
12. Rotzsche, H.: In preparation.
13. Howard, G. A., and Martin, A. J. P.: Biochem. J. 46, 532 (1950).
14. Sweeley, C. C., and Moscatelli, E. A.: J. Lipid Res. 1, 40 (1959).
15. Sweeley, C. C., Moscatelli, E. A., and Ta-Chuang Lo Chang: Anal. Chem. 33, 1860 (1961).
16. Rühlmann, K., and Michael, G.: Gas-Cromatographie 1963, Publishers Angelé, H. P., and Struppe, H. G., Akademie-Verlag Berlin 1963, p. 151.

The Role of Gas Chromatography in a Combination System for Substance Identification

G. SCHOMBURG and D. HENNEBERG

Max-Planck-Institut für Kohlenforschung, Mühlheim-Ruhr

1. Introduction

The combined application of a separation method and of an identification method allows a quick analysis of complex mixtures. The number of components and very often their structures may be determined by means of such „combinations". Reliable structural information from NMR, IR or mass spectroscopy may only be obtained from spectra of pure or strongly enriched materials. The purification of components can be done by gas chromatography, if volatility and thermal or catalytic stability of the different species allow its application.

Gas chromatography is able to achieve separations of high efficiency by means of high theoretical plate numbers, and in addition the polarity of the stationary phase may be varied. Frequently only special column types - for example open tubular columns - show the sufficient separation power desirable. It is important to realize that not only high plate numbers per meter of column are needed but that a high number of theoretical plates is available for the complete chromatographic system.

Separation power decreases strongly with increasing sample load. Especially with high-resolution-columns the amount of liquid phase within a theoretical plate is so small that even with high sample loads not enough material for spectroscopic identifications can be separated.

The amount of substance needed for the different spectra often prevents the application of the typical identification methods, if high plate numbers are to be applied. Moreover, further difficulties arise, if the separated species must be trapped from the hot eluate in which they are present in low concentrations. Also the manipulation, i.e., the transfer of the trapped samples, is sometimes difficult.

Under these aspects, on-line-combinations are of special importance especially if the separated substance is not needed in the liquid state. In this case, however, the substances hare to be isolated from the eluate. For these reasons on-line-combinations with gas chromatography are carried out in practice only mass spectroscopy. Particularly in the case of compounds not yielding informative mass spectra, additional structural information must be drawn from the gas chromatographic retention and response data.

2. Separation power and sample loads of gas chromatographic columns

The application of columns with high resolution is necessary if the species to be separated do not show appreciable differences of polarity. In this case, alteration of polarity causes no change in the relative volatility which depends on

$$\alpha_{1,2} = \frac{{}^{o}p_1 \gamma_1}{{}^{o}p_2 \gamma_2}$$

the vapor pressures of the pure components ${}^{o}p_i$ and the activity coefficient γ_i of the liquid phase. Unknown mixtures occurring, for example, in the chemistry of natural product must, therefore, be separated with columns of varying polarity in order to find the most suitable liquid phase yielding the least coincidences of peaks. With complicated mixtures containing species of strongly differing polarity new overlappings of peaks are sometimes obtained, if others are removed.

If such measurements are performed directly within a GC-MS combination the finger print information from the mass spectra of the different separated species can be used to recognize the same component within different chromatograms. A practical example for this is to be seen in Fig. 1. The different carboranes could only be recognized within the two chromatograms by measuring the molecular weights of every component. The exact correlation of peaks in the two phases was necessary for the determination of ΔI-values (KOVATS), which characterize the separated species according to their polarity. If no additional or new peaks are discovered by changing the liquid phase, the separation efficiency of the gas chromatographic system is increased in order to separate components with small differences in polarity.

The optimal sample load for a certain column system depends on the following factors:

a) Amount of liquid phase on the theoretical plate.
b) Plate number or separation efficiency.
c) Column diameter.
d) Partition coefficient.

Investigation concerning the influence of these factors have been made principally with preparative scale columns (comp. for example G. W. A. RIJNDERS [1]). Good separations with packed „analytical" columns (4 mm internal diameter, 0.5-15 weight % liquid phase on support) are obtained with sample loads of 0.01-2 mg, whereas with capillary columns of the open tubular type with 0.25 mm i. d. the possible sample loads are of the order of 0.01-1 μg per substance, i. e., 500-1000 times smaller.

Greater amounts of separated components of mixtures can be obtained by applying the following techniques:

a) Overloading of the column
(especially effective with large retention differences of the species to be separated).
b) Application of columns with greater diameters.
c) Repeated injections.

a) and b) give rise to a marked decrease in separation efficiency and c) is of course not applicable with an on-line-combination. Further difficulties occur in the trapping of separated substances of low concentration in the carrier gas.

3. The utilization of spectroscopic methods in on-line-combinations. Problems of substance requirements.

Essential points of on-line-combinations of gas chromatography and spectroscopic methods are:

a) During scanning of the spectrum only negligable or small changes of substance concentration within the carrier gas at the elution of the peak may take place. For this reason scanning times below at least 10 sec. per spectrum are desirable.

b) The sensitivity of the combined spectroscopic method has to be sufficient to obtain spectra of good quality with appreciable intensity particularly for small frequencies or masses.

The three most important methods to be considered are

mass spectroscopy,
nuclear magnetic resonance spectroscopy,
infra red spectroscopy.

3.1. Mass spectroscopy: Modern mass spectrometers with multiplier amplification allow very high scan rates. The scanning times for a normal mass spectrum in the range between mass number 20 and 400 are only 1 to 2 seconds. The sensitivity for measuring total ionization in the mass spectrometer is at least as high as that of the flame ionization detector. For these two reasons GC-MS combinations play a predominant role in the application of on-line-combination [2].

3.2. Nuclear magnetic resonance spectroscopy: Sensitivity as much as the scan rates are much too small with the classical techniques. Additionally, the sample has to be available in the liquid state (film or solution). Even with the discontinuous or off-line-combination technique, the intensity of the NMR spectrum has to be improved by application of a time averaging technique. The geometry of the probe tube which cannot be chosen independently, also prevents the measurement of a spectrum with very small samples [3].

Teranishi isolated 0.8 mg i-pulegol from the eluate of a 1000 ft 0.03" i. d. capillary column. An adequate NMR spectrum of this sample diluted in 50 μl CCl_4 was obtained after an 800 times repeated scan with a C. A. T.

3.3. Infra red spectroscopy: Scan rates of a few seconds per spectrum are only possible with specially designed instruments, for example the Beckman IR 102, which is a single beam grating instrument. The geometry of the solution [4]. The application of beam condensers allows the use of samples below ~1 mg according to intensity of absorbance of the different substances.

An on-line-technique is only possible with the vaporized sample in a heated gas cell. Spectra of adequate quality are only obtained with a reasonable path length. The high dead volume of such cells and the limited sensitivity of the IR detectors prevents the on-line-application of capillary columns. In actual practice IR spectra are only measured off-line with samples as they are obtained from analytical packed columns. Very often repeated separation and trapping of the desired peak is necessary.

R. P. W. SCOTT et al. [6] describe a combined GC-IR arrangement which works with a heated gas cell connected as a bypass to the carrier gas stream. By leading the eluate stream into the IR cell at the moment the desired „peak" is emerging from the TC detector the species is trapped together with the carrier gas in this cell. The carrier gas stream is switched to the normal exit of the chromatograph if the substance is completely trapped in the cell. The spectrum is scanned with the „standing" sample. The necessary scan rate is now independent of the elution rate of the sample. Scott calls this technique, for which 0.4 - 5 mg of trapped sample are necessary, stop-start chromatography.

On-line-combinations with gas chromatography have been realized in practice only with mass spectroscopy since only mass spectroscopy meets the three major requirements

a) High sensitivity.
b) High scan rates.
c) Measuring of the spectrum must be possible in the vapor state.

4. Combination of Gas chromatography and mass spectrometry

Considering first the gas chromatographic system, sufficient separation power and suitable choice of polarity of the stationary phase are conditions for effective combination work.

Four further points must be met in combination of gas chromatography (GC) and mass spectrometry (MS):

a) Sufficient quantity of sample.
b) Minimum bleeding of the stationary phase.
c) Minimum tailing of peaks
d) Carrier gas must be helium or hydrogen.

The amount of sample needed in the ion-source of the mass spectrometer (MS) to yield a good spectrum varies from instrument to instrument, but is of the order of magnitude of 10^{-8} g which should pass the ion source during the time for registration of the spectrum. Correspondingly, the amount of sample to be injected into the gas chromatograph (GC) depends widely on the type of column and connection of the instruments used.

The bleeding of a column causes a constant background spectrum, which may be „subtracted" from sample spectra, if it is well defined and does not overlap too greatly with that of the sample; for instance, squalane with saturated hydrocarbons. The ratio of sample to background may be improved by introducing larger sample or by lowering column temperature. In principle the mass spectrometer ion source is an integral detector in contrast to the GC detectors which are of the differentiating type giving signals only for changes of constant concentration.

For the same reason, it is better for GC-MS combination to have peaks which are only partly separated than to have peaks on a tail, as is demonstrated in Figure 2. Obviously in case A pure spectra can be achieved in front or at the rear of the partially separated peak. In case B an optimum purity of about 90% is obtainable although the separation between both peaks is twice that of case A.

Not only tailing of peaks caused by the column must be avoided. The connection system itself and the MS inlet system often give rise to additional tailing because there are large surfaces where the pressure is very low. This means that the sample molecules on the way to the ion source hit the walls of the apparatus with higher probability instead of carrier gas molecules. Adsorption and sometimes even decomposition takes place. This must be taken into account when choosing the type of connection.

The types of connection of GC and MS with and without a separator are shown in Figures 3 and 4.

If the carrier gas flow does not exceed 1-2 ml/min, the complete effluent can be admitted to the MS as shown in fig. 3 A. In this case (open tubular columns with 0.25 mm i. d.) the end of the column is introduced into the vacuum system of the MS as close as possible to the ionization chamber, to avoid collisions with the wall. This system is preferred by the authors for capillary column work because it allows complete retention of the resolution of the column [6]. For instance, with a 100 m capillary column, 0.25 mm i. d., coated with DC 200, 350,000 theoretical plates (120,000 effective theoretical plates) have been obtained.

A greater flow from the column must be split as in figure 3 C by using a capillary leak or an adjustable leak valve. This type of connection is obsolete, however, since it offers no advantages compared with the more recently introduced separators as the splitting element. Separators are devices which reduce the pressure down to values acceptable for the ion-source by using processes which result in a reduction of the amount of carrier gas and not of sample.

The two types mainly used at present are shown in figure 4 [7-9]. For trace analyses they are extremely useful. But two points should be mentioned regarding the application of separators:

Together with high resolution gas chromatographic systems a reduction of separation efficiency was reported [10] for the frit type separator. On the other hand, an example [11] of decomposition of a sample at the walls of a separator has been reported (Fig. 5). Compare also [12]. In principle the RYHAGE type separator should have advantages because the sample suffers fewer wall collisions, but no results have been reported as yet with systems of high separation efficiency or investigations of the type described by LEEMAN and McCLOSKEY [11].

Two other types of separators may be mentioned here. One uses a heated teflon tube instead of a glass frit [13]. The other works according to the opposite principle [14]. A silicon rubber diaphragm allows the sample to pass but not the carrier gas, which is insoluble in the diaphragm. This type would cause strong tailing and perhaps for this reason only off-line application is reported [15].

The enrichment factor of the various types in their applications is so great that differences become negligible. One fact that should perhaps be mentioned was pointed out by the authors early; the best separator can only be a disadvantage in cases where the effluent is introduced into the MS completely.

5. Isotope scan and fixed mass technique

The normal technique for obtaining a complet mass spectrum in about 2 sec, the fast scan method, is of limited usefulness when either the background spectrum is comparable with the spectrum of the desired compound or when two components are poorly separated. In these cases a differentiating detector, i. e., a detector capable of registering changes of concentration of a compound is desirable.

Two techniques have been used successfully which meet these conditions. The first is to operate the MS fixed at a certain mass number. Thus the MS is a specific detector. By selecting a mass which is sensitive mainly to certain compounds, these compounds are registered and other suppressed. This technique has been described by the [16-18] authors. The main advantage is that background does not exist, because the MS acts as a poor differentiating detector. Furthermore, trace components can be detected in the presence of high concentrations of major compounds if the MS is fixed at a mass number which has strong sensitivity in the trace compound spectrum, but weak sensitivity in the other. The disadvantage is that no spectrum is obtained. To obtain part of the spectral information the gas chromatographic run must be repeated on other mass numbers or a MS with 4 adjustable collectors must be used [19]. A special version is the plate-scan technique which is applicable with a MS of MATTAUCH-HERZOG type [20], where the variation of the complete spectrum versus time is obtainable.

To overcome the drawback of lack of spectral information, methods have been applied where the MS is switched alternatively between two fixed mass numbers [21] . Fig 6 shows an example.

The authors prefer a technique of recording a continuous part of the spectrum repeatedly [22]. Mass range and repetition rate can be adjusted to the individual problem. For instance, it is possible to cover 500 mass units 5 times per second. During elution of a carborane containing three boron atoms the mass region across the molecular ion was scanned about 30 times. In Fig. 7 the heights of three mass peaks corresponding to ions with $^{11}B_3$, $^{11}B_2{}^{10}B$ and $^{11}B^{10}B_2$ are given across the chromatographic peak and show a well defined separation of about 1/20 of halfwidth of the peak. This corresponds to an isotope effect on chromatographical separation of 0,05 index units per boron atom. Similarly the influence of ^{13}C has been determinde to 0.06 index units. The authors used the expression „isotope scan" for this technique, which is extremely useful to determine the peak shape of unresolved peaks. Fig. 7 represents peaks which correspond to a separation efficiency of 250,000 theoretical plates. Together with the additional separation of the chromatographic peak by the mass spectra an apparent resolution of some millions of theoretical plates has been achieved.

With this method a difference in retention has been observed between two d_2-alkanes which differed only in the position of the deuterium atoms.

6. GC-MS Identification supported by gas chromatographic retention data

It has been shown [23-28] that the retention behaviour of isomeric hydrocarbons and their simple functional derivates can be predicted using the KOVÁTS retention index system. The structural information which can be derived from retention data using open tubular columns and performing the index measurements with sufficient reproducibility is particularly valuable when applied to closed classes of compounds or mixtures of isomers. Preknowledge of the chemical origin of the separated mixtures facilitates greatly the correlation of peaks.

It is an important point that the separation of complex mixtures of isomeric hydrocarbons is nearly impossible without capillary columns. As pointed out above only mass spectra of the separated species can be obtained from the separable amounts of substance. Moreover these spectra can only be measured within an on-line GC-MS combination. Further structural informations about the separated materials, for example, from NMR or IR spectra, are not accessible.

The results mentioned below have been obtained with methylene insertion reaction (MIR) mixtures made from alkanes mainly with the intention of collecting retention data and mass spectra of isomeric hydrocarbons in order to investigate the retention behavior and the mass spectrometric fragmentation. The identification of the components of such a mixture by combined discussion of retention indices and mass spectra was supported by a wide experience of types and number of isomers, diastereomers included, arising from the MIR. The following examples are meant to illustrate the measurements which have been done in this field and a complete discussion is reported elsewhere [29].

Figure 8 and 9 show the separation of two isomeric mixtures obtained by MIR from 2-methyl-undecane and 5-methyl-undecane. The two chromatograms of every figure have been measured with capillary columns and FID in a normal gas chromatographic apparatus and with a GC-MS combination in which the capillary column end was introduced directly into the ion source of a MAT CH 4 mass spectrometer with a double ion source. It may be visually observed that there is no loss in separation within the combination.

The chromatograms from the GC - MS combination have been shortened by raising the temperature by 10 °C. The advantage of the higher temperature is the higher concentration of the isomers in the carrier gas, since separators cannot be used with this type of open tubular columns.

The correlation of peaks with the isomers to be separated from the MIR was attempted by calculation of the I^S values using branching increments, as described previously [28] (retention index in the liquid phase squalane). The use of other than non-polar stationary phases such as squalane or silicon oil DC 200 for separation and characterization of saturated hydrocarbons is not very useful because of the low solubility of these compounds in polar stationary phases.

In the case of the elution order of the 2,2-, 2,4-, 2,6- and 2,5-dimethylundecanes a safe correlation from the calculated retention data could not be obtained because of the neighbouring positions of the branched methyl groups which may be taken into consideration by introducing additional interaction increments, which could be determined accurately from the measured data. Moreover, the I^S values of the three isomers did not differ very much.

The difficulties of correlation of peaks of chromatogram II from figure 9 arise from the overlapping of too many peaks in the region between peak number 3 and 8, of which several could be expected to be diastereomers.

The Figures 10 and 11 show a comparison of measured I^S values with the values calculated simply by addition of the two branching increments. It comes out that additional interaction increments have to be introduced:

α,β-dimethyl	+ 19,5
α,γ-dimethyl	- 14,0
α,δ-dimethyl	- 7,2
α,ε-dimethyl	- 5,0
α,ζ-dimethyl	- 1,5

The reliable identification of peaks was only possible by combined use of the retention increments and of fragmentation rules for the iso-paraffins, which could be derived from the spectra, comp. D. HENNEBERG and G. SCHOMBURG [30].

Two examples of this mutual support of the two methods are given in Figures 12 and 13. Figure 12 gives a selection of mass spectra of some dimethylundecanes, whose I^S values did not show characteristic differences. The mass spectral fragmentation pattern allows a safe decision between the different possible structures, which show sufficient specifity for evaluation of branching types.

Figure 13 contains the mass spectra of three dimethylundecanes which look, at first glace, very similar whereas the I^S values show strong differences. In this case the I^S values enabled the distinction between the three possible isomers.

7. On-line-hydrogenation within the coupling part of a GC-MS-combination

Isomer mixtures of olefins are only separable with high resolution columns and by taking advantage of the polarity of special liquid phases, e. g. those containing transition metal complexes. Because of the high amounts of sample needed for NMR and IR, these methods, which could of course give decisive information about olefin structures, are not suitable.

Identification from the mass spectra is difficult or impossible, because the pattern of olefin spectra are not sufficiently characteristic and the position of the double bond within the carbon chain does not give rise to a typical spectrum. Even the position of methyl or alkyl branchings in olefins cannot be determined reliably as is possible for branched saturated hydrocarbons. Therefore the simplest procedure seems to be the GC-MS analysis of the hydrogenated sample. The hydrogenation is generally done on-line before the column with H_2 as the carrier gas and a Pt or Pd catalyst. Figure 14 shows that this method leads to a great simplification of the chromatogram, since several of the separated olefin species have the same skeleton and therefore produce one peak. It is very difficult to see from the chromatogram which saturated species was formed from which olefin species by comparison of peak areas. Thus is impossible to know which skeleton a certain olefin has.

If the catalyst is used behind the column but before the mass spectrometer the separated olefin components leaving the catalyst tube one after the other are hydrogenated and afterwards eluted into the ion source of the mass spectrometer [31, 32]. The mass spectrometer delivers from the total ionization signal the MS chromatogram, which looks completely like the olefin chromatogram although saturated hydrocarbons are indicated or measured by the mass spectrometer. The mass spectra of the saturated hydrocarbons can be identified with much higher reliability, thus giving exact skeletal information [30]. Retention indices of the saturated hydrocarbons must of course be determined by hydrogenation before the column either of- or on-line; Figure 15, example I, C_7-olefins from a codimerization of propene and butene-2. The Kovats indices I^S (squalane) of the C_7-monoolefins were well known from our own work and from the literature [33, 34].

There are several isomers which are not separable from others; for example, the peaks with the I^S values of 660.0 and 671.5. From the hydrogenation before the column only the number and the types of different carbon skeletons occurring within the mixture could be evaluated. The hydrogenation behind the column gave reliable information from the mass spectra concerning the carbon skeletons to be correlated with the various olefins. If more than two olefins are eluted with the same retention time, superimposed mass spectra of perhaps three alkanes are obtained, from which the interpretation is only possible if the spectra of the single species show differing characteristic intensities.

Figure 16 shows the chromatogram of the MIR mixture of 1-dodecene which should contain, without the cyclopropane derivative, 13 isomers of which all have been at least partially separated. The correlation of peak 1 to 7 was anticipated to be difficult, because there were always two olefins with the same carbon skeleton. An exception is the 3-Me-1-dodecene which is located amoung the other isomers, because allylic methyl substitution of the olefins always leads to a strongly lowered retention index, compare G. SCHOM-

BURG [24]. The result of the mass spectrometric alkane identification is marked by arrows above the olefin peaks.

Results: The olefin pairs 4-/9-, 5-/8- and 6-/7-methyl-1-undecene are eluted according to the elution order of the methylalkanes. The I^S values generally decrease if the methyl group migrated into the middle of the molecule, compare G. SCHOMBURG [24], A. MATUKUMA [35]. The elution order of the above pairs of olefins could not be evaluated with the same reliability. The skeletons of the olefins couls be derived for two of the three pairs from the mass spectra of the olefins obtained by GC-MS analysis without hydrogenation.

8. Conclusion

The combined use of spectroscopic methods with gas chromatography produces difficulties arising from the amount of sample necessary. With direct couplings or „on-line combinations" the isolation and manipulation (transfer) of very small samples, as they are obtained from gas chromatographic systems of high separation power, is avoided. With real on-line combinations in which the carrier gas stream is not stopped in the IR cell for example, the scan time for the spectrum must be negligibly short compared with the time during which the peak is eluted. Only mass spectrometers deliver sufficient sensitivity for general application in gas chromatography. It is a great disadvantage if the trapped sample must be brought into the liquid state before measuring the spectrum.

Retention data can be used, especially in the case of hydrocarbons, to support identification with spectroscopic methods.

GC-MS couplings can be combined with reaction gas chromatographic techniques before and behind the columns giving additional information about the structures of the separated species.

References

1. Rijnders, G. W. A.: Advances in Chromatography 3, 36 (1966).
2. McFadden, W. H.: Advances in Chromatography 4, 265 (1967).
3. Teranishi, R., et. al.: Anal. Chem. 38, 291 (1966).
4. Littlewood, B.: Cnromatographia 1, 223 (1968).
5. Scott, R. P. W., Fowlis, I. A., Welti, D., and Wilkens, T.: Gas Chromatography Rome 1966, ed. A. B. Littlewood, The Institute of Petroleum 1967, p. 318.
6. Henneberg, D., and Schomburg, G.: Z. f. Anal. Chem. 229, 359 (1967).
7. Ryhage, R.: Anal. Chem. 36, 759 (1964).
8. Stenhagen, E.: Z. f. Anal. Chem. 205, 109 (1964)
9. Watson, G. T. L., and Biemann, K.: Anal. Chem. 37, 84 (1965).
10. Völlmin, J. A., Omura, I., Seibl. J., Grob, K., and Simon, W.: Helv. Chim. A. 49, 1768 (1966).
11. Leemans, F. A. J. M., and McCloskey, J. A.: J. Am. Oil Chemists' Soc. 44, 11 (1967).

12. Kaiser, R.: Über „Tagung des Arbeitskreises für GC-MS-Kopplung" an der ETH in Zürich, Chromatographia 1, 258 (1968).
13. Lipsky, S. R., Horvaths, C. G., and McMurray, W. J.: Anal. Chen. 38, 1585 (1966).
14. Llewellyn, P., and Littlejohn, D.: Conference on Analytical Chemistry and Applied Spectroscopy, Pittsburg 1966.
15. Strube, H.: Meßtechnik 3, 63 (1968).
16. Henneberg, D.: D. Z. anal. Chem. 183, 12 (1961).
17. Henneberg, D., and Schomburg, G.: Gas Chromatography, Hamburg 1962, ed. M. van. Swaay, Butterworths, London 1962, p. 191.
18. Henneberg, D., and Schomburg, G.: Z. f. Anal. Chem. 215, 424 (1966).
19. Kaiser, R.: Chimia 21, 235 (1967).
20. Henneberg, D.: Anal. Chem. 38, 495 (1966).
21. Sweeley, C. C., Eliott, W. A., and Ryhage, R.: Anal. Chem. 38, 1549 (1966).
22. Schomburg, G., and Henneberg, D.: Chromatographia 1, 23 (1968).
23. Schomburg, G.: J. Chromatogr. 14, 157 (1964).
24. Schomburg, G.: J. Chromatogr. 23, 1 (1966).
25. Schomburg, G.: J. Chromatogr. 23, 18 (1966).
26. Schomburg, G.: Separation Science, 1, 2, and 3., 339 (1966).
27. Schomburg, G.: Anal. Chim. Acta 38, 45 (1967)
28. Schomburg, G.: Z. f. Anal. Chem. 236, 279 (1968).
29. Schomburg, G.: no published.
30. Henneberg, D., and Schomburg, G.: Advances in Mass Spectrometry Vol. 4, 1968, p. 16, Elsevier.
31. Schomburg, G., and Henneberg, D.: 7th International Symposium on Gas Chromatography and its exploitation, Copenhagen 1968.
32. Teeter, R. M., Spencer, C. F., Green, J. W., and Smithson, L. H.: J. Am. Oil Chemists' Soc. 43, 82 (1966).
33. Tourres, D. A.: J. Gas Chromatogr. 35, (1967).
34. Matukuma, A.: 7th International Symposium on Gas Chromatography and its Exploitation, Copenhagen 1968.
35. Matukuma, A.: J. Chem. Soc. (Japan), 84, 770 (1963).

Pyrolysis gas chromatography *)

W. SIMON

Department of Organic Chemistry, Swiss Federal Institute of Technology, Zurich

In pyrolysis gas chromatography, the samples are thermally fragmented in a stream of an inert carrier gas and the reaction products are passed directly into a gas chromatography column, where they are separated. There are two extreme possibilities for the application of this technique:

a) The pyrolysis gas chromatogram may be used as a „fingerprint" for the starting material.
b) Correlations between the structure of the fragments and the structure of the starting material may be obtained and used for structural elucidation.

Most of the applications described so far are of the first type (a). Because of the difficulties in the reproducibility due to the instrumental parameters, such applications are usually of local value only. Adequate long term and inter-laboratory reproducibility has not yet been achieved. The potentiality of this method is nevertheless striking. Successful differentation of micro-organismus using fingerprint pyrolysis gas liquid chromatography has been reported [1, 2], and indeed with some advantage over the classical procedure.

The successful use of pyrolysis gas chromatography in the structural elucidation of organic compounds (and in the quantitative analysis of mixtures) is possible under the following ideal conditions [3, 4]:

a) The fragmentation of the sample must be carried out at such a high dilution and the fragments so quickly stabilized that the probability of undesirable recombination reactions is small.
b) The sample must be heated to a constant pyrolysis temperature as quickly and as reproducibly as possible.

---

*) Summary of a talk given at the VI. Symposium über Gas-Chromatographie in Berlin, 14th bis 18th May 1968.

c) The pyrolysis temperature must not be too high, since uncharacteristic fragments (methane, carbon, hydrogen) may otherwise be preferentially formed.

At present, the most promising approach for the study of volatile compounds is the tubular pyrolysis reactor [5, 8, 9]. Since only a few analytical procedures are applicable on nonvolatile or slightly volatile compounds, pyrolysis gas chromatography seems to be especially attractive for application to these samples. There is general agreement that conditions a) - c) can, at present, best be met by using a filament technique in which the sample is coated, in a thin layer, onto a filament [7, 10]. This filament is then subjected to a square wave temperature time profile in a stream of an inert carrier gas [3, 4, 10, 12]. As described earlier, cylindric ferromagnetic conductors of 0.5 mm diameter may be warmed up to a constant final temperature (Curie temperature) in less than 30 milliseconds using high frequency induction heating (0.45 MHz oscillator; 1.4 KW) [3, 4, 6, 12, 13]. It has been reported that the time required for the temperature to rise to the Curie point of iron (770 °C) in the case of similar filaments is about 0.8 sec with an oscillator having on output of around 30 W [14].

When an oscillator with a certain range of frequencies and a conductor of a limited range of diameters is used, the final temperature (Curie temperature) depends only on the ferromagnetic material and is independent of any fluctuations in experimental condition [3, 4, 6]. Adequate ferromagnetic conductors covering a wide range of Curie temperatures are commercially available [3, 4, 13]. Due to the relatively high heat capacity of these conductors, the decrease in temperature after the turning off of the oscillator is slow as compared to the initial temperature rise (see fig.). Using cylindric ferromagnetic conductors of smaller diameter (0.05 mm), the cooling off will be much faster and an almost ideal temperature time profile may be obtained as shown schematically in the figure [15].

The conditions a) to c) mentioned above being met, the structure of the fragments is closely related to the structure of the original compound. Under these conditions, the method of choice for the determination of the structure of the pyrolysis products is a direct instrumental combination of thermal fragmentation, gas chromatography and mass spectrometry [16]. More detailed information about the mechanism of the fragmentation may be obtained using tracer techniques. An instrument for the simultaneous recording of the pyrolysis gas chromatograms and the pyrolysis radio gas chromatograms of ^{14}C- and ^{3}H-labelled compounds has been described earlier [6].

Under the ideal pyrolysis conditions mentioned, there is as a rule a striking parallel between the thermal fragmentation and the fragmentation induced by electron impact (mass spectrometry). This has been shown in detail in the fragmentation of phenyl-, nitrophenyl- and dinitrophenylhydrazones of aromatic aldehydes [17] as well as in a study of aromatic carboxylic acids [18] and azo compounds [19]. (See also [20]).

In the routine analysis of organic compounds, electron impact fragmentation (mass spectrometry) has certainly, in many respects, a clear advantage over thermal fragmentation whenever the sample is adequately volatile. Thermal fragmentation may even be used routinely as an analytical method for volatile samples in restricted applications. It has been shown that thermal fragmentation permits the localization of double bonds in aliphatic chains [9], whereas electron impact fragmentation usually leads to an isomerization of such double bonds. For non or slightly volatile organic compounds such as natural and synthetic polymers, pyrolysis gas chromatography is, however, becoming an attractive tool both for qualitative and quantitative analysis [10, 11, 12, 21]. Information about the structural elucidation of polymers can be obtained which would be unaccessable with comparable effort by other methods [12, 21].

References

1. Oyama, V. I., Carle, G. C.: J. of G. C. 5, 151 (1967).
2. Reiner, E.: J. of G. C. 5, 65 (1967).
3. Simon, W., Giacobbo, H.: Chem.-Ing.-Techn. 37, 709 (1965), Simon, W., Giacobbo, H.: Angew. Chem. int. Ed. 4, 938 (1965).
4. Giacobbo, H., Simon, W.: Pharm. Acta Helv. 39, 162 (1964), Giacobbo, H., Simon, W.: Festschrift zum 60. Geburtstag von Jakob Buechi (X. Perlia, Ed.). Verlag Schweizer Apothekerverein, Zurich, 1963, S. 199.
5. Dhont, J. H.: Nature 192, 747 (1961, Analyst 89, 71 (1964).
6. Simon, W.: Kriemler, P., Völlmin, J. A., Steiner, H.: J. of G. C. 5, 53 (1967).
7. Levy, R. L.: J. of G. C. 5, 107 (1967).
8. Cramers, C. A. M. G., Keulemans, A. I. M.: J. of G. C. 5, 58 (1967).
9. Levy, E. J., Paul, D. G.: J. of G. C. 5, 136 (1967).
10. Levy, R. L.: Chromatog. Rev. 8, 48 (1966).
11. Brauer, G. M.: Journal of Polymer Science: Part C 8, 3 (1965).
12. Noffz, D., Pfab, W.: Zeitschrift für analytische Chemie 228, 188 (1967).
13. Hochfrequenz-Pyrolyse, Bulletin 8.67 and 9.66. Fischer, Labortechnik, 532 Bad Godesberg, Germany.
14. Coupe, N. B., McKeown, M. C.: Pye Unicam Chromatography Bulletin 2, No. 4, 8 (1968).
15. Bühlers, Ch., Simon, W.: unpublished results.
16. Völlmin, J., Kriemler, P., Omura, I., Seibl. J., Simon, W.: Microchemical Jornal 11, 73 (1966).
17. Völlmin, J. A., Seibl, J., Simon, W.: Chromatographia, in press.
18. Kriemler, P., Völlmin, J. A., May, D. P., Simon, W.: Chromatographia, in press.
19. Völlmin, J. A., Simon, W.: Helv. Chim. Acta, in preparation.
20. Kriemler, P., Simon, W.: Review in preparation.
21. Pfab, W., Benz, W.: GDCh-Hauptversammlung, Fachgruppe Analytische Chemie, Berlin 1967.

The present work has been supported by the Schweizerischer Nationalfonds zur Foerderung der wissenschaftlichen Forschung (Research Projekt Nr. 4312).

On the Complete Evaluation of Chromatograms with the Help of Machine Computations

H. G. STRUPPE

Deutsche Akademie der Wissenschaften zu Berlin, Institut für Verfahrenstechnik der organischen Chemie, Leipzig

1. Previous efforts to evaluate the chromatograms automatically

With the first realization of the routine analysis of gas chromatograms there emerged the problem of the automatic evaluation of chromatograms. Along with the utilization of the first process-chromatographs, efforts have been made to facilitate the evaluation on the basis of reproducible sample sizes and constant qualitative composition. Here we only want to mention the line-diagram technique which consists of stopping the chart for a certain time, and the multiple-point recording technique for recording only the position of the maximum of the peaks. In the laboratory the adequate procedure has been to record fractional chromatograms with the help of integral-multipliers. Therefore, the quantitative evaluation of the chromatograms has been reduced to the task of measuring the peak-heights on the recording chart.

The digital measurement of the peak-area was done first by mechanical or electro-mechanical integrators which are basically automatized planimeters. The main drawback of these apparatus has been that they required a very good base-line, the complete separation of the peaks and during the running of the chromatogram the readings had to be taken after the completion of each peak. This drawback is eliminated by the use of the ball-disc integrator which is still the most popular type of electromechanical integrators. This device records a second trace on the chart in the form of pulses which give the integrated area of the peaks; the number of these pulses can be read later.

2. Digital-integrators, the second generation of the evaluating devices

Gas chromatographic apparatus have been known since the last years, while electronic digital-integrators have been used for the quantitative evaluation of the chromatograms in the last seven years. Following a short time during which the digital-integrators had been introduced into practice, it started a new era in the history of gas chromatography; the process and laboratory chromatographs have advanced to be secondary peripherial devices of the electronic data processing systems.

The rapid advancement in the field of electronic data processing has certainly contributed to the fact that the answers to the basic problems in this field are still changing. One thing is quite clear; the electronic data processing has opened a new field of possibilities, but it also increased our burdens. The problem begins with the recognition of the starting and terminating positions of the peaks, and the construction of the base-lines under the peaks.

2.1. Recognition of the peaks and performance of the base-line corrections

In the case of large dynamic ranges of the amplifier for the amplification of the detector signal, which is made possible by the high sensivity of the detector, when the length of the chromatograms is around 100 minutes or longer, or when automatic alterations are introduced into the separation parameters, the constancy of the base-line cannot be assumed. The analytical accuracy now does not depend on the accuracy of the determination of the shape of the peaks, but rather on the determination of the base-line, and the starting- and terminal positions of the integration of the peak area. In the simplest case one uses the pass-over of the tolerance-threshold as the starting position for the integration and terminates the latter when the next peak starts.

For the latter reason, it is more advantageous to use the pass-over of a certain slope value, as the starting position for integration, instead of using an arbitrary limit for the signal strength. As Fig. 1 illustrates, the slope threshold values are different from that of the base-line than the ordinates of the output-function [1].

In addition, the first and second derivatives of the chromatogram give the positions of the characteristic points on the peak profile, thus the maxima, minima, the inflection points, the shoulders, and the tendency of the base-line as well (see Fig. 2 and [2]). The derivatives are best formed in the analogue-phase of the measurements. The base-line can be corrected step-wise from peak to peak, or from the drift of the base-line an extrapolation can be made to the end of the next, fully separated peak.

Although by means of simple electronic circuits, special informations can be obtained with respect to the trend of the chromatographic curves, the problem of fixing the starting and terminating points of the integration still remains open.

Fig. 3 depicts a typical peak profile. The integration is started when the slope threshold J_A is exceeded (on the figure this is exaggerated to be 30°). The integration is terminated when the slope value drops below the same, now negative level. Since, however, the chromatographic peaks are generally unsymmetric, the previous method would lead to termination at higher ordinate values than those of the starting positions. In the case of step-wise baseline correction this would result in unjustified jumps. To regard the line connecting points J_A and J_{E1} as baseline would also be incorrect. Although in the case of peaks with strong tailing the method to continue integrating to the point of the same slope threshold on the rising front of the next peaks; J_{E3} has certain advantages, it does not allow baseline correction. An acceptable solution is to start integrating when the slope value exceeds the threshold limit J_A, while to terminate the integration one uses the value J_{E2} taking into account the trend of the baseline as well. Since in the case of constant seperation conditions the peaks are more flat and wide due to diffusion effects, the increased width of the peaks leads to an increased systematic error of the peak area determination. This happens because the flatter peaks are recognized later. Where this effect cannot be avoided by means of programmed alteration of the separation conditions, it would be necessary to provide for a programmed alteration of the slopesensitivity.

2.2. Basic principles and possibilities for evaluating devices

Although there are no theoretical difficulties regarding the meaning of the quantities derived from the chromatograms, the commercially available evaluating devices have not yet fully solved the problems of the complete evaluation of the chromatographic cur-ves.

The basic quantities of the chromatograms are the following:

1. the retention times of the individual peaks an the deadtime of the column, and
2. the height, width and area of the individual peaks, where any of these three quantities is approximately determined by the other two ones.

Only certain types of integrators make it possible to integrate the peak profiles and se-cure the adequate accuracy of the digital recording of the retention times. The evaluating devices do not perform the ratio formation of the retention time and the dead-time (i. e., the formation of the ratio using the total retention time of a non-sorbed component), this operation necessitates the use of an additional computer. In addition the existing evalua-ting devices give the noncharacteristic total retention time, instead of indicating the dead-time and restarting timing.

All digital integrators are characterized by the fact that their purpose is to perform quantitative analyse. In the case of the routine quantitative determinations, it is sufficient indeed to give the total retention time or the peak number in order to have an orientation in the chromatogram and to assign the area values to the components. This is, however, no longer sufficient when the qualitative composition changes somewhat from sample to sample. Now, however, the laboratory work requires such evaluating devices which are not only routine analyzers but which can provide extra facilities in addition to evaluating the peak areas.

The computer-compatible output of the contemporary integrators make it possible to per-form the calculation of mass-percents, quantitative correction factors and all sorts of calculations which are based on quantitative results. The retention indices can be com-puted for the purpose of identification. The computer, however, cannot calculate the se-parability, the necessary theoretical plate number and the degree of separation of the overlapped peaks. It would be desirable if the digital integrator could calculate the height of the maxima, the height and position of the minima, in addition to the retention time and the integrated area. A further development would be possible in the way of the evaluation of the chromatograms via the not very frequently used formula given by JAMES and MARTIN [3] for the determination of the theoretical plate number n:

$$n = 2\pi \left(\frac{t_{dr} \cdot h_{max}}{A}\right)^2$$

t_{dr} is the total retention time, h_{max} is the peak-height in the maximum, and A is the peak-area in the units of t_{dr} and h_{max}

A similar relationship could be used for the R resolution:

$$R_{1,2} = \sqrt{\frac{\pi}{2}} \frac{t_{dr2} - t_{dr1}}{\frac{A_2}{h_{max2}} + \frac{A_1}{h_{max1}}} = 1{,}253 \frac{t_{dr2} - t_{dr1}}{\frac{A_2}{h_{max2}} + \frac{A_1}{h_{max\,1}}}$$

where $b_W = \sqrt{\frac{2}{\pi}} \cdot \frac{A}{h_{max}}$ (b_W: peak widht at the inflexion points)

Here we are not going into details about the computation possibilities in connection with the above formulae. In a way the same possibilities emerge as in the case of the direct contact with a computer.

3. Analogue-digital converter as the only fore-unit of the computer

3.1. The on-line operation gas chromatograph computer

Besides the direct coupling of the real-time digital integrators to the gas chromatographs, and the application of the intermediate storage devices prior to the digital integrators, there has recently been much discussion about, and the utilization has begun of, the direct processing systems. We dispense with the expensive digital integrators and try to delegate as many tasks as possible to the computer which is anyway at our disposal.

The first such devices were described by KAROHL [4] in as early as 1963 for IBM 7080 computers.

The time dependent analogue signal of the detector is fed into a voltage-frequency converter for digitalization and stored on magnetic tape for further use in the computer, or fed into a digital voltmeter, as analogue-digital conversion unit, and used as input in prescribed intervals to either a storage device or to the computer.

As it has been discussed by KNECHTEL [5] in the case of the C 090-10 digital computer, the reading-in can be effected 25 times per second in on-line mode of operation. It is, therefore, enough if the first chromatogram peak to be evaluated has a half-width of at least a quarter of a second. The computer is not utilized efficiently, due to the storage of the called-in values that are obtained from the converter at, say, every 40 msec, not even in the case when the first integration steps are effected right after calling-in. Al -

most the entire computation time is left for the processing of the other data. A timer calls in at predetermined time intervals a subprogram for the storage of the data, through a circuit-breaking channel. The application of several such circuit-breaking channels renders it possible to process the data of several gas chromatographs by the same computer. Since the analogue-digital converter is neither used efficiently in the case of a few msec-s necessary for a measurement, therefore several gas chromatographs can be processed through the same digital-analogue converter by the computer. (In the case of certain types of problems, as naturally the case of logical decisions that can be derived from the quantitative composition of the sample to be analyzed, programs can already be found in program libraries.) Some problems remain to be solved in connection with the characterization of the peak-asymmetry and the overlapping of inadequately separated peaks since there are not generally accepted and applied calculational methods for these purposes. Only one thing is clear; the cost of such calculations is formidable except when machine computational methods are used.

3.2. Remarks about data reduction

While the digital integrator summarizes the information content of a 20 minutes long, 10 component chromatogram in the form of 20 numbers, in the case of on-line coupling for the example given above according to the 40 msec scanning interval the analogue signal is being resolved into 30.000 digital ordinate values. In the latter case there would be too much information fed into the computer as compared to the necessary volume of information for the exact evaluation of the chromatogram. For this reason the question as to how much information is needed for the satisfactory description of the peak-profile has been extensively studied. In the case of an undistorted Gauss curve the value of the maximum and the position of one inflexion point would be enough. Unfortunately the gas chromatographical peaks are never identical with the exact Gauss curve. Any increase in the data enhances the precision of the determination of the profile, but, on the other hand, it necessitates more storage space.

Especially in the case of on-line devices it is possible to determine each peak profile by 500-200 digital values at reasonable cost.

The mere summation of the densely packed ordinates already gives the area beneath the curve an adequate accuracy. From these values the following quantities can be obtained: area, peak height, peak width and the position of the inflexion points.

In the case of the on-line devices the great number of the input informations necessitates the provisional reservation of a few thousand storage locations. In the case of the off-line operation the storing of the digital values is an expensive and problematic aspect.

When the chromatograms are represented in the form of punched tapes, the scanning speed depends on the peripherial punching device. There are certain limits to the construction of narrow peaks from 20 or more values.

While in on-line operation all 40 msec intervals can be scanned, in the case of a mechanical tape punching device the interval that can be scanned must be greater than 500 msec.

Table 1 shows the consequences of this latter aspect.

If we take it as starting assumption that 15 ordinate values are enough for the construction of the peak profile then in the case of a scanning interval of 0,5 sec the following results are obtained.

Table 1. Determination of the peak profile from 15 or more ordinate values in the case of a scanning interval of 0,5 sec.

Separation efficiency of the column given by	$n = 5{,}54 \left(\frac{t_{dr}}{b_H} \right)^2$	Necessary total retention time of the narrowest peak
n = 1 600		t_{dr} = 1,25 min
n = 10 000		t_{dr} = 3,13 min
n = 90 000		t_{dr} = 9,36 min

By a scanning interval of 0,5 sec, all those peaks can be evaluated whose total retention time is lenger than 1,25 min. In the case of capillary columns not even the retention time value of 9,36 ist of critical importance since the adequate efficiency is normally obtained for a column length 100 m and deadtime 7 min.

For fast analyses and low boiling components in short capillary columns the above scanning interval is, of course, not enough. On the other hand it should also be considered that the necessary punched tape length can be as much as between 20 and 100 m.

Table 2. Length of the number columns and the punched tapes in the case of digitally recorded chromatograms for a scanning interval of 0,5 sec

	Duration of analysis t_A = 20 min	Duration of analysis t_A = 120 min
Number of scanned values	2400	14 400
Length of the number column on the tape printer	10,0 m	60,5 m
Length of punched tape for binary representation in five-hole system	18,4 m	110,4 m
Number of values for the last chromatographic peak in the case of		
n = 1 600	240	1440
n = 10 000	96	576
n = 90 000	32	192

When a constant scanning interval is used there are rather long data tapes to be processed. Is it, however, necessary to describe the last peak by a few hundred ordinate values ? Surely, it is not necessary. Then the scanning interval should be gradually or continually altered by means of a given program, which of course should be known to the computer which determines the time coordinate only from number of the input ordinate values. The input of the corresponding ordinate and abscissa values and the scanning of the chromatograms in intervals that are irregular and depend on the magnitude of the ordinate values, does not seem advisable due to increased computational costs in the case of gas chromatography.

In addition the simultaneous transfer of ordinate and abscissa values would lead to the doubling of the technically feasible minimal scanning interval length, which is a drawback that is far from being compensated by the achieved shortening of the punched tape. The programmed alteration of the scanning interval in the specific case of gas chromatography entails a logarithmical speed suppression. (In the case of giving the correct parameters the first and last chromatogram peaks would be described by the same number of values. The selection of correct parameters does of course presume the knowledge of the actual separation efficiency.)

This system has, however, a drawback: it is not possible to input several chromatograms simultaneously through the same analoguedigital converter, since the scanning intervals could overlap. It seems to be a practical solution to the problem to lengthen the output scanning interval after 100, 200, 300 and 400 values to 1, 2, 3 and 4 sec, resp. In this case a two-hour chromatogram could be described by only 1647 ordinate values instead of 14,400 ones. For parallel analogue-digital converters it would be possible furthermore to divide the scanning intervals for other chromatograms using arbitrary starting time.

Now, since we have just shown a method for preliminary data reduction in the following we are going to deal with data processing within the computer.

4. Data processing in the first programme stage

In the case of a series of ordinate values the computer should be able to determine the position and, in given cases, the trend of the baseline, it has to detect the starting peak position, it should be able to reduce the peak ordinate values by the value of the extrapolated baseline, and finally it should recognize the determinating peak position, the shoulders and minima. The summation of the reduced ordinate values yields the peak area. The number of the input ordinate values up to the actual peak maximum determines the total retention time.

In the case of giving the substance-specific factors and the identification of the chromatographic peaks, the computer can print the quantitative analysis results in mass % units for each component, provided the separation is ideal. This evaluation gives results which cannot justify the application of an electronic computer, and, on the other hand, this result of the evaluation is the least satisfactory from the point of those demands that are posed by:

the practical routine analyses of overlapped peaks,
the identification of peaks,

the choice of adequate working parameters in the case
of continually changing separation problems, and
the most different methodical and theoretical works.

Therefore, we have done at the „Institut für Verfahrenstechnik der organischen Chemie, Leipzig" research as to how to solve the above problems by an electronic computer in a two stage programme.

4.1. Apparatus

The necessary equipment consists of an analogue-digital converter (type 4013, VEB Funkwerk Erfurt) which can be connected directly to the amplifier output, but also to a following potentiometer of the self-balancing recorder (Fig. 4). By an external timer a scanning process according to a rhythm of 0,6, 1.2, 1.8, 2.4, 3.0 sec is started. The actual voltage value is transferred, through adequate converters, to a tape printer and a tape punching device. A step number will also be put in front of the values printed. It is used only for control purposes. Then the punched tape with the voltage values can be read into the computer. This values are received in the core storage device.

4.2. Computing operations in the first program stage

In the first program stage, the statistical moments of the individual peak profiles as given by GRUBNER and KUCERA [6] are computed, printed, and simultaneously punched on a result-tape together with different characteristic data put in. The calculation of the statistical moments is carried out on the basis of the following formulae:

$$h \sum a = A$$

$$\frac{h \sum a \cdot l}{\sum a} + h \cdot s = \mu_1^0$$

$$\frac{h^2 \sum a \cdot l^2}{\sum a} - \left(\frac{h \sum al}{a}\right)^2 = \mu_2$$

$$\frac{h^3 \sum al^3}{\sum a} - 3 \left(\frac{h \sum al}{\sum a}\right) \cdot \left(\frac{h^2 \sum al^3}{\sum a}\right) + 2 \left(\frac{h \sum al}{\sum a}\right)^3 = \mu_3$$

where

h: the scanning interval, in sec,
a: the ordinate value reduced with respect to the baseline, in mV,
l: the step number of the ordinate values after the peak start,
s: the step number of the peak start from the beginning of analysis,
μ_1^0: the first moment related to the beginning of analysis,
μ_2: the second central moment,
μ_3: the third central moment.

It can be seen that the calculation is relatively simple and computing by means of a desk computer would be difficult only because of the great number of summations. The problematical questions of this first program stage are the same as in using digital integrators. The possible trend of the baseline must be digitally corrected, momentary virtual signals (voltage surges) must be suppressed, and the beginning and termination of the peaks must be recognised in time. The feasible procedure has already been discussed in the literature in connection with digital integrators and has also been programmed for different computer types.

4.3. Advantages of using moments

While the manual evaluation of chromatograms had to be based on easily available quantities as the total retention time, peak width, height or area, the use of digital measurement and computing technique renders it possible to determine the peak profile more exactly. Gas chromatographic peaks could be approximated by Gauss curves only under special conditions. Alway distorted, asymmetric profiles occur in practice. Starting from describing diffusion processes in the absorption column, GRUBNER and KUČERA [6] arrived at a mathematical model of gas chromatography in which a certain peak asymmetry is normal. From their theory, it follows that not the total retention time of the peak height, but the center of gravity of the peak related to the beginning of analysis is the characteristic quantity. On account of the peak asymmetry the location of the peak maximum is somewhat shifted towards the center of gravity of the profile. The location of the center of gravity of the peak is nothing else as the first moment. The width of an asymmetric profile is also expressed incorrectly by the width at a given height. What does it mean: width at the height of the inflexion points? In the case of asymmetric non-Gaussian-curves neither the location of the inflexion points, nor the two inflexion points at the same height are known. Hence, the peak width as the standard deviation related to the mean value has to be given by the second (central) moment.

While the moments have no relevance for the determination of the areas of separated peaks, they can yield good approximations of the profiles via moments of higher order in the case of overlapped peaks and shoulders. But, to day we have no adequate experiences as to how the characteristics of asymmetric peak profiles may be described by statistical moments. It is, however, quite sure that not the moments themselves but only their combinations, i. e., their relative values, allow clear statements. Thus, for example, the quotient $(\mu_1^0)^2/\mu_2$ equals n, the number of theoretical plates. However, of the peak has a tailing, the number of theoretical plates becomes much smaller if calculated on the basis of the moments instead of the total retention time and the peak width. The following table illustrates this behavior. Here the numbers of theoretical plates are shown as calculated either from the total retention time and the peak width at the half height, or from the total retention time, peak area, and peak height, or, finally, on the basis of the moments.

The difference of the values for n can be explained by the fact that μ_2 as the square of standard deviation of a distorted asymmetric peak is greater than the square of the value which is determined at the alleged inflexion points as standard deviation. Only in the case

Table 3. Comparison of numbers of theoretical plates determined from the peak width b, the peak area A, and the peak height h_{max} with those calculated from statistical moments.

	for n-hexane	for n-octane
$n = 5{,}54 \left(\frac{t_{dr}}{b_H}\right)^2$	795	1005
$n = 2 \left(\frac{t_{dr} \cdot h_{max}}{A}\right)^2$	780	963
$n = \left(\frac{\mu_1^o}{\mu_2}\right)^2$	655	682

(Conditions of analysis: column 20% apiezon L on sterchamol;
carries gas: 6,6 cm/sec, nitrogen; column temperature: 125 °C)

of an ideal Gaussian curve, the latter value can be interpreted as a measure for the standard deviation. Though the interpretation of the moments as well as of the quantities derived from them must still be made clear, yet the use of relative moments of the first order instead of the relative retention r shows the advantage of statistical moments. The first moment gives the location of the center of gravity of the peak profile; moreover, it is obtained with a greater accuracy than the retention time. So the retentions derived from the moments will lead to a more exact identification.

5. The complete evaluation of chromatograms in the second program stage

After the first program stage, the computer had produced a result-tape containing the moments of the individual peaks an their areas, it can be decided which further informations in the chromatogram should be sought. The result-tape of the first program stage is then, at the same time, the data-input for the second program stage, in which the relevant gas-chromatographic characteristics and the result of the quantitative analysis will be computed from the moments.

A division into a reduction of data (calculation of moments from the ordinate values) and an evaluation of moments (calculation of chromatographic characteristics from the moments), resp. seems to be advantageous from certain points of view. Thus the problems of obtaining the ordinates in the first program stage are quite different from the computing instructions necessary for the second program stage. Especially for the qualitative and quantitative evaluation of the peaks it is advantageous if the computing instructions will be fixed on the basis of the chromatogram determined digitally and evaluated according to the moments. Especially when a series of investigations or the comparison of more chromatograms is the matter in question, the divided calculation of gas-chromatographic characteristics is to be favored. The second program stage can then be shortened considerably corresponding to the statement of the problem.

The second program stage contains a series of calculations to be solved independently, i. e. computing operations

1. for the quantitative analysis,
2. for the calculating the flow of the carrier gas,
3. for the qualitative analysis,
4. for evaluating the column.

The calculations to be carried out are fixed by a list of operation numbers and input informations.

In the case of quantitative analysis there are different operation numbers which call in the corresponding subroutines for the 100%-method (giving the mass or mol percents of every component), for the absolute calibration, for the evaluation with an internal standard, or for the mixing method, and carry out the calculation. This happens after putting in the necessary conditional informations as correction factors, molecular weights, and data for the calibration materials or the comparison of chromatograms. Three further operation numbers call in subroutines computing

The linear velocity u of the carrier gas,
the kr^4-value as the measure of the free cross section or
the free gas volume V_g in the column, and
the flux F_0 of the carrier gas in ml/min.

In the case of the quantitative evaluation the relative retentions determined by the moments are calculated for all the peaks and/or für the retention indices of every component, and are related to arbitrary reference peaks; thereafter the V_g-values and their logarithms are computed.

It is not intended in the second program stage to compare the retention indices determined with tabulated (i. e. stored) values from the literature or earlier experiments. In the case of a great number of values to be stored, this last stop of the automized identification is performed only in laboratories with a limited material supply.

Performing a basic column evaluation belongs to treating scientific measurement results with respect to a good planning of experiments. Thus among the individual operation numbers operations are carried out, such as the determination of dead-time retentions, of the least retention of neighbouring peaks giving the necessary effective number of theoretical plates as well, of the greatest effective number of theoretical plates in the chromatogram, of the separation and the time-related separation of given peaks, etc.

6. Final remarks

In the way described above all the evaluation labor spent today on a chromatogram can be transferred to an electronic data processing system. Thus, the possibility of an overall evaluation of chromatograms emerges; such an evaluation cannot generally be achieved now because of the lack of manpower and time, consequently important final conclusions must often be drawn and decisions must often be reached quite empirically, on the basis of insufficient experimental data.

On the other hand, using an electronic data processing system is very expensive indeed, and we cannot say to day to what extent this kind of evaluation of chromatograms will prove well, taking into account the economic aspects too.

Future predictions in this respect would be difficult. All the same, the following circumstances should be mentioned:
Until the existing medium and large data processing systems will not be fully exploited in three shifts, the computing centres can also be concerned with gas chromatography. Efficiency and reliability of the intermediate terms will determine to which extent gas chromatograms can be evaluated. The necessary computer programs can easily be compiled or found in program libraries. It is, however, well known, that this way is not always the most expedient one. For universal electronic computer the cost of programming and the computing times are considerable while the computing operations are rather simple. In the case of increasing use of the computer an effort is made to cut down a number of such tasks.

Since the digital integrators can hardly fulfil the present and future demands on evaluation devices, the modern form of chromatogram evaluation is realised via special small computers. Already to-day data processing systems are commercially available which have in the computer part only what is needed in gas chromatography. Especially an analogue part will considerably simplify the following digital computing, this analogue part producing the derivatives and the characteristic points of the chromatogram. As the advertisement proves:

> „If you have more than three gas chromatographs you need a GC-data processing system",

the firms hope for the wide use of such small devices. Obtaining such apparatus would be economical for bigger GC-laboratories only, as these devices are still more expensive than three gas chromatographs together, even if they are constructed from serial component parts for spectral evaluation devices and general computer systems. Even further on a great number of routine problems will be solved only with the help of digital integrators without any coupling to a computer. To this end, however, the integrators must still be developed. Besides the retention time and the peak area, the data of the peak width or the peak height are desired in order that the integrators may be applied beyond qualitative analysis as well. Both such integrators and using special small computers or coupling gas chromatography to central data processing systems will lead us to some new knowledge through extensive chromatogram evaluations. This knowledge should also help us in gaining time for the actual experimental works.

Acknowledgments

The author takes pleasure in thanking Dr. habil. G. GEPPERT, Leipzig, under whose guidance the data processing system described was constructed, for his many valuable comments on the subject, and Dr. GRUBNER, Prague, for his interest and encouragement with respect to the evaluation of statistical moments of gas chromatographic peaks, on the occasion of the author's stay in his laboratory.

References

1. Baumann, F., and Tao, F.: J. of Gas Chrom. 5, 621 (1967).
2. Karohl, J. G.: J. of Gas Chrom. 5, 627 (1967).
3. James, A. T., and Martin, A. J. P.: Analyst. 77, 915 (1952).
4. Karohl, J. G.: Paper on the 7th Conference on Analytical Chemistry in Nuclear Technology, Catlinburg, Tenn., Oct. 1963, Abstr. 46
5. Knechtel, G.: Comment, H. 1 (1967).
6. Grubner, O., and Kučera, E.: Gas-Chromatographie 1965, ed. by H. G. Struppe, Akademie-Verlag, Berlin 1966, p. A 13

Preparative Gas Chromatography

M. VERZELE

Laboratory of Organic Chemistry State University of Ghent

1. Introduction

In gas chromatography with analytical purposes the sample size varies from fractions of a microgram (with flame ionization, electron capture or ß ray detector etc.) to mg amounts (hot wire detector, etc.). The separated substances are not collected.

When collection is carried out with the purpose to do further experiments with the collected material, the field of preparative gas chromatography is entered. When the „further experiment" is recording of a spectrum (U.V., I.R. or M.S.) or perhaps a colour test, the sample size to be collected can be very small and is in the mg range. Preparative gas chromatography uses therefore sample sizes from this value onwards and up to 10-100 ml. Undoubtedly preparative gas chromatography will soon enter the industrial field with even much larger sample sizes.

With such variation in sample size, it is evident that the technical problems are also very variable and that a single solution can not be advanced.

To this variation in sample size must be added the variation in the difficulty of the separations. A deliberate technical choice has therefore to be made at the start. Because of the possibilities of instrumentation it turns out that this choice can be narrowed to four different techniques which can be distinguished more or less clearly from each other.

1. Separations of not too complex mixtures in sample sizes in the mg. range. Here, normal analytical gas chromatography can be used and the only problem is the collection of the separated components.
2. Separations of mixtures containing a large number of components with a rather large boiling range; sample size from 10 µl to a few ml. The main purpose of this sort of separation is characterization of the substances.
3. Separations on larger amounts of mixtures containing only a few components; sample size from 100 µl to 20 ml. When larger amounts of a multicomponent

mixture like in 2 have to be separated, it is better to fractionate the complex mixture into several few component mixtures and to use technique 3.

4. The same as under 3 but on much larger scale.

These points will be detailed in this paper which is based mainly on studies from the author's laboratory. Preparative gas chromatography is an extension of analytical gas chromatography and it will be assumed that the reader is familiar with the theoretical and technical fundamentals of this instrumental method.

2. Theoretical considerations relevant to practical preparative gas chromatography

2.1. M. A. S. S.

When the sample size of a single substance is gradually increased it is well known that the peak band width also increases. A two component mixture that can be separated with small samples will, with larger and larger samples in the end show peak overlap and only partial separation. There is thus a well defined „Maximum Allowable Sample Size" or M. A. S. S. for that mixture on that particular column which can just be separated.

This M. A. S. S. depends on the nearness of the two peaks from each other or in other words on their relative retention or α-value, since $\alpha = \frac{V'_A}{V'_B}$, where V'_A and V'_B are the adjusted retention volumes (or time or distances on the chromatogram). „Adjusted" means measured from the air pear or its equivalent [1]. The α value reflects the difficulty of the separation of A and B.

For a multicomponent mixture, the pair of peaks closest together or with the lowest a value will determine the M. A. S. S. of the whole sample. Only the lowest a value of the mixture has in this case to be considered.

This M. A. S. S. is not only dependent on the difficulty of the separation (1) but also on the separation power of the column (2) and on the time that this separation power is applied (3).

(1) The difficulty of a separation is reflected as stated by its α-value. This is a measure of the relative volatility of the relative volatility of the components and also of the selectivity of the stationary phase for that particular mixture.

$$\text{Indeed } \alpha = \frac{V'_A}{V'_B} = \frac{\gamma_A \cdot p^o_A}{\gamma_B \cdot p^o_B} = \frac{k_A}{k_B} = \frac{K_A}{K_B} \text{ where}$$

where V'_A and V'_B are adjusted retention volumes (or time or distances as measured on the chromatogram)

γ_A and γ_B are activity coefficients

p^o_B and p^oA are vapour pressures

k_A and k_B are partition ratios

K_A and K_B are partition coefficients.

On another stationary phase, another γ can give a higher α value and therefore a larger M. A. S. S. Even small α value increases have a large influence and it is therefore necessary to try several stationary phases for optimalization of this factor. The α values are easily determined on analytical size columns.

The highest α values still needing gas chromatographic techniques are around 1.5 to 2. For higher values separation by distillation is probably possible or even easier. The lowest α values which can still give reasonable results are around 1.09 to 1.1. Lower values should be avoided as much as possible by a judicious choice of the stationary phase. Indeed it is nearly always possible with only 2 or 3 stationary phases of sufficient polarity difference to find one, giving at least $\alpha = 1.10$. This exploitation of different activity coefficients (γ) on different stationary phases is only possible with mixtures containing only a few components (cis/trans isomer mixtures for example). With more complex mixtures there are so many α values to consider that at least one pair of the components will always happen to give a low α value, regardless of the nature of the stationary phase. In choosing the stationary phase for multicomponent mixtures, general polarity is therefore the only factor to consider (polar substances on polar stationary phases and vice versa).

(2) The separation power of a preparative column is best expressed in plate numbers (N). Although this has many drawbacks, if only because of the variations of N with temperature, k value, stationary phase etc., it still is a useful comparison basis. In analytical gas chromatography plate numbers are usually calculated with the equation

$$N = 16 \left(\frac{V_R}{Y}\right)^2 \qquad (2)$$

where V_R is the retention volume of the peak (or time or distance on the chromatogram) measured from start to peak maximum

Y is the band width at the peak slope tangent intercepts with the base line.

The equation is a simplification of the more correct one which is

$$N = \left(\frac{V_R}{Y}\right)^2 \left(\frac{1}{k} + 1\right) \qquad (3)$$

Since k is usually large in analytical work the simplification is warranted. In preparative gas chromatography on larger columns, the time element often forces us to use smaller k values and then the factor $1/k + 1$ is no longer negligible and it is better to use equation 3 to calculate plate numbers. In another form equation 3 becomes

$$N = 16 \cdot \frac{V_R \cdot V'_R}{Y^2} \qquad (4)$$

and this is the more practical form.

With the increased sample sizes used in preparative gas chromatography, band width Y is also increased and N values calculated with these peaks reflect only the separation power of the column for that particular sample size.

The plate number calculated for the peaks of preparative scale samples is therefore called „preparative scale plate number" and is symbolised by N'. These N' values decrease with increasing sample size, which means that N' and sample size must always be given together. The evolution of N' for a 6 m x 9 mm coiled column is shown in Table 1.

Table 1. Preparative scale plate number or N' value for samples of cumene at 145°. Column 6 m x 9 mm. Chromosorb W 30-60 mesh - SE-30 silicone gum 25 % w/w with 200 ml H_2/min.

Sample size	N'
10 µl	2340
50 µl	1380
100 µl	710
200 µl	400

(3) The time of separation is determined mainly by temperature and is expressible as a partition ratio k. This is the ratio of the retention volumes (time or distance) of the peak considered and the air peak. Low k values mean that the substances are swept rapidly through the column and it is obvious that in this case the M. A. S. S. is smaller than could be found for the same mixture at lower temperatures and therefore higher k values. This k value should not be larger than about 20, otherwise the separations last too long, but should also not be smaller than about 2-3.

2.2. The Use of $\alpha \cdot k$ and N'.

In analytical scale gas chromatography the following equation (5) relates α, k and N

$$N = 16 (\alpha/\alpha - 1)^2 (1/k_2 + 1)^2 \tag{5}$$

This equation is used to find the necessary column length to separate a mixture with given α and k. Since k is between 3 and 20, this equation can be simplified to

$$N = 16 (\alpha/\alpha - 1)^2 \tag{6}$$

In preparative scale gas chromatography equation (6) will not be used to find column length, but to find the maximum allowable sample size (M. A. S. S.) of a problem.

A sample calculation will serve to illustrate this. Consider the 6 m x 9 mm column of table I and a limit duration of the separation around 45 minutes, which is equivalent to a partition ratio in the range of 20. With such a partition ratio the column will have its maximum separation power and the second term in equation (5) can be neglected. For an α value of 1.25 (this is the case for the trans-cis decalin mixture at about 160°) equation (6) leads to 400 required plates. From table 1 it can then be deduced, that the column gives that preparative scale plate number for sample sizes of 200 µl. With a 1/1 decalin mixture this means then that 400 µl is the maximum allowable sample size of this mixture (M. A. S. S.) which will still give complete separation. This is found to be the case experimentally and both peaks give about 400 plates.

Most mixtures contain unequal amounts or a larger number of components. In this case the two components in the mixture giving the lowest α values are used to calculate N. Of these two components the percentage (x) in the mixture of the one present in largest amount is calculated. We then have for this general case

$$M.A.S.S. = 100\,(SS(N'))/x \qquad (7)$$

where SS(N') is the sample size of a single substance giving the calculated N value on the preparative scale column to be used. Consider for example a five component mixture with lowest $\alpha = 1.09$ between second and third peak, the second peak being the largest and representing 20% of the total mixture. Column no. 1 of Table 5 could separate a M. A. S. S. of 100 x 500 ml/20 = 2.5 ml.

From the foregoing is should be clear that the M. A. S. S. is variable and is determined by the α value and the complexity of the mixture. N' values necessary then for all possible α values are calculated easily enough and again neglecting the second term of equation (5) we obtain Table 2.

Table 2. Plate numbers necessary to separate mixtures with the α values as indicated.

α value	N' values	α value	N' values
1.03	17496	1.12	1384
1.04	10816	1.13	1221
1.05	7056	1.15	940
1.06	4960	1.20	576
1.07	3745	1.25	400
1.08	2915	1.30	300
1.09	2342	1.40	195
1.10	1936	1.50	144
1.11	1632	2.00	64

From Table 2 it is clear why α values below 1.1. should be avoided. The separation power of the column has to be very high and Table 1 shows that only low M. A. S. S. will be possible.

A separation with α = 1.10 needs already around 2000 plates and this for preparative scale samples. Considering the unavoidable drop in plate number with larger samples, this means that the plate number of preparative gas chromatography columns for small samples must be substantially higher than 2000 plates. This can only be achieved with rather long columns.

3. Columns

3.1. Column Design and Variations of N' values

It is obvious that the M. A. S. S. should be as large as possible and that this is a very important point in preparative scale gas chromatography. The simplest and originally the most often tried way to obtain large M. A. S. S. values is to increase the column diameter. This has however an adverse effect on band width and it is well known that because of this, separations which are for example easily possible on a 2 m x 4 mm column with a 1 μl sample cannot be obtained on a 2 m x 4 cm column with 100 μl samples (same increase in column cross section and in sample size).
The repetitive injections of small samples on relatively smaller columns is therefore a useful approach. This technique is the basis of commercial instruments, the best known is the Varian Aerograph Autoprep 700. When really large samples are needed this technique, although automated, takes too much time.
Another solution is the simultaneous use of a number of parallel columns and this idea has also been exploited in commercial instruments like the Beckmann Megachrom. A set of exactly identical columns, necessary for the success of a multiple parallel column instrument, is not easy to obtain and this technique has largely been abandoned nowadays.

A recent solution in preparative scale gas chromatography is the continuous flow technique of DINELLI, POLEZZO and TARAMASSO [2] where a cylindrical stack of columns revolve before the injection port. This continuous operation resembles somewhat the liquid-liquid continuous separation of S. S. D. developed in this laboratory [3]. Technically it must be much more difficult than a more traditional approach and the future will have to show what this technique can do.

To return to columns with increased diameter, the band widening factor has been the subject of many investigations. While earlier efforts were mainly directed at better and more uniform packing, later on, attention went to regulate the flow of gas artificially through the column by devices which are now known as „flow homogenizers" [4, 5]. Positive results have only been reached recently and wider bore columns of normal length (2 m) provided with such flow homogenizers are probably the only answer to industrial scale gas chromatography [6].

It is however more correct to say that larger M. A. S. S, on these increased diameter columns is the result of increased column volume rather than of increased column diameter. Now, column volume cannot only be increased by using wider bore columns, but also by using longer columns of a more conventional gas chromatographic diameter. Both types of column: short wide or long narrow present advantages and disadvantages. From a purely technical standpoint it can be said that

short wide columns are: faster, more, expensive, consume more gas (He!) and can be used at slightly lower temperatures. They <u>must</u> be operated at high gas rates;

long narrow columns are: slower, less expensive, consume less gas and have to be used at slightly higher temperatures. They <u>can</u> be operated at low gas rates.

These points are however of secondary importance to the M. A. S. S. and this as far as the column is concerned will depend on the N' values.

N' values decrease with increasing sample size but the rate of decrease is determinde by a number of factors. The influence of the nature of the carrier gas, of the mesh size of the support and of the amount of stationary phase will be discussed further. Another factor is column, shape. We can indeed as explained put the same amount of support material to work in a short wide bore column or in a long narrow bore column. N' values for such columns are given in Table 3.

Table 3. N' values for

a) short wide bore column filled with Chromosorb A 30-40 mesh coated with Carbowax 20 - M 5 g/100 ml. Cumene samples at 140° with 1 l H_2/min. Column volume 12 l.
b) long narrow bore column (19 sections of 100 cm long with tapered ends) filled with Chromosorb A 10-20 mesh coated with SE-30 silicone gum - 5 g/100 ml. Benzene samples at 80° with 600 ml H_2/min. Total volume of column: 8.8 l.

Sample size	N' values	
	2 m x 7.5 cm column	19 m x 3 cm column
100 µl	177	2900
200 µl	149	2390
500 µl	132	1632
1 ml	183	760
2 ml	161	480
5 ml	125	225
10 ml	125	96

The long narrow column shows large variations of N'; it can handle all problems with sample sizes which vary enormously but which are large enough for a great number of investigations. The short wide column shows practically no variation of N'; it cannot handle difficult problems, but allows very large samples of the easier problems. From Table 3 it could be deduced that long narrow bore columns are to be preferred in all cases. This is however not so because the introduction of „flow homogenizers" [5] can improve the N' values markedly. Such a column with dimensions of 2 m x 10 cm equipped with flow homogenizers showed according to the published chromatograms about 600 plates for ml samples [7]. The short-wide bore column of Table 3 also contained flow homogenizers, of our own design, but obviously, although we tried a large number of models and combinations, they were not very successful.

It remains true then that longer narrow columns can handle a broader range of problems (from α 1.08 and up with samples from 0,1 to 20 ml) while short wide columns are more indicated when really large volumes are needed (from 1 ml and up with α at least 1.20). This then for columns with dimensions as in Table 3. From the foregoing it is evident that for every preparative gas chromatographic column N' values should be determined as in Table 3. These values permit one to calculate M. A. S. S. and in general to compare different columns.

3.2. Choice of column Dimensions

Optimalization is needed, but this should not go too far and in column choice versatility should be considered. This is especially so in a research laboratory. Therefore, as stated before, long narrow columns are to be preferred. Further reasons for this choice are then based on considerations about collection and gas rates. In laboratories where samples are used for identification and for exploratory chemical purposes, the needed sample size can be narrowed to the range 10 μl to 10 ml. To make the separated substances easily available they should be collected in a vessel of about 1 ml to 25 ml volume. Larger volumes of the collectors will lead to losses for obvious reasons. In a collecting device of 1 to 25 ml volume the gas rate should be rather low otherwise good recovery will be impossible. While some substances can be collected adequately with gas rates up to 1 l/min in the collectors under consideration, other substances need gas rates as low as 25 ml/min (see further). Although the optimum for the gas rate in preparative scale work is not so pronounced as in analytical gas chromatography, gas rates of 25 to 1000 ml/min can only be applied with equally good result in a rather narrow column. Since on the other hand high sample capacity can only be obtained with larger volumes of column filling material, rather long and narrow columns will be needed.

4. Practical Considerations

A preparative gas chromatograph contains the same elements as a normal chromatograph and an extra collection device. The sample is injected (1) and travels into a column (2) filled with a support (3) coated with a stationary phase (4). The sample is pushed through the column with a carrier gas (5) into a detector (6) and is collected in fractions (7). At the same time the chromatogram is recorded (8).

4.1. Injection

In analytical gas chromatography the sample is so small that vaporisation is not a big problem. This is not so for preparative gas chromatography and for samples in the ml range complete vaporization before allowing the sample into the column, is a major difficulty. The simplest solution is to avoid this vaporization and to use on column injection. This technique gives excellent results (Fig. 1).

To obtain Fig. 1 injection time in the injector block was varied from a fraction of a second to much longer periods. The results as shown are the best obtainable. That on column injection is superior is not surprising, it should indeed best approach the ideal of putting the sample as a plug on the column.

A drawback of on column injection of large samples is that the stationary phase could be stripped rather quickly from the first part of the column. This stripping possibility was investigated by on column injecting one hundred (100) times 5 ml of a mixture containing benzene, toluene, ethylbenzene and isopropylbenzene in a 20 m x 9 mm coiled column. The support was coated with SE-30 silicone gum. After the 100 injections the column was emptied in small portions and the amount of stationary phase in each portion was determined by extraction and weighing. The results are that stripping is negligible and that the stationary phase distribution has not changed. This could be expected since chromatographic results before and after the injections were practically the same. The separating power of the column for small samples dropped slightly after a few 5 ml injections but remained afterwards remarkably constant as can be seen in Table 4. For large samples the „preparative scale plate number" changed practically not at all, taking into account the bad reproducibility of plate number determinations.

Table 4. Analytical and preparative scale plate numbers for 20 m x 9 mm coiled glass column in function of column „age" as explained in text

Plate numbers for benzene, toluene, ethylbenzene, cumene

Sample size	Number of 5 ml injections at 165°			
	first injection	after 30 injections	after 60 injections	after 100 injections
50 μl	952	792	796	680
	1180	905	1032	841
	1471	1098	1220	1098
	1530	1154	1368	1126
1 ml	64	63	62	70
	164	149	142	172
	346	279	309	317
	576	443	468	491

It must of course be remembered that this experiment is for SE-30 an that a more soluble stationary phase may give different results. Anyhow, stripping of a few inches of a column 10 m or more in length must be of little effect. Not so on a short wide bore column; here stripping a few inches or channelling in the first part of the column certainly must be felt rapidly. Vaporization is then necessary in this case.

In any case injection of large volatile samples into a heated column can cause a rather large increase of pressure. Backflash of part of the sample into the cold carrier gas lines must of course be avoided. Normally this is achieved by installing a „back flash valve". However, in our experience even the best back flash valves tend to become dirty

and thus to fail after some time. Therefore the back flash valves are better replaced by a needle valve.
This is closed just before injection and reopened about 1 minute later. The effect of this operation can hardly be seen on the chromatograms obtained with the large volume columns under consideration.

With on column injection, an injector heater is not needed. On our instruments the injection port is only provided with a small analytical type heater for work with very small samples. All examples of separations shown in this paper were obtained with on column injection.

4.2. Columns

4.2.1. Column material:

Columns are usually made of metal. Copper has to be avoided because it is easily coated with an oxide layer which catalyzes decomposition of organic substances. Stainless steel is to be preferred but this is rather expensive. Aluminum columns are generally best suited. Some halogenated substances requiring high working temperatures cannot be chromatographed on aluminium columns. Decomposition is smallest in glass columns, but these of course break rather easily. An attractive advantage of glass columns is that it can be seen if filling has been achieved as should be, or also if the stationary phase has charred by overheating.

4.2.2. Shape of Columns:

All types of columns (streight, in sections, coiled, etc.) can give good results and N' values for a number of different columns are shown in Table 5.

Column 1: ovalized(see further) coiled aluminum column 21 m x (1,2 x 2.5 cm) Chromosorb A 20-30 mesh Carbowax 20 M. 5 g/100 ml support or 10% w/w. Cumene samples at 145°. 600 ml H_2/min.

Column 2: straight round stainless steel column 19 m x 3 cm in 19 parallel sections of 100 cm with tapered ends. Chromosorb A 10-20 mesh, SE-30.5 g/100 ml support or 10% w/w Benzene samples at 80° 600 ml H_2/min.

Column 3: ovalized coiled aluminum column 15 m x (1.2 x 2.5 cm). Chromosorb A 20-30 mesh, SE-30 5 g/100 ml support or 10% w/w, cumene samples at 145° - 500 ml H_2/min.

Column 4: coiled glass column 75 m x 9 mm. Chromosorb G 10-20 mesh SE-30 5 g/100 ml support or 10% w/w. Cumene samples at 145° 400 ml H_2/min.

Column 5: coiled glass column 20 m x 9 mm. Chromosorb W 30-60 mesh, SE-30 5 g/100 ml support or 25% w/w. Cumene samples at 145° 200 ml H_2/min.

Column 6: the same as column 1 but only 6 m in length.
Coil diameter for columns 1, 3 and 4 is 30 cm, for columns 5 and 6 it is 12 cm.

Table 5. The N' column shows the plates reached by the indicated samples

Column:	1	2	3	4	5	6
N'	21 m x (1.5 x 2.5) cm cm	19 x 3 m cm	15 m x (1.2 x 2.4) cm cm	75 x 0.9 m cm	20 x 0.9 m cm	6 x 0.9 m cm
17496					10 µl	
10816						
7056					50 µl	
4960						
3745	10 µl	10 µl	10 µl			
2915	100 µl	100 µl	50 µl		100 µl	
2342	500 µl		100 µl	100 µl		10 µl
1936			200 µl	200 µl		
1632	1 ml	500 µl	500 µl	500 µl	200 µl	
1384						50 µl
1211						
940	2 ml		1 ml	1 ml		
		1 ml			500 µl	100 µl
576			2 ml			
		2 ml		2 ml		
400	5 ml					200 µl
300		5 ml	5 ml		1 ml	
195				5 ml		
144						
64						

The columns have to fit in the oven of the chromatograph. Coiling is the best way to achieve compactness and to fit large volume long columns in as small an oven as possible. With column diameters above a few mm, coiling produces an ovalization of the column section. This ovalisation is a beneficial factor, since, it turns out experimentally that coiled ovalized columns show about 10% more plates than equivalent non ovalized columns. Ovalized columns have been advocated before [8]. The reason for this better performance must be better temperature equilibration and minimalization of the „race track effect". By this is meant that gas molecules at the outer wall of a coiled column have to travel a longer distance than those at the inner wall. This causes band broadening and is the more apparent coil diameter is smaller. Finally, long metal columns and certainly glass columns must be supported in a frame.

4.2.3. Filling the columns:

Relatively shorter metal columns can be filled upright and coiled afterwards. Longer columns and glass columns have to be coiled and then filled. Even with extremely long columns like no. 4 of Table 5 this is not so difficult and only a question of patience. The coated support is simply poured through a funnel in the upstanding column and is forced in by water pump suction and with the aid of a vibrator applied to the column frame. For large volume long columns and with some supports, filling can take several hours. This is also the case for glass columns where great care has to be taken. The columns can be emptied by a reverse procedure and we have frequently filled and emptied the same 75 m x 0.9 cm coiled glass column. The most important point in column filling is that this should be carried out in such a way as to have the best possible permeability.

Firstly, the support material should therefore be free of fine particles or powder which can be formed during the coating procedure or by frequent handling of the support in general. Sieving is not recommended here since this can also produce powder. A blow process is better. To do this the support is passed through an inclined glass tube. The inclination angle will determine the rate at which the support is falling through the tube. The support is made to fall over a distance of about 10-15 cm on to a clean paper sheet while air is blown through it. Powder will collect on another spot of the paper sheet than the main support bulk.

Secondly, the column should be filled as lightly as possible. Indeed, in smaller bore columns the packing density effect seems relatively unimportant. For a 6 m x 9 mm glass column (Autoprep 700) filled with Chromosorb W 60/80 mesh and coated with SE-30 we could variate the packing weight between 68 and 81 g, the column still being completely filled for the lower value. With these packing densities and for trans-cis decalin, the preparative scale plate numbers at 160^0 and with 200 ml/min H_2 as gas rate, were remarkably similar, so were in fact the analytical plate number values. There was however a big difference in retention times because of the difference in inlet to outlet pressure ratio (compressibility factor). The loosely packed column is nearly twice as „fast" as the most densely packed column. It seems therefore that for smaller bore preparative gas chromatographic columns, maximum packing density should not be sought for, but that to the contrary lightly packed columns should be used because of the beneficial influence on analysis time.

Thirdly, for a precoiled column it is not irrelevant which end is chosen for filling in the support. Columns can indeed be filled from either the injection point end or from the detector point end. The highest packing density will be produced at the column end opposite to the one chosen for filling. As the highest density should be at the inlet port where we have the high inlet pressure, it is clear that columns should be filled using the detector end part. Comparative experiments, filling column no. 1 of Table 5 in the two possible ways, showed that the detector end filled column was 1.5 as fast as the injector end filled column. Under an inlet pressure of 700 mm Hg the H_2 gas rates were respectively 630 ml/min and 430 ml/min.

An example, demonstrating this importance of the way the column is filled, is shown in Fig. 2. This separation of cis and trans 3,4-dimethylcyclohexanone has α = 1.15 on Carbowax 20 M. The 21 m ovalized column no. 1 of Table V separated 4 ml at a time as could be deduced for a 50/50 mixture with this a from Table 5. The interesting point here is however that with the same outlet gas rate, A (which is the chromatogram on the column filled from the detector end side) is about 1.5 times as fast as B (which is the chromatogram on the column filled from the injector end side). Elution time for A was 80 minutes, for B it was 120 minutes. Each time a column is filled, it has to be conditioned overnight (250^0 for SE-30 - 190^0 for Carbowax 20 M) regardless if the filling is freshly prepared or has already been used in another column.

4.3. Support

4.3.1. Material:

For analytical gas chromatography the quality of the support is usually chosen as high as possible.

In preparative scale gas chromatography much larger volumes of costly support material are needed and there could be a tendency to be less severe in its choice. This is wrong and the contrary line should be followed. Indeed, in preparative gas chromatography temperatures are usually slightly higher and the time the substance remains at this temperature longer than in analytical gas chromatography. Thermal decomposition of the chromatographed substances and of stationary phases is therefore a major difficulty of preparative gas chromatography. This thermal decomposition is catalyzed by the support material and high quality supports are indeed least active in this respect. In fact the following „activity sequence" can be advanced for the materials we had the occasion to study [9]; glass beads < gas Chrom P ~ Chromosorb W < Chromosorb G < Chromosorb A Chromosorb P < Sterchamol.

Another sequence can then be based on the capacity (size of M. A. S. S. for equal volume of support with the same amount of coating and in the same mesh size). According to our experience this sequence looks about like: Chromosorb P > Chromosorb A ~ Sterchamol > Chromosorb W > Gas Chrom P > Chromosorb G > glass beads.

Capacity difference are however not so very big that this point should be determining in choosing a support. For Chromosorb A capacity is only about 2.5 times larger than for glass beads. Catalytic activity is therefore the most important factor and the materials to be recommended are: glass beads, Gas Chrom P, Chromosorb W and G or any similar material of a different brand.

Glass beads (density ≈ 2) give small capacity and low resolution. Their use will be discussed more extensively further.

Gas chrom P and Chromosorb W (Johns Manville, Applied Science - State College Pennsylvania) (density 0.22) cannot be obtained in grain size coarser than 30-60 mesh. Therefore they can only be used in not too long columns otherwise the pressure drop becomes exsessive. They are best suited for columns like no. 5 and 6 Table 5. Chromosorb W has slightly larger capacity than Gas Chrom P while the latter gives slightly better results with analytical separations [10].

Chromosorb G (Johns Manville) (density 0.55) can be obtained in any mesh size. It gives a lower M. A. S. S. but is so hard that powder formation is no problem.

Chromosorb A (Johns Manville) (density 0.45) can also be obtained in any mesh size. It gives better separations and therefore larger M. A. S. S. than Chromosorb G. It is more destructive toward samples and is somewhat brittle.

Chromosorb P (Johns Manville) (density 0.45) is very similar to Chromosorb A but is even more active for substance destruction.

Sterchamol (Sterchamol Werke - Dortmund - Germany) is a low cost support but it is very active.

4.3.2. Support mesh size:

In analytical gas chromatography, band widths are determined by such factors as gas rate, gas nature, support particle size, stationary phase percentage viscosity, etc. In preparative gas chromatography the large sample size is the most important factor determining band width. Therefore the other factors just mentioned are not so critical. Particle size can thus be increased without reducing the effectiveness of the column very much. This is demonstrated by the experimental results shown in Fig. 3.

The advantage of coarse support materials is that they allow the use of very long columns without increasing the pressure drop appreciably. Although small there is still an influence of the mesh size on column performance with these coarse supports. This is shown in the next Table 6. This shows the results of comparative experiments with Chromosorb A in mesh sizes 10-20, 20-30 and 30-40 in a 75 m x 9 mm coiled glass column.

The stationary phase was SE-30 5 g/100 ml of support - 10% w/w. Filling the column took equally long for the three supports (about 4 hours); emptying took only about 1.5 to 2 hours. The largest inlet pressure which was accepted in this glass column was 3 kg/cm^2 and with this pressure, hydrogen gas flow through the 20-30 and 30-40 mesh support columns was rather similar around 250 and 150 ml/min respectively. This was not as high as could be hoped for. The retention time for the air peak was about 20 to 30 min and the analyses lasted very long. With the 10-20 mesh size support results were much better. Hydrogen gas flows as high as 200 ml/min could be obtained with only 1 kg/cm^2 inlet pressure. The retention time for the air peak was 15 min. Comparative results of the plate numbers for these columns are found in Table 6.

Table 6. Plate numbers for 75 m x 8 mm coiled glass column.
The temperature was 140° and the partition ratio was about 5.

	200 ml H_2/min			500 ml H_2/min
				Support mesh size
i-propylbenzene	30-40	20-30	10-20	10-20
25 µl	12416	7120	5024	4736
250 µl	2176	1944	2304	2048
1.25 ml	496	416	560	496
2.5 ml	240	256	229	208
5 ml	172	170	157	115

Absolutely identical conditions cannot be realized. While the flow rate was about the same for the first three rows of Table 6, the pressure drop was much larger for the first two rows (about 3 kg/cm^2 inlet pressure) than for the 10-20 mesh support of the third row (1.2 kg/cm^2 inlet pressure). Deviations in Table 6 from the expected values must be explained by this. For identical inlet pressures the gas rate was much larger for the 20 10-20 mesh support (fourth row). Table 6 shows again that the ratio of plate numbers for fine grain versus course grain supports is higher for analytical samples than for preparative samples.

A number of separations were carried out with the columns of Table 6. With decalin at about 190° and with 200 ml H_2/min as gas flow the largest sample giving practically complete separation of the cis and trans isomer is 7.5 ml on the 30-40 mesh column and 6 ml on the 20-30 mesh column. On the 10-20 mesh column it is only 4 ml. Nevertheless the 10-20 mesh support is the one to be recommended in these exceptionally long columns.

It has nearly the same separation power of the other supports but presents the advantage of being much faster. With 3 kg inlet pressure (gas rate 400-500 ml H_2/min) and at 190° it separates 4 ml decalin mixture in 45 minutes, while the other columns needed more than two hours.

Although these very long columns are interesting, better and faster results can be obtained with the same volume of support material in slightly enlarged columns of only 10 to 20 m length (see the columns of Table 5 and the separation of Fig. 2).

In such columns the pressure drop is not so servere and the mesh size may be 20-30 or even 30-40. These last mesh sizes are therefore the recommended ones for long and large volume columns.

4.4. The stationary phase

4.4.1. Nature of the stationary phase:

The number of stationary phases in analytical gas chromatography is still increasing and even alarmingly so. The job of having to choose is already complicated enough. In preparative scale gas chromatography the choice is however much simpler because it is economically unfeasible to have a large number of large volume columns. It is therefore obvious that „universal" stationary phases with a high maximum allowable operating temperature (M. A. O. T.) will be most useful.

We have used mostly SE-30 silicone gum, in 2 to 10 l volume columns, carbowax 20 M, Apiezon, Versamid and polytetramethylene oxide (P. T. M. O.). In smaller volume columns (6 m x 9 mm) a greater choice is possible and the large number of examples of separations published by K. DEEMICK [11] can provide a guide line in this respect. SE-30 is non-polar stationary phase and can be used up to temperatures of 300°.

Apiezon L is an alternative non-polar column. It can be used up to 250° but is rather expensive.

Carbowax 20 M or polyethylene glycol is a polar stationary phase and can be used up to 210°.

PTMO or polytetramethylene oxide was made in our laboratory from tetrahydrofuran. It is also a polar phase, but spreads the substances in a different pattern over the chromatogram than carbowax. It can be used up to 230°.

Versamid is a polyamide and thus a polar stationary phase. It can be used up to 250°. Versamid does not always come in the same quality. Some samples are more stable than others.

The maximum allowable operating temperatures (M. A. O. T.) mentioned above are only an indication. Variations are possible on different support materials. Also, the figures are for a katharometer (hot wire) detector. With a flame ionization detector for example, M. A. O. T. may be lower.

4.4.2. Coating percentage:

The amount of stationary phase on the support is usually given on a weight/weight basis. The different supports have however variable densities as indicated above and therefore giving the coating as a weight/weight percentage is not the best way. On glass beads for example this w/w percentage is normally below 1 % and this might be thought to be much too low to permit preparative separations. The density of glass beads is however high (2.0) compared to for example, Chromosorb W (0.22). This means that comparable column lengths, one filled with glass beads with 1 % w/w stationary phase and one with chromosorb with 10% w/w stationary phase, will contain about the same amount of stationary phase. For this reason it is preferable to express the coating amount on a weight/volume basis. For example 2 g stationary phase/100 ml Chromosorb W is about 10% w/w while 2 g stationary phase/100 ml glass beads is only about 1 % w/w.

What is then the coating weight/volume ?

To verify if this parameter is critical a 6 m x 9 mm glass column filled with Gas-Chrom P as support was tested successively with 1, 2, 3, 4, 5, 6 and 7 g SE-30 silicone gum/ 100 ml of support (± 5, 10, 15, 20, 25, 30 and 35% w/w) in an Autoprep 700 instrument. Test substances were iso-octane and cyclohexane. While there is a maximum in the plate number around 4-5 g/100 ml coating with analytical size samples as is well known, it turns out that the preparative scale plate number is still slightly increasing at 7 g/100 ml coating with the larger samples.

It could be concluded that liquid loadings should be further increased for preparative scale work, but this is not so. Larger liquid loadings increase the separation time and especially increase the needed temperature. This is often detrimental to the capacity of a column, since α values generally diminish with higher temperatures. In fact, since the difference is not so very great between 2 and 7 g/100 ml coating, we can use the lower coating amounts with advantage.

Similar experiments with high boiling mixtures confirn the above results.

There is, however, another and even more important reason why in general a low coating weight/volume support should be adopted. With homologous series, separation results mainly from differences in boiling points, activity coefficients are similar and in this case, increasing the amount of coating increases the M. A. S. S. Separations of homologous series are, however, easy (large α) and even with a low coating weight/volume support M. A. S. S. will be relatively large.

With most isomer mixtures, boiling points are more similar and separation results mainly from differences in polarity. The polarity of the support material plays a definite role here and with thick coatings the support influence cannot come through like it does with thinner coating layers. These assertions, proved by the results presented in Table 7, have been confirmed repeatedly.

Table 7. The indicated samples were separated in the sizes given on 4 l long coiled glass columns filled with Chromosorb A coated with silicone gum. The carrier gas was hydrogen at 500 ml/min

M. A. S. S. on SE-30		Mixture	Column Tempe-rature
5 g/100 ml or 10% w/w	12.5 g/100 ml or 25% w/w		
10	25	benzene, toluene, ethylbenzene, cumene	140°
4	8	methylpropylketone, methyl-iso-butylketone, cyclopentanone, methyl-isopentylketone	122°
15	30	ethylacetate, ethylpropionate, ethyl-isobutyrate	100°
1	1	the four isomeric butanols	100°
21	35	dichloromethane, chloroform and tetra-chloromethane	65°
5	10	methanol propanol	90°
1,5	0,3	cis and trans ß-decalone	230°
4	1	cis and trans decalin	190°

Instead of compromising it is better to choose a low coating weight/volume support. This will give highest M. A. S. S. for difficult isomer separation, while the M. A. S. S. for easier homologue separations will be sufficiently high anyway. Except for special applications a costing of 5 g/100 ml support can thus be recommended.

The influence of the support material on the separation is also shown by the fact that α values of isomer pairs change with different support materials all other conditions remaining equal. This must be due as said to residual adsorption and it is normal that this is more conspicious with lower coating amounts (see Table 7).

Such α value changes due to the support material are then:

1) cis and trans ß-hydrindanone have
   α = 1.25 on Cnromosorb A 5 g/100 ml SE-30
   α = 1.30 on Chromosorb W 12.5 g/100 ml SE-30
2) cis and trans ß-decalone have
   α = 1.23 on Chromosorb G 5 g/100 ml SE-30
   α = 1.19 on Chromosorb A 5 g/100 ml SE-30
3) of 2.4.5-trimethyl-2-phenyldioxolane there are three isomers (I, II and III):
   α I/II is 1.16; α II/III is 1.08 on Chromosorb G 5 g/100 ml SE-30
   α I/II is 1.16; α II/III is 1.10 on Chromosorb W 12.5 g/100 ml SE-30
4) 4) cis and trans 2,4-dimethyldioxolane
   α = 1.14 on Chromosorb W - Carbowax 20 M - 5 g/100 ml
   α = 1.08 on Chromosorb G - Carbowax 20 M - 5 g/100 ml

These α values were calculated from results for small samples and for approximately identical temperatures. A drop from α = 1.14 to 1.08 as in example 4 reduces the M. A. S. S. very much.

4.5. Carrier Gas

4.5.1. Carrier gas and detection:

In analytical gas chromatography with katharometer detection, hydrogen or helium are the best carrier gases. The use of nitrogen results in a considerable loss of sensitivity, and the non-linearity of the detection together with possible peak reversals are additional drawbacks with this gas. In preparative scale gas chromatography, however, gas consumption is an important factor. Helium is ideal, but the price is high. Hydrogen is dangerous, although it may be pointed out here that an air-hydrogen mixture only becomes inflammable at hydrogen concentrations of more than 4%. This means that in a fairly large room (5 x 6 x 4 m) an entire cylinder ($\pm$5 m^3) may be emptied without danger of explosion. Undoubtedly, however, in preparative gas chromatography the use of nitrogen should be considered.

With nitrogen and katharometer detection [13] the troublesome effect of peak inversion can occur. BOHRMEN and PURNELL [13] have shown that the signal of a katharometer detector is influenced by several factors. The thermal conductivity factor heats the katharometer when a substance passes and this gives a „positive" deflection. The heat capacity of the substances, however, also plays a part and has a cooling effect. If the heat capacity factor is the more important, peak inversion occurs. High gas flow rates, spiral filaments and high temperatures of filament and detector oven favor peak inversion.

With hydrogen and helium, thermal conductivity far outweighs the other factors, but with nitrogen this is not the case and peak inversion can occur easily with this gas. Commercial katharometer blocks are usually designed for analytical purposes and have as small a volume as possible. The use of such a detector block for preparative purposes with nitrogen at high flow rates produces just the conditions where peak inversion is most easily observed. This is, for example, the case for the Varian Aerograph Autoprep 700 preparative gas chromatographic unit. The chromatograph has also analytical possibilities and in fact the detector block is the same as in analytical chromatographs produced by this firm. Inversion then occurs easily. This seemingly very annoying phenomenon can, however, be guided in such a way that inversion is the normal state of affaire. This is achieved by setting the detector oven at a high temperature (250 ^{0}) and by using high flow rates and high detector currents. Most substances then show completely reversed peaks with a strong signal over the whole possible concentration range. All that is necessary then is to switch the polarity of the recorder.

This is for example the procedure to be followed using the instrument mentioned above for all concentrations of acetone, ethyl acetate and benzene, substances for which badly shaped peaks and inversion trouble have been reported frequently in the literature. The highest signal for a reversed peak can be obtained with a nitrogen gas flow of 250 ml and a detector current of 200 mA. The efficiency of the columns is, however, impaired by such a high gas rate using nitrogen and the high katharometer currents considerably shorten the detector filament life time. The same detector current will heat a katharometer wire to a much higher temperature in a nitrogen stream than in a hydrogen or helium stream. With more normal conditions e.g. gas rate 200 ml and a detector current of 150 mA, the response is only slightly less and the separation is good.

An important point with all detectors is the linearity of the response to sample load. With helium and hydrogen the response of a katharometer detector is linear for very small sample loads and shows decreasing signals for the large samples used in preparative chromatography. The reverse is true for the inverted peaks obtained with nitrogen. For small samples there is no linearity and the response first falls off with increasing sample, while remaining linear for larger sample loads. With a flow rate of 300 ml and 150 mA as bridge current the sensitivity of the detection in hydrogen and nitrogen for large samples is nearly the same.

This is shown in Fig. 4. To obtain an easily readable diagram, all measured peak areas were recalculated as peaks for 2 ml an the sample scale ordinate was compressed for larger sample loads.

The absolute value of the surfaces cannot be compared in Fig. 4 because measurements were obtained on different recorders and with a different flow rate and detector current. The scales were adapted to show that for really large samples the surface area is about the same for the normal peaks in hydrogen and the inverted peaks in nitrogen. For small samples then the inverted peak area with nitrogen is larger, but the peaks are lower and broader than for the positive peaks obtained with hydrogen. This is a normal result with regard to the higher non-linear response in nitrogen (response is lower at peak maximum).

Although as shown it is possible to work with nitrogen on a katharometer detection instrument, this cannot be recommended, since quick breakdown of the detector is unavoidable. With a hot wire detector hydrogen or helium are thus far the best.

With a flame ionization detector (together with a splitter of course - only a fraction of the separated substance will be burned) nitrogen can be used without great difficulty. We have little experience with this technique, but several commercial instruments use this approach (Varian Aerograph 705 and 712, F & M model 776 for example) and the predictable difficulties with the splitter seem to be solved satisfactorily.

4.5.2. Carrier gas and efficiency:

With a katharometer, detection is not linear and the peaks drawn on the chromatogram are not the true reflection of the real peak coming out of the chromatograph. This distortion is different for hydrogen (not linear for large concentration) and nitrogen (not linear for small concentration). Because of this, comparison of N' values obtained with the two gases is deceiving. Experiments with large samples indicate that the difference in N' value is negligible. Here again, sample size is the most important factor determining band width and it is only normal that carrier gas nature is relatively unimportant as far as N' values go. For more details about this point see reference [14].

4.5.3. Carrier gas and speed of separation:

Hydrogen gets more easily through a column than nitrogen and therefore the inlet pressure with hydrogen can be much lower than with nitrogen. Because of this, columns operated with hydrogen are faster. With smaller preparative scale gas chromatographic columns like for example a 6 m x 9 mm column, the time difference is not so important, but with larger columns this difference becomes large anough to be taken into account.

4.5.4. Carrier gas rate:

A normal analytical column has a volume of about 5 to 25 ml. Gas rates used in such columns vary between 20 to 100 ml/min. A small preparative scale gas chromatographic column of 6 m x 9 mm for example contains about 500 ml material. Optimal gas rates are however not increased by a factor of 100 but are only around 200 - 300 ml/min. This can be deduced From Fig. 5.

Fig. 5 shows that the optimum gas rates for large samples is about the same as for small samples. As could be expected gas rate is again not as critical in preparative as in analytical gas chromatography. With 5 l to 10 l volume in long narrow columns (columns no. 1, 2, 3, 4 and 5 of Table 5) normal carrier gas rates vary between 200 to 1000 ml/min. With short wide columns the gas rate is more critical and has to be increased approximately following the section surface increase. A 2 m x 1000 mm column has a cross sectional area which is about 250 times larger than for analytical columns. Normal gas rates are then also about 5 to 10 l/min.

As conclusion to this paragraph about carrier gas, hydrogen and helium are excellent with all instruments. Nitrogen can be used on instruments with a flame ionization detector; the time for separation will then be longer.

4.6. Detector

This point has already been treated partly in the preceeding paragraph about carrier gas because these two have to be discussed together.

In preparative gas chromatography the sensitivity of the detector may be thought to be rather unimportant. With the really large samples of rather easy problems this is indeed so and strong attenuation of the signal will assure very good base line stability. With more difficult samples the situation is not the same. Firstly with the larger gas rates of preparative gas chromatography, the detector sensitivity and stability is already poorer than with the smaller gas rates of analytical gas chromatography. Because of this reduced stability attenuation has to be used.

Secondly, the limit amount that can be detected with a katharometer in analytical work is 0.01 μl and with high k values about 10 times that much can even present difficulties. Since a concentration in measured and is not an absolute amount and since the volume of our large preparative gas chromatography columns (5-10 l) is 200 to 1000 larger, the limit that can be detected here is 10 to 100 μl. Trial separations will have to be carried out with such sample sizes and because condensation from dilute vapor is difficult the sample may be lost. This can be a reason to prefer flame ionization detection (FID) for preparative scale gas chromatography. With FID a splitter has to be used and only a few percent are normally burned. In some cases up to 10-20% of the sample has to be burned because of sensitivity considerations.

With katharometer detection, really large samples especially of high boiling substances can cause kicking of the pen. This can be so strong that the chromatogram becomes useless. An example is shown in reference 12. A special large volume detector may be necessary or alternatively a bypass has to be drilled in the detector block [12]. In this way pen kicking is avoided and the sensitivity of the detector is practically unchanged.

4.7. Substance recovery

Recovery of the separated substances in preparative gas chromatography is a problem for which there is no single answer. It can be very good but it can also be very bad. Low boiling substances give only partical condensation because their vapor pressure, even at low temperatures, is still high or simply because sufficient cooling of the vapor loaded carrier gas is impossible to obtain rapidly enough. High boiling substances can produce aerosols which are also easily swept away by the carrier gas. Various methods have been advanced to solve this problem, such as electrical precipitation [15, 16]; centrifugation of the collector bottle during chromatography [17]; the use of gradient cooled collectors to prevent aerosol formation [18, 19]; filling the collecting bottles with a glass wool trap [20].

None of these methods seems to have found general acceptance. It should also be noted that only a simple collector bottle can be practical since it soon turns out in a laboratory doing preparative gas chromatography that a very number of these collector bottles are needed. The form of the collector bottles for work as depicted in this paper is not very important. We have tried several models, but were unable to improve upon the commercially available Autoprep 700 collectors [21].

The efficiency of these collector traps is often improved markedly by filling their upper-part as tightly as possible with glass wool. The first drops of recovered substance form a liquid layer on the glass wool and this acts as a stripping column. The glass wool scrubber dols not always improves recovery. Its resistance to gas flow can cause leaks at the connection between the column exit and the collection bottle. This is indeed a delicate point to which the necessary attention must be given. Another reason is that clogging can occur and is unavoidable with solidifying substances. This is for example the case with chloroform collected at very low temperatures. In general we use more collectors without than with glass wool scrubbers. Recovery is then determined by a number of factors as discussed further.

4.7.1. Volatility of the substances:

This point is obvious. As a comparison example follow the recoveries for 200 µl of ether, cyclohexane, iso-octane and trans-decalin, chromatographed with a gas rate of 200 ml H_2/min at their respective boiling points and collected at -20°. Results were: 0, 39, 45 and 88% recovery respectively. The effect of further cooling on the recovery of ether was studied. 200 µl of ether was chromatographed isothermally at 70° otherwise like above with cooling of the bottles at 0, -8, -14, -25, -65 and -80°. Recovery was respectively 0, 2, 17, 24, 83 and 89%. Adequate cooling is thus very important for the recovery in preparative scale gas chromatography.

4.7.2. Concentration:

The concentration of the substances in the gas phase plays an important role in the recovery, which increases with increased concentration. This concentration is determined by column parameters such as the volume of mobile phase in the column and the separation efficiency of the column (more plates per meter gives narrower bands and higher concentrations). For a given column, however, these are not very variable. Other factors directly related to the concentration are the amount of material put on the column, and more particularly the temperature of elution large samples chromatographed at a high temperature will show the highest concentration in the eluate. Unfortunately however, the most interesting are nearly always the rarest and most complicated chemical mixtures which are therefore difficult to separate, so that in this case only small samples and long retention times (low temperatures) can be used. The recovery percentage will be low although it is just with such mixtures that high recovery is most interesting. That the sample size has a direct influence on the concentration of the eluted substances is obvious, but the influence of the temperature of elution is not so obvious and is shown by the results given in Table 8.

High separation effect, long retention times and high temperature at elution can be obtained together by temperature programming the separation. This is however not always possible and not so easy as isothermal operation.

Table 8. Gives the mean value of five determinations. The collecting bottles were tared and the collected sample weight was compared with the weight of the injected sample as delivered by a syringe. Other conditions were: on column injection on Autoprep 700; gas rate 200 ml/min H_2; collecting bottles cooled in ice water

Percentage recovery of trans-decalin as a function of elution temperature (isothermal chromatography).

Quantity (µl)	Column temperature					
	200°	180°	160°	140°	120°	100°
200	93.1	87.8	72.7	71.3	66.7	47.0
100			69.0			

An alternative is to run the chromatogram at a lower temperature for the desired time, stop the flow of carrier gas, increase the temperature of the column oven and then restore the gas flow. In this way the substances are chromatographed long enough to be separated and are eluted at a high temperature. That recovery is better by this technique is shown by the figures for 200 µl ether, cyclohexane, iso-octane and trans-decalin which were run in the conditions as described before, but with an additional temperature increase of 70° with the gas flow stopped just before elution. The recovery was respectively 14.4, 42.4, 61,5 and 95,4%, which is decidedly higher than without temperature increase.

4.7.3. Gas flow rate:

High speeds of the gas flow through the collecting trap must be detrimental to the recovery. One way to ascertain this is by diminishing the gas flow rate just before elution of a peak. In a recovery experiment on 200 µl ether, cyclohexane, iso-octane and transdecalin, as described above, the temperature was increased by programming and the gas flow was reduced to 25-30 ml/min just before elution of the peaks. This was very easily effected without changing any control setting of the instrument, by inserting a suitable bleeding capillary through the injection gasket at the appropriate moment.

The recovery results were now respectively 31.8, 73.4, 84.6 and 97.0%. While such a drastic gas rate reduction would have a marked influence on baseline stability in normal analyses with high sensitivity, this is not the case in preparative work, since the sensitivity of the instruments is much reduced with regard to the high sample load. Comparative gas chromatograms on decalin are shown in Fig. 6.

As explained before in this paper, with preparative scale samples, such reduction of the gas rate is possible in narrow columns without harmful effects to the separation. This is once more shown by the n' values in Table 9.

So far, the experiments in this paragraph about recovery were carried out with rather small columns (m x 9 mm) and thus small samples for which recovery is essentially difficult. For larger columns with larger samples, recovery is much better.

Table 9. N' values for 500 ml benzene at 80^0 (k = 5).

Gas rate in ml/min	Column 2 of table 5	Column 6 of table 5
250	1345	1447
500	1455	1497
750	1304	-
1000	1416	1460
1250	1555	1427
1750	1579	-
2400	-	1505

With aerosol forming substances, increasing elution temperature or reducing the gas rate does not always help. For substances of low volatilizy (and this is mostly the case with these aerosol formers) adsorption on silica gel can bring a solution. This is shown in the separation of Fig. 7 of the substances of Chart 1 [22].

Collection for five separations with a 150 μl sample solution was obtained by blowing the effluent gas on 2 g silica gel (10-20 mesh) in a glass tube of 12 x 40 mm inserted in the chromatograph outlet. The tubes were eluted with ether and gave 133 mg of 1, 78 mg of 2 and 173 mg of 3. Substance 2 is isomeric with 3 but its structure is not known. The same silica gel tubes can be used repeatedly. Although this point was not checked specifically we think collection was quantitative.

Collection of mg samples for mass spectra or infrared spectra for example can often be achieved very simply. For solids and high boiling liquids insertion of a capillary in the chromatograph outlet is very often enough. An example of this is shown in Fig. 8. A large number of such deuterated enol ethers and analogues were thus purified for spectral investigation.

With more volatile substances, leading the chromatograph effluent through a U tube cooled in liquid air is usually sufficiently effective. An example of a separation where collection was achieved in this way is shown in Fig. 10 (see further).

4.8. Recording of preparative separations

The recorder must be as sensitive as in analytical scale gas chromatography (e.g. 1 mV). As the preparative scale separations last long it is best to provide a very slow paper movement of the recorder. Most examples shown were obtained with a paper speed of 1 inch/ 10 min.

5. Apparatus construction

Excellent instruments, applying in fact most of the principles discussed in this paper, are commercially available. A complete discussion of the construction of our instruments seems therefore unnecessary. A schematic drawing is shown in Fig. 9.

The oven is 1.2 m long and has a diameter of about 40 cm. The heating is at one side of the tubular oven. By regulating the input of fresh air a temperature gradient can be created or not. Up to 150-200° the temperature can be held remarkably constant throughout the oven. Because fresh air is blown continuously in the oven, eventual build up of carrier gas through leaks (hydrogen!) is impossible.

The detector is coupled to the electrical circuits of an Autoprep 700. Flipping a switch permits use of the 6 m x 9 mm column of this latter instrument. Using hydrogen as carrier gas and with 150 mA as katharometer heating current the detector hot wires have a variable life time of about 6 months to a year. With nitrogen and at higher current intensities this life time may be much shorter. The coiled ovalized columns can be made cheaply out of aluminium (columns 1 and 3 of tables 5). The Following procedure was used: the tubes (6 m length x 24 mm diameter) were flattened until the smallest diameter was 15 mm, filled with sand, coiled mechanically to a coil diameter of 30 cm, emptied, sections joined by soldering up to the desired length, checked for gas leaks, washed with water, acetone and dried, filled with the coated support by vibration and placed in the oven described above. Coiled and ovalized glass columns of the same dimensions can also be made. To do this a 1 m x 30 cm cylinder of 2 mm sheet steel was mounted horizontally so that it could be rotated very slowly, with a reducing grar box. Pyrex glass tubes of 1.5 m length are coiled under the flame on the cylinder and welded to each other until the desired column length is obtained. With tubes of 1.5 m x 8-9 mm opening and 10-11 mm outside diameter this can give up to a 75 m x 9 mm column [23]. With tubes of 2 cm opening the coiled column has an ovalized section, but this is a favorable factor as discussed above. These glass columns are conditioned against stress by baking overnight in an oven at 500°. After this the column is taken off the steel cylinder and mounted on a frame. The length wise rods of this frame, actually supporting the glass coils are provided with hooks to immobilize the column.

To facilitate connection to and removal of glass columns from the gas inlet and outlet of the chromatograph, a 30 cm piece of stainless steel capillary with 1.5 cm opening is fixed to the column ends. As an alternative, leakproof connections can also be obtained with fittings in injection septum rubber. This is better than teflon fittings which are too hard and can only be used once.

6. Examples of separations

In the introduction of this paper it has been stated that about four different preparative scale gas chromatographic techniques can be distinguished, this according to the sort of mixture and to the amount of substance that is needed.

6.1. Not too complex mixtures - spectrometric analysis

A normal analytical gas chromatograph with katharometer detection and a column of 1 to 3 m x 4 to 6 mm will do the job easily. Examples are found in Figs. 1 and 8 and another is given in the following Fig. 10.

The sample size of Fig. 10 is already not negligible although the column is small. Collection of the substances of this separation was mentioned in the paragraph on recovery.

6.2. Complex mixtures - spectrometric analysis

Fox mixtures containing a large number of components (10 or more) α values below 1.1 are unavoidable and the column will need an exceptionally high separating power. In this case the boiling range will also be rather large and programming is indicated. A characteristic of preparative scale separations is that they last longer than analytical separations. Therefore the programming rate must be slower and will normally be in the range of 0.5 to 2°/min. Suitable columns for this sort of problems have following characteristics: 20 m x 9 mm filled with Chromosorb W 30-60 mesh and coated with 5 g/100 ml. Such an column with Apiezon L as stationary phase produced 20,000 plates with 10 μl samples. See column V of Table V. An example of such a separation is shown in Fig. 11. Polytetramethylene oxide is used because it is a „universal" phase, giving equally good results for substances having very different polarity. Polytetramethylene oxide spreads the peaks over the chromatogram in a different way than Carbowax for example. It is therefore useful as an alternative polar phase and it presents the advantage of being slightly more stable to temperature effects than Carbowax 20 M. The 20 m x 9 mm columns in question, filled with a 30-60 mesh material show poor permeability. The analyses last long and the separation of Fig. 11 for example lasted 4 hrs.

Such columns are also indicated when the search for the best stationary phase fails to produce α values of at least 1.1.

This is for example the case for the cis-trans isomer mixtures of 4-methyl-2-alkyl-dioxolanes with alkyl equal to iso-propyl and n-butyl. The α-values on Carbowax 20 M and on polytetramethylene oxide (PTMO) were similar and between 1.07 and 1.09. A large volume 20 m x (1.5 x 2.5 cm) column filled with Chromosorb G 20-30 mesh with Carbowax 20 M failed to separate even 100 μl of the mixture. The maximum N' value for this column was only 1750 and for α 1.09 at least 2300 plates are necessary. A 20 m x 9 mm coiled glass column (similar to column 5 of table 5) filled with Chromosorb W 30-60 mesh and coated with PTMO separates a 100 μl sample completely, with N' values for the peaks between 3000 and 5000. M. A. S. S. on this column for this problem is then between 150 and 200 μl.

6.3. Simple mixtures - larger amounts

Columns 10 to 20 m in length and with a section of about 2 cm are to be used in this case. With coiled columns, ovalization of the section is best. The support can have a mesh size

Table 10

No.	Mixture	α min	MASS in ml	Total separated amount in ml	Column	t^o	Support mesh 20-30	Stat. Phase g/ml	Gas rate H_2
	Meta dioxanes cis/trans				Aluminium 21 m x 1.5-2.5	100°	Chrom. A	Carbowax 20 M 5/100	700
1	meta dioxane benzene		10	150					
2	4-methyl-6-iso-propyl	1.36	8	25	id.	140°	id.	id.	600
3	4,6-di-n-propyl	1.25	6	30	id.	140°	id.	id.	600
4	4-methyl-6-t-butyl	1.30	6	20	id.	140°	id.	id.	600
5	4,6-dimethyl	1.35	8	8	Aluminium 15 m x 1.5-4.0	100°	Chrom.G	SE-30 5/100	1000
6	4,5-tetramethylene	2	2	15	id.	100°	id.	id.	600
	+ 1,3-diolacetate	1.32							
	+ 2 other peaks	1.22							
7	4,6-di-iso-butyl	1.1	0.2	0.2	id.	170°	id.	id.	700
8	4,5-pentamethylene	1.11	0.5	8	Aluminium 21 m x 1.5-2.5	170°	Chrom. G.	Carbowax 20 M 5/100	600
9	4,6-si-i-propyl	1.25	4	28	id.	150°	id.	id.	700
10	4-methyl-6-n-butyl	1.3	5.5	12	id.	150°	id.	id.	700
11	4,6-diethyl	1.3	2.5	12	id.	170°	id.	id.	600
12	4-methyl-6-iso-butyl	1.31	6	15	id.	170°	id.	id.	600
13	4-methyl-6-ethyl	1.38	10.5	10.5	id.	130°	id.	id.	600
14	4-methyl-6-n-Pr	1.36	7.5	21	id.	135°	id.	id.	700
15	4-methyl-6-sec-butyl	1.36	5	42	id.	165°	id.	id.	700
16	di-n-butyl-4,6	1.14	2	18.5	id.	180°	id.	id.	700
17	di-sec-butyl-4,6	1.12	0.5	19	id.	180°	id.	id.	700
18	5-ethyl-5,5-diethyl	2.3	3.8	3.8	id.	150o	id.	id.	700

of 20-30 or 30-40. An example of such a separation, illustrating further technical details is shown in Fig. 2.

The α value for these cis and trans 3,4-dimethyl-cyclohexanones is 1.15. With equation (6) or from Table 2 it is deduced that 940 theoretical plates are needed. According to Table 5 this column still produces 940 plates with a 2 ml sample. Since the mixture contains about equal amounts of both isomers, MASS = 100·2/50 = 4 ml as calculated with equation (7).

Other examples, showing the M. A. S. S. values in function of α and conditions in general are found in Table 10.
A similar series of cis-trans dioxolane mixtures was also separated. It is interesting to note that the dioxolanes could not be chromatographed on Chromosorb A but indeed so on Chromosorb G.

This sort of preparative gas chromatography allows the separation of sufficient material for subsequent chemical work. The general procedure to do this is illustrated in the example of Fig. 12 which is about the separation on column no. 1 of Table 5 of a mixture of dimers obtained from 2-methyl-2-pentene and 2-methyl-1-pentene mixture. The values of the problem are first investigated on different analytical columns available. A preparative column is chosen according to the result (Carbowax 20 M is often most useful) and a test chromatogram is run at a temperature about 20 to 50° higher than the temperature used for the analytical separation (trace A in Fig. 12 obtained at 130°).
With this result a final temperature is chosen and a trial separation is run (trace B in Fig. 12 at 110°). Analytical results A and B allow one to deduce approximately the M.A.S.S. which is here only 1 ml since the smallest α value is 1.08-1.09.

A trace of one of the 5 separations with 1 ml which we did is then shown in C of Fig. 12; also indicated on C are the fractions of the composite peaks collected as 1, 2, 3 and 4 (total recovery for 1, 2, 3 and 4 was 60%). Finally an analysis was run of each collected fraction on the same column. The injections were carried out every 10 min as shown in part D of Fig. 12. Peak 1 is too broad and is still a mixture. Peaks 2, 3 and 4 are sufficiently pure for further work although this did not seem so predictable from trace C. This is however something which we have experienced repeatedly, namely that the result is better than could be hoped for from the aspect of the chromatograms. Subsequent I. R. and N. M. R. analyses showed that the substances are cis and trans 4,6,6-trimethyl-3-nonene and 3-ethyl-2,4,4-trimethyl-1-heptane.

The type of column under discussion in this paragraph is also the one to be chosen for enrichment of trace amounts to a more amenable concentration. Very large volumes of complex mixtures are then injected (considering the earlier experiments about stripping, an SE-30 column seems best indicated) and although the chromatogram shows scarcely separation, collection is carried out around the retention time of the desired trace component. The enriched mixture can then be treated in a normal way.

A recovery problem with such a column is then found in the reaction mixture obtained by exchanging the proton of chloroform by deuterium. The deuteroform was contaminated with 0.25% of 1.1-dichloroethane. This impurity made the deuteroform unusable as solvent for N. M. R. spectroscopy. The boiling point difference is only 4 ^{o}C and fractional distillation was unsuccessful. Preparative scale gas chromatography on column 1 of Table 5 was possible with samples of 40 ml. At a gas rate of 500 ml H_2/min, recovery was bad. Clogging of the glass wool scrubber wos partly the cause. By reducing the gas rate at the moment of peak elution to 100 ml H_2/min and in an Autoprep 700 collection bottle of 25 ml capacity without any scrubber, the recovery goes up to 80-90%. Even the impurity could be completely collected with an elution speed of only 25 ml H_2/min. In this way the 100 mg present in one 40 ml shot of the mixture was condensed an its nature was ascertained by N. N. R. spectroscopy. Elution of the deuteroform peak in this way however lasts several hours !

Another example of recovery is cyclohexane 80% of which was recovered under the following conditions: column 20 m x (1.5 x 2.5 cm) Chromosorb G 20-30 mesh with Carbowax wax 20 M. 2 ml sample with 800 ml H_2/min. at 80°, condensed in a 5 ml Autoprep 700 collector: recovery 81%, 5 ml sample with 800 ml H_2/min at 80°, condensed in A 25 ml Autoprep 700 collector: recovery 77%. The collectors were cooled at -40°; with lower gas rates recovery is over 90%.

6.4. Very large samples

Scaling up the technique described in this paper by a factor of 2 or 3 seemsquite possible, although unnecessary for research laboratory purposes. Scaling up by a factor 10 to 100or even more however needs another approach. This would bring gas chromatography in the field of industrial scale separations. The introduction of „flow homogenizers" seems to be a valid answer. In section C_1 on column design this point has already been discussed. We have little experience with this technique and refer the reader to the literature [5, 6, 7].

7. Further points of interest

7.1. Use of glass beads in glass columns

The possibility of catalysed thermal destruction has been referred to repeatedly in this paper and it has also been stated that glass is one of the best material tanse to avoid this. Coating amounts on glass beads have also been mentioned and practical amounts turn out to be 1 to 2 g stationary phase per 100 ml of glass support material. Coating glass beads with this last amount of an oily stationary phase gives a rather sticky material, very difficult to fill into a column (although certainly not impossible). Stationary phases which are solid at room temperature present no great difficulties and Carbowax 20 M on glass beads thus gives a satisfactory column filling. Different glass beads sizes were tested. The finer the beads the greater the separation power of the column, but also the greater the pressure drop. A satisfactory compromise turned out to be glass beads with a diameter of 0.8 mm (Ballotini shot lead glass beads no. 11 obtained from JENCONS-HEMEL Hempstead-England). Similar experiments were carried out with glass wool. This was cut in a mixer until the longest strands were about 4-5 mm. The resulting material was purified by frequent decantation and by washing with strong acid, soap water und chloroform. Coating with stationary phase was carried out as usual. Filling columns with such coated glass wool is much more difficult than with coated glass beads. The separation power of the glass wool columns is better than for the glass beads columns, but the pressure drop is excessive. Therefore we continued this line, using exclusively glass beads with 0.8 mm diameter.

About 6 l of such glass beads coated with 2 g/100 ml Carbowax 20 M was filled in a 20 m x 2 cm coiled glass column. The glass column was made from 1.5 m glass tubes with 2.0 cm opening as described in more detail before.

To facilitate connection to and removal of this column from the gas inlet and outlet, a 30 cm piece of stainless steel capillary with 1.5 mm opening was fixed to its ends. The permeability of the column was good, giving 500 ml H_2/min with 1250 mm mercury inlet pressure at temperatures up to 200°. The separation power of this column for small samples is much below the results obtainable with conventional supports. For large samples it however compares well with the earlier columns developed in this laboratory for large scale preparative gas chromatography (see Table 11).

Table 11. All results obtained with cumene as sample at 145° (k ~ 5) and with 500-700 ml H_2/min. Column 1: 0.8 mm glass beads coated with 2 g Carbowax 20 M/100 ml support in 20 m x 1.2-2.4 cm coiled glass column. Column 2 : 15 m x 1.5-2.5 cm, filled with Chromosorb A 20/30 mesh coated with 5 g SE-30/100 ml support. Column 3 : 21 m x 1.5-2.5 cm filled with Chromosorb A 20/30 mesh coated with 5 g Carbowax 20 M/100 ml support.

N' values

Sample size	Glass Beads	Chromosorb A	
	Column 1 Volume 6.2 l.	Column 2 Volume 5.1 l.	Column 3 Volume 7.7 l.
100 µl	508	2626	3075
200 µl	542	1913	3301
500 µl	541	1803	2584
1 ml	456	1093	1731
2 ml	314	673	976
5 ml	174	270	373
10 ml	99	110	154

An example of a separation on the glass beads columns is shown in Fig. 13. The mixture was obtained by alkaline treatment of dichloromethyl-iso-butylketone.

Only peaks 6, 7 and 8 were identified, being $(CH_3)_2CH-CH=CHCOOMe$, $(CH_3)_2CHCH_2COCHCl_2$ and $(CH_3)_2-CHCH(CH_2Cl)COOMe$ respectively. Similar mixtures obtained from other dichloromethyl alkylketones were also separated on this column. Purification of n-butyliodide and n-butylbromide by preparative gas chromatography was only possible on glass beads as support material. A series of deuterated n-butylhalides was thus isolated from the reaction mixture, using a 1.5 m x 9 mm column and with SE-30 as stationary phase.

The trend in our laboratories is to use more and more such glass beads in glass columns.

7.2. Retention times

Reproducible retention times can only be obtained with small samples. With the larger samples of preparative scale gas chromatography, retention times are not reproducible an and the peaks can show typical deviations. Early peaks (low k) tail while late peaks (high k) show leading (see Fig. 7 and 8). In between peaks can be perfectly symmetrical. Tailing and leading of peaks is often discussed in function of non linear partition isotherms of the substances. As inferred above, however, with larger samples all early peaks tail an all late peaks lead, regardless of substance nature. Typical changes also occur in peak front retention. This is shown in the next Table 12.

Table 12. 20 m x 9 mm coiled glass column. 15-20 mesh support with SE-30 silicone gum. Gas rate 200 ml H_2/min - 165°

Sample size	Peak front retention in mm for:			
in µl	Benzene	Toluene	Ethylbenzene	Cumene
50	62	89	127	165
100	62	89	129	168
2200	61	91	135	176
500	57	89	135	179
1000	55	89	136	181
2000	53	88	140	190

Reproducibility is only achieved with small samples.

For low partition ratios peak front retention decreases with increasing sample size and for high partition factors it increases. In between values showing no change are possible (toluene). Peak front retention with increasing sample size is influenced by several counteracting factors. It

a) decreases because of exponential sample introduction. This is most strongly felt for substances with small partition ratio (higher temperatures), low gas flow rates on short columns with small pressure drop (coarse support);

b) increases because of increased resistance to gas flow rate with large samples. This is most strongly felt with high gas flow rates and in longer columns with high pressure drop (small grained support). Another reason for peak front retention increase is found in temperature effects. Evaporation at the rear end of a first band cools the column locally and this retards a second substance. High partition factors should enhance these temperature influences and this is indeed so. This effect is general and also occurs on analytical size columns with larger samples. As far as practical results go, these effects are not important [22].

7.3. Miscellaneous

7.3.1. Temperature:

Mixtures with only a few compounds are best separated isothermally. For complex mixtures temperature programming may be needed. We have also studied „chromathermography" or the technique where a temperature gradient is applied along the column but this has only a minor influence in preparative scale gas chromatography [9].

7.3.2. Pressure:

Using about 5-7 kg/cm^2 inlet pressure and 4-6 kg/cm^2 outlet pressure, does not increase the capacity of preparative scale gas chromatography in a significant way. The most notable difference with normal work is an increase in separation time.

Reversedly with reduced pressure (pumping over a restrictor at the column outlet, p_i = 600 mm H_g, p_0 = 200 mm Hg) a reduction in separation time was noted. To reestablish the same retention time as for normal conditions the oven temperature can be reduced by about 20 to 50°. This can have some importance in connection with stationary phase stability, but the technical difficulties are big and we do not see much interest in this approach.

7.3.3. Column ovalization:

We have also tried ovalization in the axis ratio of 5/1 (flat column) and 4/2.5 thus greatly increasing the column volume. The results were not significantly better than for the 1.5-2.5 ovalized columns mentioned in Table 5.

7.3.4. Catalytic effects outside the column:

The separation of cis and trans 2,4-dimethyldioxolane gives a normal chromatogram with α = 1.08 on Chromosorb G - Carbowax 20 M. A „multiple process" [11] separation of these two isomers however gave products which showed the same composition on analytical gas chromatography as the starting mixture. The reason of this isomerization of the pure compounds after the separation, was an acid fouled detector. Washing the detector with tetralin and bicarbonate water removed the acidic contamination and subsequent separations proceeded in a normal way.

8. Conclusion

Problems in preparative gas chromatography are related to: speed, resolution, recovery and decomposition of stationary phase and chromatographed substances. One of these aims can only be obtained at the expense of the others. Compromising is therefore necessary. Ways to do this according to the problem have been advanced in this paper. The column is the heart of preparative gas chromatography. Best results can be expected with column length/diameter ratios between 300 and ?000. For a research laboratory column volumes of 25 ml, 2 l and 5-7.5 l are indicated. In the smaller volume columns support with mesh size between 30-60 and for the larger volume columns between 20-40 is recommended. The coating percentage is best around 5 g per 100 ml of support.

For each column a table similar to Tables 1, 3, 5, 6 and 11 should be determined. Separation problems are first investigated on analytical columns to determine α values and sample composition. The α values lead to the necessary plate number of the separation with

$$N = 16\left(\frac{\alpha}{\alpha - 1}\right)^2 \qquad (6)$$

In the table just mentioned the sample size which gives this plate number S. S. (N') can then be found. The mixture sample which can then be separated on the maximum allowable sample size (M. A. S. S.) is 100.SS(N')/x where x is the percentage of the compound for which the total amount may not exceed SS(N') in the injected sample. In the multitude of problems to be solved by preparative gas chromatography in the research laboratory, we can distinguish

1. 10-200 μl of mixtures containing only a few compounds and with all α values above 1.1. For this a normal analytical column and isothermal operation is sufficient.
2. 100 μl - 2 ml of multicomponent mixture with large boiling range. Temperature programming is indicated and a column of the type 20 m x 1 cm filled with a 30-60 mesh support.
3. 100-20 ml of mixtures containing only a few compounds and with all α values above 1.1. For such separations a column of the type 20 m x (1.5 x 2.5) filled with a coarse support (20-30 or 30-40 mesh) is best. Good cooling of the collector bottles is essential for recovery. For very volatile substances the gas rate must be reduced during collection (this is possible during the separation, with the narrow bore column advocated above). For aerosol forming substances adsorption on coarse silica gel is very often a good solution.

References

1. Nogare, S. D., and Juvet, R. S.: Gas Liquid Chromatography, Intersc. Publishers (1962).
2. Dinelli, D., Polezzo, S., and Taramasso, M.: J. Chromatog. 7, 477 (1962). J. Gas Chromatog., 2, no. 5, 150 (1964).
3. a) Alderweireldt, F.: Analytical Chemistry, 33, 1920 (1961).
   b) Alderweireldt, F., and Verzele, M.: Bull. Soc. Chim. Belg., 70, 703 (1961).
4. Frison, G.: J. Chromatog., 6, 243 (1961).
5. Debbrecht, F.; Found M. Company technical paper, no. 28 (1966).
6. Chemical & Engineering News; 43, 46 (1965). Chemical Week; March 20 (1965).
7. Facts and Methods, F. & M. company publication vol. 6, no. 5 (1965).
8. Taft, E.: Pittsburgh Conference on Analytical Chemistry, March 1964 Varian Aerograph Documentation, W-116.
9. Verzele, M., Cauwenberghe, K. Van, and Bouche, J.: J. Gas Chromatog. in press.
10. Verzele, M., Bouche, J., De Bruyn, A., and Verstappe, M.: J. Chromatogr. 18, 253 (1965).
11. Dimick, K. P.: G. C. preparative separations, Varian Aerograph laboratory manual (1966).
12. Verzele, M.: J. Gas Chromatog. 4, 180 (1966).
13. Bohemen, I., and Purnell, J. H.: J. Appl. Chem. (London) 8, 433 (1958).
14. Verzele, M.: J. Chromatog., 15, 482 (1964).
15. Kratz, P., Jacobs, M., and Mitzner, B.: Analyst, 84, 671 (1959).
16. Thompson, A.: J. Chromatog., 6, 454 (1961).
17. Wehrli, A., and Kováts, E.: J. Chromatog. 3, 313 (1960).
18. Teranishi, R., Corse, J., Day, J., and Jennings, W.: J. Chromatog. 9, 244 (1962).
19. Stevens, R., and Mold, J.: J. Chromatog., 10, 398 (1963).
20. Dimick, K., and Taft, E.: J. Gas Chromatog., 1, 7 (1963).
21. Verzele, M.: J. Chromatog., 13, 377 (1964).
22. Morisur, J., Furth, B., and Kossanyi, J.: Bull. Soc. Chim. Fr. (1967) in press.
23. Verzele, M.: J. Chromatog. 19, 504 (1965).

Gas solid Chromatography on Organic Crystals - A Study of New Adsorbents Avalaible for the Analysis of Complex Organic Compounds by Gas-Solid Chromatography

C. VIDAL-MADJAR and G. GUIOCHON

Ecole Polytechnique, Laboratoire du Professeur L. JACQUE, Paris

1. Introduction

The bright potentialities of gas solid chromatography, especially its great selectivity and high efficiency at large carrier gas flow rates have been emphasized by SCOTT and PHILLIPS [1], KISELEV [2], HALÁSZ and HORVÁTH [3] and GIDDINGS [4] among others in the last few years. These authors have shown that a successful adsorbent should have a very homogeneous surface, without micropores or high adsorption energy sites to achieve great selectivity and fast mass transfer in the absorbed state.

Until quite recently there was practically only one adsorbent, graphitized thermal carbon black to fulfill this requirement [5, 6]: KISELEV and YASHIN [7], HALÁSZ and HORVÁTH 3 have shown how this adsorbent has an unusual selectivity and allows many separations which are impossible by gas-liquid chromatography to be obtained easily the retention of any compound depending only on the geometrical structure of its molecule [8, 9, 10, 11]. However the existence of only one adsorbent suitable for the analysis of non-gaseous organic compounds contrasts with the versatility of gas-liquid chromatography using several hundred stationary phases. A new method to generate a great number of adsorbents with homogeneous surfaces and various selectivities would be very useful.

To decrease to too large adsorption enthalpy on graphitized carbon black and to obtain different adsorbents with various selectivities, HALÁSZ and coworkers [12, 13] add small amounts of liquid compounds which act as surface modifiers and lead to the intermediate method, gas-adsorption layer chromatography, which has some unique properties.

Another method of preparing a full range of new adsorbents suitable for gas solid chromatography is to use crystals of organic compounds with high melting points. Some papers have been previously published in this field but it seems that the potentialities of this method have not yet been fully realized.

SCOTT [14] has used benzophenone deposited on Chromosorb; ALTENAU and MERRITT

[15] have coated Chromosorb with copper amine complexes. The elution zone of polar compounds tails badly on these adsorbents because of the presence of high energy adsorption sites on the surface of Chromosorb, as we have shown in the case of benzophenone coated on this support [16, 17]. BURKE and ROGERS [18] have recently shown that there is a great difference between the adsorption properties of the pure isomeric phthalic acids and those of the adsorbents obtained by coating them on Chromosorb which is probably caused by the influence of the solid surface of the support on the orientation of the Molecules of the acids. Thus the use of Chromosorb or other diatomaceous earths as support material for solid organic adsorbents should be avoided. ALTENAU and ROGERS [19, 20] used some copper amine complexes and HEVERAN and ROGERS [21] several pure phthalic derivatives, directly as column packings and obtained small efficiencies, most propably because of the very small specific surface of the ground crystals.

Anthraquinone coated on graphitized carbon black [22] allows some very interesting chromatographic separations. However this compound has a thermal stability to be widely used. Metallic phthalocyanines used as organic adsorbents, coated on graphitized carbon black offer a much brighter prospect since they are thermally very stable [23]. In a previous paper we have shown some of the unusual chromatographic properties of this kind of support, using copper phthalocyanine deposited on graphitized carbon black. The efficiency is very high and the elution zones are symmetrical even with polar compounds like phenols [24]. New specificities as exemplified by the separation of anthracene from phenanthrene on a one meter long column [25] have been found; they are very different from those obtained in gas liquid chromatography.

PECSOK and VARY [26] have used some metallic phthalocyanines as a suspension in a silicone oil deposited on Chromosorb T; they have shown that each metallic phthalocyanine has its own specificity towards some of the main organic functions. However, this method cannot demonstrate clearly the affinity of the phthalocyanine because of the influence of the liquid phase.

Many new adsorbent surfaces can be obtained by coating various metallic phthalocyanines on graphitized carbon black. The specificity observed for each phthalocyanine is different. This specificity and the high efficiency obtained for all the columns had allowed us to achieve unusual and difficult separations.

In this paper, however, we have mainly studied the adsorption properties of these new surfaces. Coating the phthalocyanines on graphitized carbon black allows one at the same time to obtain highly efficient columns because of the sufficiently large surface area and the non selective adsorption properties of this support, and to determine exactly the adsorptive behavior of each phthalocyanine because there is no effect of the liquid phase, especially no competition and dilution, whereas the effect of the support is kept small as shall be shown later.

2. Experimental

The chromatographic equipment used was made in our laboratory: it has a flame ionization detector and a splitting system which allows the injection of small sample amounts (less than 0,1 μg). All the columns are made of 2 mm i. d. stainless stell tubing. The metallic phthalocyanines were prepared by adding an alcoholic solution of the anhydrous metallic chloride to a solution of dilithium phthalocyanine in absolute alcohol. As the anhydrous nickel chloride is not soluble in ethyl alcohol, we have used nickel chloride hydrate ($NiCl_2$, 6 H_2O); this is propably the origin of the unusual adsorption properties of nickel phthalocyanine. Metal-free phthalocyanine has been prepared by adding slightly acidified water to the alcoholic solution of dilithium phthalocyanine. The dilithium phthalocyanine is prepared by the procedure described by BARRETT, FRYE and LINSTEAD [27].

As soon as the phthalocyanine starts to precipitate the necessary amount of graphitized thermal carbon black is added to the solution (STERLING, M. T., CABOT, C.). The mixture is stirred for one hour, until the precipitation is complete. The solid is then washed with ethyl alcohol to eliminate the excess of dilithium phthalocyanine. It is then washed with distillated water dried and sieved. The 100-200 μm fraction is retained. The weight of solid packing used in a one meter long column is roughly 2.5 g. Very reproducible chromatographic results have always been obtained when the procedure described here is used for the preparation of the adsorbents.

To elimate the oxygen which might oxidize the organic solid phase at high temperature, the hydrogen used as carrier gas is purified by flowing through a bed of an oxidation catalyst and then through a Molecular Sieve 5 A trap; with such a carrier gas it is possible to work several days at 300 °C without changing the column performances. At a temperature of 280 °C the column stability is indefinite. Before work begins the columns are heated at 220 °C for 24 hours with hydrogen as carrier gas.

For every solute studied we have determined the retention volume per unit of surface area (it is given in cm). The specific surface area of each adsorbent has been measured using the B. E. T. method. They were always around 9 m^2/g. The differential enthalpy of adsorption is obtained from the slope of a plot of log k' vs. $1/T$ (where k' is the column capacity factor and T is the absolute temperature). From these two results we have calculated the differential entropy of adsorptions ΔS (using the procedure suggested by SCOTT [14]; the adsorbed standard state is the one defined by DE BOER [28] in which the average distance between the adsorbed molecules is the same as it is in the gaseous three dimensional standard state at the same temperature.

3. Results

The retention volumes per unit surface area of the adsorbent, the differential adsorption enthalpies and entropies have been measured for a number of organic compounds representative of the main functions : alkanes, alkenes, aromatic hydrocarbons, ethers, ketones alcohols, amines, phenols etc... (Cf. Tables 1 to 6) on the adsorbents obtained by coating 5 % of each of the following phthalocyanines: hydrogen, copper, cobalt, zinc and nickel on graphitized carbon black. Some previous systematic experiments have shown that if the adsorption enthalpy decreases with increasing percentage of phthalocyanine between 1 % and 3 %, from 3 % up to 15 % (we have not tried higher percentages) the adsorption enthalpies and the relative retentions are practically unchanged.

Systematic studies on the adsorbents made by coating 5 % of the organic solid compounds on graphitized carbon black will allow us to compare the reactivities of the different phthalocyanines. The adsorption enthalpy on pure graphitized carbon black (STERLING, M. T.) is quoted in tables 1 to 6 every time it has been measured by KISELEV and his coworkers [6].

It may be seen in table I that the adsorption enthalpies of the hydrocarbons, except benzene, on the various phthalocyanines are always 1 to 2 kcal/mole lower then the values obtained with graphitized carbon black. The same differences have been observed with anthraquinone deposited on graphitized carbon black [22]. This decrease in the adsorption enthalpy has a very useful practical consequence because it makes it possible to work at a lower temperature than that necessary with graphitized carbon black alone.

Table 1. Adsorption of normal alkanes

Solute		nC_5	nC_6	nC_7	nC_8	nC_9	nC_{10}	nC_{11}
T (^oK)		370	370	370	435	435	435	435
Sterling MT	ΔH kcal.mole^{-1}	8.9	10.4	12.5	13.4	14.8	16.7	18.0
Ph	log U_a	$\bar{1}.05$	$\bar{1}.51$	$\bar{1}.93$	$\bar{1}.44$	$\bar{1}.77$	0.08	0.39
	ΔS cal.mole^{-1}/°	6.9	8.2	9.5	10.3	11.3	12.1	12.9
	ΔS cal.mole^{-1}/°	11.6	13.0	14.5	15.0	15.7	16.1	16.6
PhCu	log U_A	$\bar{1}.11$	$\bar{1}.55$	$\bar{1}.99$	$\bar{1}.47$	1.79	0.11	0.41
	ΔH kcal. mole^{-1}	7.5	8.8	9.9	10.8	11.7	12.6	13.8
	ΔS cal.mole^{-1}/°	13.0	14.3	15.4	15.9	16.6	17.0	18.4
PhCo	log U_A	$\bar{1}.12$	$\bar{1}.57$	0.04	$\bar{1}.52$	$\bar{1}.85$	0.16	0.47
	ΔH kcal. mole^{-1}	7.5	8.7	9.7	10.6	11.6	12.4	13.3
	ΔS cal. mole^{-1}/°	13.0	14.0	14.6	15.2	15.9	16.3	17.0
PhZn	log U_A	$\bar{1}.18$	$\bar{1}.64$	0.09	$\bar{1}.54$	$\bar{1}.89$	0.22	0.54
	ΔH kcal.mole^{-1}	7.7	9.2	10.3	11.3	12.3	13.3	14.3
	ΔS cal.mole^{-1}/°	13.2	15.1	15.9	16.8	17.5	18.2	18.8
PhNi	log U_A	$\bar{1},13$	$\bar{1},58$	0.00	$\bar{1}.48$	$\bar{1}.81$	0,12	0.41
	ΔH kcal.mole^{-1}	7.8	9.1	10.1	11.1	12.1	13.0	13.8
	ΔS cal.mole^{-1}/°	13.8	15.1	15.9	16.6	17.2	17.9	18.4

The same retention time is observed at a 50° to 100° lower temperature. However, the selectivity is sometimes poorer. So for some difficult separations, a lower percentage (1 %) of phthalocyanine had to be chosen.

Table 1 shows the results obtained with the saturated hydrocarbons. The metallic phthalocyanines have a slightly larger adsorption enthalpy than the metal-free phthalocyanine. The maximum difference is 1 kcal/mole with nickel phthalocyanine. These results show the non-specificity of the phthalocyanines towards saturated hydrocarbons, which is in agreement with KISELEV's theory [2], since these molecules can give only non specific interactions.

Table 2. Adsorption of Various C_6 Hydrocarbons

Solute		Cyclohexane	n-Hexane	Hexene-2 cis	Hexene-2 trans	Ø
T (^{o}K)		370	370	370	370	370
Sterling MT	ΔH kcal.mole^{-1}	8.7	10.4	9.7	10.2	9.8
	log U_A	$\bar{1}.22$	$\bar{1}.51$	$\bar{1}.39$	$\bar{1}.45$	$\bar{1}.33$
Ph	ΔH kcal.mole^{-1}	7.2	8.2	8.2	8.2	8.3
	ΔS cal.mole^{-1}/^{o}	11.6	13.0	13.5	13.2	14.1
	log U_A	$\bar{1}.32$	$\bar{1}.55$	$\bar{1}.49$	$\bar{1}.52$	$\bar{1}.55$
PhCu	ΔH kcal mole^{-1}	7.3	8.8	8.6	8.6	10.1
	ΔS cal.mole^{-1}/^{o}	11.3	14.3	14.3	14.1	14.8
	log U_A	$\bar{1}.14$	$\bar{1}.57$	$\bar{1}.59$	$\bar{1}.57$	$\bar{1}.85$
PhCo	ΔH kcal.mole^{-1}	7,6	8,7	9,0	8,7	11,2
	ΔS cal.mole^{-1}/^{o}	12.2	14.0	14.9	14.0	19.5
	log U_A	$\bar{1}.38$	$\bar{1}.64$	$\bar{1}.64$	$\bar{1}.64$	$\bar{1}.86$
PhZn	ΔH kcal.mole^{-1}	7.9	9.2	9.2	9.2	11.1
	ΔS cal.mole^{-1}/^{o}	12.7	15.1	15.1	15.1	19.4
	log U_A	$\bar{1}.18$	$\bar{1}.58$	$\bar{1}.54$	$\bar{1}.56$	$\bar{1}.50$
PhNi	ΔH kcal.mole^{-1}	8.0	9.1	9.1	9.1	9.1
	ΔS cal.mole^{-1}/^{o}	13.3	15.1	15.4	15.1	15.4

Table 2 allows comparison of the retention volumes, the enthalpies and the entropies of adsorption of 6 carbon atoms hydrocarbons of different structure, possessing π electrons or not. For these compounds it is observed that the phthalocyanines have an increasing affinity in the following order: metal-free, nickel, copper, cobalt and zinc phthalocyanines. The elution order of the cis and trans isomers of hexene 2 is the same with metal-free, nickel and copper phthalocyanine and with pure graphitized carbon black. The elution order is reversed with cobalt and zinc phthalocyanines. The chromatograms of Figure 1 show the separations obtained with metal-free phthalocyanine (Figure 1 a) and cobalt phthalocyanine (Figure 1b). In the latter case, with the more specific cobalt phthalocyanine benzene is strongly retained while the elution order of the cis and trans hexene 2 isomers is reversed. This order is the same as the one found in gas liquid chromatography.

Data on the adsorption of aromatic hydrocarbons by the various phthalocyanines are given in Table 3. The order of increasing affinity is the same as the one observed above for molecules having π electrons.

The different specificities observed according to the adsorbent used show that it is possible to resolve the three xylene isomers by choosing the most appropriate type of phthalocyanine. The elution order of the aromatic hydrocarbons is the same with graphitized carbon black and with metal-free or nickel phthalocyanine. However the resolution of meta xylene from para and ortho xylene is not so good with one of these two phthalocyanines than with the carbon black surface, undoubtedly because of the decrease of the adsorption enthalpy and of the slight affinity of these adsorbents towards aromatic hydrocarbons, which leads to the begining of a separation between para and ortho xylene (Figure 2a). Figure 2b shows the separation of the aromatic hydrocarbons on copper phthalocyanine. The affinity of this adsorbent towards molecules having π electrons leads to very good resolution of the three xylene isomers which are not, however, resolved from n-propyl-benzene. With cobalt and zinc phthalocyanines which are still more specific, it is not possible to resolve meta xylene from para xylene. On the other hand, both isomers are well resolved from ortho xylene (Figure 2c). However if the percentage of cobalt phthalocyanine is lowered to 1.5 %, the adsorption properties of the modified support are intermediate between those of the graphitized carbon black and those of the organic solid compound, the adsorption enthalpy is increased and it is possible to achieve a better selectivity. As shown in the chromatogram of figure 2d, all the aromatic hydrocarbons in the mixture from benzene to n-propyl benzene are resolved.

The separation shown in Figure 2d is intermediate between that obtained with metal-free phthalocyanine (Figure 2a) and that obtained with copper phthalocyanine (Figure 2b). These results show that the affinity of these adsorbents towards substituted aromatic hydrocarbons is larger when the substituents are methyl groups.

Table 4 gives the results concerning the adsorbtion by the different phthalocyanines of molecules having oxygen atoms with unshared p-electrons but no OH groups. The specificity of the adsorbents may be estimated by comparing the adsorption enthalpy of these compounds with that of the saturated hydrocarbons which have the same number of heavy atoms, for example diethyl ether and n-pentane, methyl propionate and n-hexane, dipropyl ether and n-heptane. The order of increasing affinity of the phthalocyanines is the same as the one found formerly towards aromatic hydrocarbons, except for nickel phthalocyanine: Ph $<$ PhCu $<$ PhCo $<$ PhZn. In this case, zinc phthalocyanine is much more reactive than cobalt phthalocyanine. Only nickel phthalocyanine behaves differently. It has a weak affinity towards the molecules which have π electrons, while it is the most active adsorbent towards the oxygenated compounds. The affinity of nickel phthalocyanine for any compound depends largely on the value of its dipole moment.

Table 3. Adsorption of aromatic hydrocarbons

Solute		Ø	ØCH$_3$	ØC$_2$H$_5$	Øi(C$_3$H$_7$)	mX	pX	oX	Øn(C$_3$H$_7$)	1,3,5 T.M.B.	1,2,4 T.M.B.	1,2,3 T.M.B.
T (^{o}K)		370	435	435	435	435	435	435	435	435	435	435
Sterling MT	ΔH kcal.mole^{-1}	9.8	11.6	12.7	13.2				14.3			
Ph	log U$_A$	$\bar{1}$.33	$\bar{1}$.15	$\bar{1}$.33	$\bar{1}$.48	$\bar{1}$.49	$\bar{1}$.52	$\bar{1}$.52	$\bar{1}$.59	$\bar{1}$.86	$\bar{1}$.93	$\bar{1}$.98
	∠ H kcal.mole^{-1}	8.3	9.2	10.0	10.3	10.7	11.3	11.3	11.3	12.7	12.7	12.7
	ΔS cal.mole^{-1}/^{o}	14.1	14.0	14.7	14.7	15.6	16.8	16.8	1.66	18.4	18.2	18.2
PhCu	log U$_A$	$\bar{1}$.55	$\bar{1}$.23	$\bar{1}$.46	$\bar{1}$.59	$\bar{1}$.68	$\bar{1}$.70	$\bar{1}$.72	11.75	0.08	0.17	0.24
	ΔH kcal.mole^{-1}	10.1	10.6	11.1	11.3	11.8	12.0	12.3	11.8	13.5	13.7	13.9
	ΔS cal.mole^{-1}/^{o}	14.8	16.5	16.5	16.5	17.2	17.5	18.2	16.8	19.3	19.3	19.6
PhCo	log U$_A$	$\bar{1}$.85	$\bar{1}$.48	$\bar{1}$.68	$\bar{1}$.76	$\bar{1}$.94	$\bar{1}$.94	0.03	$\bar{1}$.96	0.31	0.43	0.50
	ΔH kcal.mole^{-1}	11.2	11.4	11.6	11.7	12.9	12.9	13.6	12.6	13.8	14.0	14.3
	ΔS cal.mole^{-1}/^{o}	19.5	17.2	16.8	16.6	18.7	18.7	19.8	17.7	18.9	18.9	19.1
PhZn	log U$_A$	$\bar{1}$.86	$\bar{1}$.48	$\bar{1}$.71	$\bar{1}$.79	$\bar{1}$.97	$\bar{1}$.97	0.05	$\bar{1}$.98	0.34	0.47	0.57
	ΔH kcal.mole^{-1}	11.1	11.5	11.7	11.7	13.3	13.3	13.6	12.3	14.1	14.3	14.5
	ΔS cal mole^{-1}/^{o}	19.4	17.7	16.8	16.6	19.3	19.3	19.5	17.0	19.4	19.4	19.4
PhNi	log U$_A$	$\bar{1}$.50	$\bar{1}$.14	$\bar{1}$.39	$\bar{1}$.54	$\bar{1}$.56	$\bar{1}$.60	$\bar{1}$.60	$\bar{1}$.68	$\bar{1}$.94	0.03	0.08
	ΔH kcal.mole^{-1}	9.1	10.1	10.6	11.3	11.5	11.5	11.5	12.3	12.8	13.1	13.1
	ΔS cal.mole^{-1}/^{o}	15.4	15.9	15.9	16.8	17.0	16.8	16.8	18.4	18.4	18.6	18.6

Table 4. Adsorption of various oxygenated derivatives

Solute		CH_3-CO-CH_3	$(C_2H_5)_2O$	nC_5	$C_2H_5COOCH_3$	nC_6	$(C_3H_7)_2O$	nC_7
μ		2.89	1.15		1.71		1.30	
T(^oK)		370	370	370	370	370	370	370
Sterling MT	ΔH kcal.mole^{-1}	8.3	8.8	8.9		10.4		12.5
Ph	log U_A	$\bar{2}.83$	$\bar{2}.94$	$\bar{1}.05$	$\bar{1}.26$	$\bar{1}.51$	$\bar{1}.73$	$\bar{1}.93$
	ΔH kcal.mole^{-1}	7.1	6.8	6.9	8.6	8.2	9.0	9.5
	ΔS cal.mole$^{-1}/^o$	13.2	11.9	11.6	15.1	13.0	14.0	14.5
PhCu	log U_A	$\bar{2}.95$	$\bar{1}.04$	$\bar{1}.11$	$\bar{1}.39$	$\bar{1}.55$	$\bar{1}.81$	$\bar{1}.99$
	ΔH kcal.mole^{-1}	7.7	7.5	7.5	9.3	8.8	10.1	9.9
	ΔS cal.mole$^{-1}/^o$	14.3	13.2	13.0	16.5	14.3	16.7	15.4
PhCo	log U_A	$\bar{1}.37$	$\bar{1}.36$	$\bar{1}.12$	$\bar{1}.80$	$\bar{1}.57$	0.10	0.04
	ΔH kcal.mole^{-1}	10.8	10.0	7.5	11.6	8.7	12.3	9.7
	ΔS cal.mole$^{-1}/^o$	20.5	18.6	13.0	20.8	14.0	21.3	14.1
PhZn	log U_A	0.00	0.12	$\bar{1}.18$	0.48	$\bar{1}.64$	0.79	0.09
	ΔH kcal.mole^{-1}	13.5	13.1	7.7	13.5	9.2	14.9	10.3
	ΔS cal.mole$^{-1}/^o$	25.2	23.5	13.2	23.0	15.1	25.1	15.9
PhNi	log U_A	$\bar{1}.38$	$\bar{1}.21$	$\bar{1}.13$	$\bar{1}.77$	$\bar{1}.58$	$\bar{1}.92$	0.00
	ΔH kcal.mole^{-1}	13.1	10.8	7.8	15.9	9.1	13.8	10.1
	ΔS cal.mole$^{-1}/^o$	26.8	21.3	13.8	32.7	15.1	26.2	15.9

The influence of the value of the dipole moment is also noticeable with the other phthalocyanines : the adsorption enthalpy of acetone, which has a strong dipole moment, is always higher than those of diethyl ether and n-pentane although a cetone has only 4 heavy atoms. This result is the opposite of the one observed on a really non specific adsorbent like graphitized thermal carbon black, on which the adsorption enthalpy of acetone is lower than those of diethyl ether and n-pentane and comparable to that of n-butane.

On the most active adsorbents, that is on nickel and zinc phthalocyanines, other very polar compounds like alcohols, phenols and pyridine are strongly retained and their adsorption isotherms are not linear, in the range of sample size we have used (the retention time decreases by 50% when the amount of sample is increased from 0.1 μg to 10 μg).

Table 5. Adsorption of nitrogen derivatives

Solute		Pyridine	Aniline	N-Me aniline	N, N di Me aniline
μ		2.2	1.48	1.67	1.57
T (^oK)		435	435	435	435
Sterling MT	ΔH kcal.mole^{-1}	10.0	13.1		14.0
Ph	log U_A	$\bar{2}.87$	$\bar{1}.37$	$\bar{1}.71$	$\bar{1}.96$
	ΔH kcal.mole^{-1}	8.3	9.1	10.1	11.3
	ΔS cal.mole^{-1}/^{o}	12.9	12.4	13.1	14.7
PhCu	log U_A	$\bar{2}.93$	$\bar{1}.56$	$\bar{1}.94$	0.25
	ΔH kcal.mole^{-1}	10.1	11.3	13.1	14.1
	ΔS cal. mole^{-1}/^{o}	16.5	16.6	19.1	19.8
PhCo	log U_A	0.39	0.14	0.58	0.49
	ΔH kcal. mole^{-1}	16.8	15.0	16.7	15.7
	ΔS cal.mole^{-1}/^{o}	25.5	22.5	24.4	22.5
PhZn	log U_A		0.56	$\bar{1}.00$	0.90
	ΔH kcal.mole^{-1}		15.6	16.6	16.1
	ΔS cal. mole^{-1}/^{o}		21.8	22.1	21.4
PhNi	log U_A		0.78	0.13	0.09
	ΔH kcal.mole^{-1}		21.1	16.6	13.7
	ΔS cal.mole^{-1}/^{o}		33.3	26.2	19.8

Table 5 shows the results obtained with some nitrogen compounds : pyridine, aniline, N-méthylaniline, N, N-diméthylaniline. Here again, the affinity increases from metal-free phthalocyanine to zinc phthalocyanine. With metal-free or copper phthalocyanine, the elution order of the nitrogen compounds is the same as the one observed with graphitized carbon black. On the contrary with cobalt or zinc phthalocyanine, the elution order depends heavily on the value of the dipole moment of the molecules studied and the adsorption enthalpy of the four compounds studied here increases in the order of their dipole moment although pyridine has only six heavy atoms. Cobalt phthalocyanine shows very important, unusual properties : a great affinity for nitrogen compounds, especially aromatics, and a comparatively low specificity for highly polar molecules. This will allow to use it in gas-solid chromatography to achieve specific separations with symmetrical peaks. The chromatogram of Figure 3a shows the analysis of a mixture of aniline, N-methylaniline, N, N-diméthylaniline and the three isomers of toluidine. The complete separation of these compounds is not possible with metal-free or copper phthalocyanine which are non specific adsorbents towards these molecules. The chromatogram of Figure 3b shows the separation on cobalt phthalocyanine of a mixture of pyridine, picolines and lutidines. The 3, 4 and 3,5 isomers of lutidine are also resolved with symmetrical peaks but their retention times are much higher than that of the picoline isomers and they are not shown on the figure. This separation is only possible with cobalt phthalocyanine, the selectivity of which towards nitrogen compounds is very high while still giving symmetrical peaks.

Nickel phthalocyanine has a behaviour which is much different from the other phthalocyanines. For example, the adsorption enthalpy of aniline is very large, 21 kcal/mole. The N-methylaniline also is strongly adsorbed, but its adsorption enthalpy is lower than that of aniline. With both compounds the adsorption isotherms are not linear and the retention time decreases with increasing sample size therefore, the values given in table 5 are extrapolated for zero sample size.

Only N, N-dimethylaniline is eluted with symmetrical peaks; the adsorption enthalpy of this compound is the same within 10 % on copper, cobalt, zinc and nickel phthalocyanines. The anomalous behaviour of nickel phthalocyanine could be explained by the presence on its surface of water molecules which could form hydrogen bonds with aniline and N-methylaniline. These water molecules may originate during the precipitation of nickel phthalocyanine because we had to use nickel chloride hydrate since the dehydrated chloride is not soluble in organic solvents.

Table 6 shows the retention volumes, the differential adsorption enthalpies and entropies of some alcohols and phenols on metal-free, copper and cobalt phthalocyanine. The adsorption enthalpy of alcohols is always 2 to 3 kcal/mole higher than that of the corresponding saturated hydrocarbons. The adsorption entropy is about the same on all the phthalocyanines used here. These results show that these phthalocyanines exhibit some selectivity towards the alcohol funtion, probably because of the unshared p-electrons of the oxygen atoms. However, this affinity is not important enough to produce a heterogeneous surface and symmetrical peaks are obtained for all alcohols when one the three phthalocyanines quoted table 6 is used.

Table 6. Adsorption of hydroxylated compounds

Solute	$T(^{o}K)$	Ph			Ph Cu			Ph Co		
		$\log U_A$	ΔH kcal.mole^{-1}	ΔS cal.mole^{-1}/o	$\log U_A$	ΔH kcal mole^{-1}	ΔS cal. mole^{-1}/o	$\log U_A$	ΔH kcal. mole^{-1}	ΔS cal. mole^{-1}/o
$nC_5H_{11}OH$	435	$\bar{2}.99$	11.3	19.3				$\bar{1}.29$	11.7	18.8
$nC_6H_{13}OH$	435	$\bar{1}.38$	11.9	18.9	$\bar{1}.42$	12.4	19.8	$\bar{1}.60$	12.2	18.4
$nC_7H_{15}OH$	435	$\bar{1}.60$	12.3	18.6	$\bar{1}.73$	13.1	20.0	$\bar{1}.80$	12.8	18.8
$nC_8H_{17}OH$	435	0.02	13.0	18.4	0.07	13.6	19.6	0.17	13.1	17.9
Phenol	435	$\bar{1}.39$	10.9	16.6	$\bar{1}.51$	11.4	17.0	$\bar{1}.60$	11.5	16.8
2 Me Phenol	435	$\bar{1}.67$	11.5	16.6	$\bar{1}.81$	12.3	17.7	$\bar{1}.90$	12.5	17.7
3 Me Phenol } 4 Me Phenol }	435	$\bar{1}.73$	11.5	16.3	$\bar{1}.88$	12.3	17.5	0.00	12.7	17.7
2,6 di Me Phenol	435	$\bar{1}.92$	11.8	16.1	0.14	13.5	19.1	0.25	13.8	19.1
2,5 di Me Phenol } 2,4 di Me Phenol }	435	0.04	12.7	17.7	0.24	13.9	19.5	0.36	14.5	20.2
2,3 di Me Phenol	435	0.04	12.7	17.7	0.29	14.3	20.2	0.45	14.9	20.7
3,5 di Me Phenol	435	0.10	13.0	18.0	0.29	14.3	20.2	0.45	14.9	20.7
3,4 die Me Phenol	435	0.16	13.3	18.4	0.37	14.7	20.7	0.50	15.1	20.9
2 Eth Phenol	435	$\bar{1}.86$	12.0	16.8	0.00	12.8	18.1	0.13	13.2	18.4
3 Eth Phenol 4 Eth Phenol	435	$\bar{1}.96$	12.7	18.0	0.11	13.3	18.7	0.22	13.4	18.4

With nickel or zinc phthalocyanine which are much more reactive toward polar compounds, as mentionned previously, the adsorption isotherms of alcohols are not linear and the retention time changes with the amount of sample injected in the range of 0.1-10 μg. The same result ist obtained with phenols. The adsorption enthalpy and the retention volumes are almost the same for phenols and for the aromatic hydrocarbon obtained by changing the hydroxyl group into a methyl group. The difference in the adsorption enthalpy of these two related compounds is never more than 1 kcal/mole [24]. As we have shown in a previous paper dealing with adsorption of phenols on copper phthalocyanine [24], the retention pattern of methyl and polymethylphenols on metal-free, copper and cobalt phthalocyanines is the same as the one observed on an non-polar liquid phases i. e., their elution order is the order of increasing carbon atoms number of the compounds. However, some differences exist in the elution order of the phenols having the same molecular weight since the retention on the phthalocyanines depends also on the geometrical structure of the molecules. Meta cresol is, for instance, eluted before para cresol on the adsorbents studied, while an non-polar liquid phase para cresol is eluted first. Other inversions in the elution order are observed with 2, 3 and 3, 5 dimethyl phenols and with ethyl phenols.

The adsorption properties of uncoated graphitized carbon black and those of the adsorbent obtained by coating 1 % metal free phthalocyanine on the same black can be compared on figure 4 which shows the analysis of the same terpene mixture as the one analysed by PETOV and SHCHERBAKOVA on pure Sterling M. T. [10]

The efficiency and the permeability of the column used here for this analysis are much better than those of a pure Sterling M. T. column because the phthalocyanine acts as a binding agent of the fine carbon black particles. The reduction in the adsorption enthalpy generally observed when organic solids are coated on graphitized carbon black allows one to achieve the separation at a much lower temperature for example, carvone is eluted here in 25 minutes at 157 ^{o}C with a four meter long column (Figure 4a) instead of 40 minutes at 180 ^{o}C with a 1. 2. meter long column [10].

The specificity of the adsorbent is relatively low because Sterling M. T. is coated with only 1 % of metal free phthalocyanine. Nevertheless the relative retention of polar compounds is larger on this adsorbent than on pure graphitized carbon black (see the migration of camphor and fenchone peaks). The surface, however, is homogeneous enough to allow elution of terpene alcohols with symmetrical peaks (Figure 4b).

The elution order of the terpenes of a given group shows that their retention depends only on their geometrical structure (Figure 5). Menthol, neomenthol and menthone have the same retention time, howeves, they are retained longer than isomenthol, neoisomenthol, neoisomenthone, howeves longer which are themselves eluted in the same time. Menthone and neomenthone are components of the mixture shown in the chromatogram (Figure 4), with the same experimental conditions as for Figure 5. The broadening of the common peak of menthol, neomenthol shows the begining of a separation.

4. Conclusion

This work shows that it is possible to obtain new adsorbents by coating graphitized thermal carbon black with different phthalocyanines. Surfaces of great homogeneity can be obtained with highly reproducible properties. The surfaces obtained with some phthalocyanines are even more homogeneous than the surface of graphitized carbon black; for example, phenols are eluted with symmetrical peaks on some phthalocyanines whereas they are not on pure graphitized carbon black. This result is important for theoretical purposes as well as for analytical work.

The achievement of new homogeneous surfaces of a known chemical structure is very important for the theory of adsorption. In this case the crystalline structure of the phthalocyanines is such that the surface will be predominantly made of the planes on which phthalocyanines molecules are lain.

The potential of adsorption may then be calculated using the method described by KISELEV and his coworkers [11]. The adsorption enthalpies measured on the adsorbent made by coating 5 % phthalocyanine on graphitized carbon black do correspond to the adsorption on the phthalocyanine surface with only negligible contribution from the carbon black support. This is supported first by the fact that the adsorption enthalpies are the same whatever the amount of phthalocyanine added to the carbon black above 3 % and up to 15 % (we did not try higher amounts). Furthermore the values of the potential of adsorption on metal-free phthalocyanine were computed by KOUZNETSOV who is working in our laboratory this year. These theoretical values agree fairly well with the experimental data.

The analytical results are important as well. The fact, that homogeneous surfaces and high efficiencies are obtained is clearly not restricted to the use of phthalocyanines, many other compounds and especially organic compounds could be used. The specificity depends on the compound used and especially for phthalocyanines on the affinity towards π and unshared p-electrons. This affinity increases from metal-free to copper and cobalt phthalocyanine. As is observed with pure graphitized carbon black, however, the retention on phthalocyanines depends to a great extend on the geometrical structure of the molecules : the elution of normal and branched alkanes is the same whether graphitized carbon black is coated with phthalocyanines or not. With metal-free or copper phthalocyanines, only the polar compounds are selectively retained. With cobalt phthalocyanine the most important factor determining the retention is the ability of the molecule to give specific interaction. Cobalt phthalocyanine could become very important in analytical practice because it is a specific adsorbent towards the aromatic ring and a non specific adsorbent towards molecules containing positively charged groups. This allows selective resolution of many organic compounds and especially of aromatic amines and pyridine derivatives.

The usefulness of these new adsorbents in high temperature gas chromatography is exemplified in Figure 6 and 7 which show the separation of some polynuclear aromatic hydrocarbons and the separation of two isomeric steroids, etiocholanolone and androsterone. The great differences in the retention times of these two isomers may be explained by their molecular structure, etiocholanolone is the cis-isomer while androsterone is the trans-isomer. Their separation depends only on their geometrical structure. The easy separation of etiocholanolone from androsterone and that of anthracene from phenanthrene [25] show how gas solid chromatography offers selectivities which are impossible to reach by gas liquid chromatography.

References

1. Scott, C. G., Phillips, C. S. G.: Gas Chromatography 1964 A. Goldup ed. The Institute of Petreoleum, London, 1965 p. 266.
2. Kiselev, A. V.: ibidem p. 238.
3. Halasz, I., Horvath, C.: Anal. Chem., 36, 1178 (1964)
4. Giddings, J. C.; Methodes Physiques d'Analyse, 2, 13 (1966).
5. Kiselev, A. V.: Adv. Chromatography, Vol. 4, ed. J. C. Giddings, R. A. Keller; Dekker, N. Y., 1967, p.113.
6. Kiselev, A. V., Yashin, Y. I.: Gas Adsorption Chromatography, Nauka Press, Moscow, 1967.
7. Kiselev, A. V., Yashin, Y. I.: Gas Chromatographie 1965, G. H. Struppe ed., Akademie-Verlag, Berlin, 1965, supplément p. 43.
8. Avugul, N. N., Isirikyan, A. A., Kiselev, A. V., Lygina, I. A., Poshkus, D. P.: Izv. Akad. Nauk SSSR Otd. Khim. Nauk, 1314 (1957).
9. Avgul, N. N., Kiselev, A. V., Lygina, I. A., Poshkus, D. P.: ibidem, 1196 (1959).
10. Petov, G . M., Shcherbakova, K. D.: Gas Chromatography 1966, The Institute of Petroleum, London, 1967 p. 50.
11. Kiselev, A. V., Kouznetsov, A. V., Filatova, I V., Shcherbakova, K. D.: Zh. Fiz. Khim. to be published.
12. Halász, I., Heine, E., Horvath, C., Sternagel, H. G.: Brennstoff Chemie 44, 387 (1963).
13. Schneider, W., Bruderreck, H., Halász, I.: Anal. Chem. 36, 1533 (1964).
14. Scott, C. G.: Gas Chromatography 1962, ed. M. Van Swaay, Butterworth, London, 1963 p. 36.
15. Altenau, A. G., Merritt, C.: J. Gas Chrom., 5, 30 (1967).
16. Vidal-Madjar, C., Guiochon, G.: Bull. Soc. France, 1966, 1096.
17. Vidal-Madjar, C., Guiochon, G.: J. Phys. Chem., 71, 4031 (1967).
18. Burke, M. F., Rogers, L. B.: J. Gas Chrom., 6, 75 (1968).
19. Rogers, L. B., Altenau, A. G.: Anal. Chem., 35, 915 (1963).
20. Altenau, A. G., Rogers, L. B.: ibidem, 36, 1726 (1964), 37, 1432 (1965).
21. Heveran, J. E., Rogers, L. B.: J. Chromatog., 25, 213 (1966).
22. Vidal-Madjar, C., Guiochon, G.: Separation Science, 2, 155 (1967).
23. Moser, F. H., Thomas, A. L., Phthalocyanine Compounds, ACS Monograph 157, Reinhold Publishing Corp., N. Y., 1963.
24. Vidal-Madjar, C., Guiochon, G.: C. R. Acad. Sciences, Paris, 265, 26 (1967).
25. Vidal-Madjar, C., Guiochon, G.: Nature, 215, 1372 (1967).
26. Pecsok, R. L., Vary, E. M.: Anal. Chem., 39, 289 (1967).

27. Barrett, P. A., Frye, D. A., Linstead, R. P. J.: J. Chem. Soc., 1938, 1157.

28. De Boer, J. H.: The Dynamical Character of adsorption, Clarendon Press, Oxford, 1953.

The Thermodynamics of Adsorption and Solution and Nonideality of the Gas Phase of some Volatile Metal Chlorides

F. M. ZADO and R. S. JUVET

Department of Solid State, Institute „Rudjer Bošković", Zagreb, Yugoslavia, and Department of Chemistry and Chemical Engineering, University of Illinois, Urbana, Illinois

The gas chromatographic separation of metal chlorides boiling above 200 °C represents a rather involved experimental as well as theoretical problem as shown in earlier communications [1, 2]. Although the elution of lower boiling metal chlorides has now been performed by several workers [1-6], such difficulties as the corrosive properties of metal chloride vapors and the choice of suitable column packing materials have undoubtedly contributed to a lag in the investigation of these very interesting systems.

In this paper studies are reported of CsCl and RbCl as solid adsorbents and as components of fused-salt liquid phases for the gas chromatographic elution of $SbCl_3$ and $MoCl_5$. The nonideality of metal chloride vapors will also be briefly discussed and the second virial coefficients of a series of metal chlorides, tabulated.

1. Experimental

Adsorption and solution studies of $SbCl_3$ were carried out in a home-made gas chromatograph using a Gow-Mac Model 625 gas density detector and an F. and M. SI-4 solid sampler as injection device.

Nitrogen, dried by means of Type 5A Molecular Sieve, was used as carrier gas. Columns for adsorption studies were made of borosilicate glass, 9 mm o. d. by 45 cm in length, packed with 80-100 mesh RbCl or CsCl and dried for several hours at 300 °C in a stream of dry nitrogen. Experimental conditions for gas-liquid partition chromatography with fused salts as liquid phases were similar to those reported earlier [2].

2. Results and Discussion

The partitioning properties of a series of molten chloride eutectic liquid phases have previously been characterized [2]. Contrary to organic systems where liquid phases may be divided on the basis of polarity, inorganic fused salt liquid phases are best classified as to their ability to enter into complex formation with the volatile metal chloride solutes. The retentive property of a liquid phase depends to large measure upon the available chloride ion concentration in the melt.

Antimony trichloride readily forms a series of chloro-complexes with alkali metal chlorides [7, 8, 9, 10], but thermodynamic data are not available on these systems. Since fused salt liquid phases are often multicomponent systems, it is generally not possible to derive directly from partition gas chromatographic measurements quantitative thermodynamic information on complex formation between the solute and the individual components of the melt. In order to derive more reliable data on the stability of antimony chloro-complexes with RbCl and CsCl, a technique involving overloading of the column was applied in order to obtain well-defined tabletop peaks as shown in Figure 1. Since the tabletop peak height is proportional to the vapor pressure of the metal chloride over its dissociating complex [11], it follows from the CLAPEYRON equation that:

$$\frac{d \log h}{d\, 1/T} = \frac{H_{assoc.}}{2.303\, R} \qquad (1)$$

where h and $H_{assoc.}$ are the tabletop peak height and the molal heat ot association of the complex formed, respectively.

In Table 1 thermodynamic results derived from this technique are compared with those obtained through both conventional adsorption peak retention measurements and GLC measurements on eutectic melts.

Table 1. Heat of Interaction of $SbCl_3$ with Various Alkali Metal Chloride Substrates

Interacting Metal Chloride	Tabletop Method $H_{assoc.}$ kcal/m.	GSC Method H_a, kcal/m.	GLC, Eutectic melt, H_s, kcal/m.
CsCl	-26.4	-31.2	-31.6 (in CsCl-AgCl) H_m, -21.2 kcal/m.
RbCl	-22.4	-	-22.9 (in RbCl-AgCl) H_m, -12.5 kcal/m.

Heat of mixing, H_m, is defined as the sum of the heat of solution plus the heat of vaporization. Preliminary experiments using solid AgCl as column packing have shown that $SbCl_3$ was eluted as „air peak" from a 4 ft x 6 mm glass column, indicating no marked specific interaction in this system.

Inspection of the phase diagrams of the CsCl-$SbCl_3$ system and the RbCl-$SbCl_3$ system [10] indicates the existance of two complexes in both systems and higher stability of the cesium complexes in comparison to the rubidium ones. Incongruent melting points are observed in the RbCl-$SbCl_3$ system for the two complexes RbCl · 2$SbCl_3$ and 3 RbCl · $SbCl_3$ while well-defined melting points are found in the CsCl-$SbCl_3$ system for the two complexes, 3 CsCl · 2$SbCl_3$ and 3CsCl · $SbCl_3$. In view of these facts it is obvious that it is difficult to derive heats of decomposition for a particular antimony chloro-complex, especially in the case of cesium. At a working temperature of about 330 °C for the RbCl-$SbCl_3$ system, it would appear that the heat derived by means of the tabletop method is most likely related to the Rb_3SbCl_6 complex alone. It is interesting to note here the considerably higher heat of adsorption obtained for $SbCl_3$ on CsCl when measured in the conventional manner using small sample sizes of about 0.1 mg. This heat of chemisorption, of course, cannot be simply interpreted in terms of heat of complex formation of any particular cesium antimony chloro-complex owing to the possibility of two complexes being formed simultaneously. The result of -21.2 kcal/m calculated for the heat of mixing of $SbCl_3$ in the AgCl-CsCl eutectic would appear to offer more information than the heat of solution of -31.6 kcal/m. It must be recalled that CsCl-AgCl eutectic is not a simple melt but a mixture of a cesium silver chloro-complex and an excess of free cesium chloride. The apparent heat of solution measured gas chromatographically actually contains terms for the heat of mixing of the solute in the binary mixture and a heat of vaporization while the heat of mixing itself contains terms for only the heat of solution plus the heat of complexation it is interesting to note the rather surprisingly small difference between the values for RbCl and CsCl interaction energies with $SbCl_3$ derived via the tabletop method in spite of rather large differences in the corresponding heats of mixing of $SbCl_3$ in RbCl-AgCl and CsCl-AgCl eutectics. These two liquid phases, of course, represent completely different environments.

While the gas chromatographic elution of antimony trichloride is easily accomplished, the elution of molybdenum pentachloride is much more involved. The literature on the chemical properties of $MoCl_5$ is somewhat discouraging because of the possibility of disproportionation of this material at elevated temperatures into less-volatile chlorides [12], extremely high reactivity particularly with traces of oxygen, and a series of possible interactions with higher alkali metal chlorides [13] leading to the reduction of Mo(V) to Mo(IV), thus possibly preventing the elution of $MoCl_5$. However, it was found that $MoCl_5$ is well retained on a RbCl-AgCl eutectic liquid phase with a specific retention volume of 61 ml/g at 332 °C (65° above its boiling point). Although somewhat tailing, the general peak shape of the $MoCl_5$ does not suggest any of the above serious effects.

The most interesting elution behavior of $MoCl_5$ was the extremely low retention observed on a TlCl-$InCl_3$ eutectic liquid phase as shown in Figure 2. A complete discussion of the solute-solvent interactions for this system will be presented elsewhere. $NbCl_5$ and $TaCl_5$, on the other hand, are well retained and separated on the same liquid phase [2]. The order of elution of these solutes indicate that highly specific interactions are taking place during the elution since all three solutes are eluted in the reverse order of their volatility. Since the heat of solution of $MoCl_5$ on RbCl-AgCl liquid phase has not as yet been determined, there is insufficient evidence to prove conclusively that the retention of $MoCl_5$ on this liquid phase is due to complex formation in this melt, although a high concentration of free chloride ions does exist. On the other hand, the relatively low concentration of available chloride ions in the TlCl-$InCl_3$ liquid phase, where Tl_2InCl_5 and $InCl_3$ exist at the eutectic composition, favors low retention of $MoCl_5$.

The nonideality of the gas phase of metal chlorides was suspected as being an important correction factor in deriving solution thermodynamic data. A thorough literature review has revealed a complete lack of data on compressibility measurements of not only metal chlorides but of other volatile inorganic compounds as well, From the comprehensive study of the saturated vapor density and pressure data, in the evaluation of critical parameters of the metal chlorides carried out by NIELSON and coworkers [14] second virial coefficients may be derived. The well-known virial form of the BERTHELOT equation of state was used to derive values for the second virial coefficient of these materials. The values of second virial coefficient were also calculated using experimental saturated vapor density and pressure data. These data as well as the value of the BOYLE temperature (where the second virial coefficient becomes zero) and other pertinent parameters are listed in Table 2.

Table 2. Theoretical and Experimental Values of Second Virial Coefficient of Volatile Metal Chlorides

Compound	Mol. Weight	T (°K)	P atm	T_{BOYLE} (°K)	-B (BERTHELOT) ml/mole	$-B_{exp.}$ ml/mole
BCl_3	117.2	452	38.2	1525	873	-
BF_3	67.18	261	49.2	880	137	174
$AsCl_3$	181.3	654	57.7	2208	867	628
$SbCl_3$	228.1	794	58.7	2680	886	1745
CCl_4	154	546	45.0	2207	978	920
$SiCl_4$	170	507	37.1	1711	899	2675
$GeCl_4$	214.4	552	38.0	1863	759	1433
$SnCl_4$	260.5	592	36.9	1998	1164	1192
$TiCl_4$	189.7	638	45.7	2153	1018	3193
$ZrCl_4$	233	777	58.8	2619	584	575
$HfCl_4$	320.3	725	58.9	·2446	495	824
$VOCl_3$	173.3	636	-	2146	-	1493
$NbCl_5$	270.17	807	46	2723	1343	-
$TaCl_5$	358.2	767	43	2588	1225	-
$HgCl_2$	271.5	972	114	3280	532	701
$MoCl_5$	273.2	850	51.8	2868	1255	-
WCl_6	396.6	923	49.1	3115	1319	-
$WOCl_4$	341.7	782	52.0	2639	1077	-
$ReOCl_4$	344.0	782	48.5	2637	1152	-

Disagreement in the values obtained for the second virial coefficient by these two methods is most likely due to the high uncertainty in the saturated vapor density data, the values of which were obtained mostly from orthobaric densities. Undoubtedly, these values of the second virial coefficient are less dependable than those obtained by the BERTHELOT equation and should require further evaluations. These data, although not very precise, are useful in further solution thermodynamics studies. Since metal chlorides are more or less symmetrical, dense molecules with a slightly negative outer shell of chloride atoms, it is not surprising that even heaviest metal chlorides (e. g., WCl_6, Mol. Wt., 396.6; $B_{BERTHELOT}$ = -1319 ,l/mole) have second virial coefficients two to three times lower than those of nonpolar C_6-C_8 hydrocarbons.

This is not unexpected result in view of structural characteristics of both types of molecules. Relatively strong dispersion forces acting among elongated hydrocarbon molecules are making the second virial coefficient large and proportional to the molecular weight, or more correctly, to the incremental dispersion force for each additional CH_2 group. On the other hand, regardless of molecular weight, local charge distribution in the molecules of metal chlorides creates additional repulsion forces among these molecules lowering the value of second virial coefficients.

Although correction of the activity coefficient for the nonideality of the gas phase through the equation [15].

$$\log \gamma_o = \log \frac{RT}{Mp_o V_g T} + \frac{(B_{22} - V_1)\,(p - p_o)}{2.303\ RT} \qquad (2)$$

would not be expected to be large, the values of the uncorrected and corrected activity coefficients of some metal chlorides as derived from literature data [2, 3] are presented in Table 3.

Table 3. Activity Coefficients, Corrected and Uncorrected, of Metal Chlorides in Some Organic and Inorganic Liquid Phases

Solute	Liquid Phase	T (^oC)	-B (BERTHELOT)	P_c	γ	γ_o
$SbCl_3$	$NaFeCl_4$	470	886	224	2.6	2.5
$SnCl_4$	Squalane	423	1192	1505	0.81	0.84
$TiCl_4$	Squalane	423	1014	1062	0.55	0.56
$NbCl_5$	Squalane	473	1665	219	0.31	0.30

It should be noted that this correction is not larger than 3% and could be ignored except in the most careful work.

3. Acknowledgment

We wish to thank the U. S. National Science Foundation for support of a portion of this work under grant NSF GP 5151X.

References

1. Juvet, R. S., and Wachi, F. M.: Anal. Chem., 32, 290 (1960).
2. Zado, F. M., and Juvet, R. S.: In „Gas Chromatography, 1966" A. B. Littlewood, Ed., Elsevier, New York, 1967, pp. 283-295.

3. Keller, R. A.: J. Chromatog., 5, 225 (1961).
4. Sie, S. T., Bleumer, J. P. A., and Rynders, G. W. A: Separation Sci., 1, 41 (1966).
5. Wilke, J., Losse, A., and Cackmann, H.: J. Chromatog., 18, 482 (1966).
6. Stumpp, E.: Angew. Chem., 79, 385 (1967).
7. Denison, J. E., and Freund, H.; Anal. Chem., 37, 1766 (1965).
8. Ducloux, E. H.: Microchemie, 2, 108 (1924).
9. Kendall, J., Crittenden, E. D., and Miller, H. K.: J. Amer. Chem. Soc., 45, 963 (1923).
10. Plyuschev, V. E., Stepina, S. B., Zimina, G. V., and Zhilyakov, V. G.: Tsvetn. Metal, 7, 112 (1964).
11. du Plesis, L. Q.: J. Gas Chromatog. 1/II/6 (1963).
12. Saeki, Y., and Matsuzaki, R.: Denki Kagaku, 33, 155 (1965).
13. Edwards, A. J., Peacock, R. D., and Said, A.: J. Chem. Soc. 4643 (1962).
14. Nielson, L. A.: and coworkers in a series of papers published in Zh. Neorg. Khim. (Russ. J. Inorg. Chem).
15. Adlard, E. R., Khan, M. A., and Whitham, B. T.: in „Gas Chromatography, 1960", R. P. W. Scott, Ed., Butterworths, London, 1961, pp. 251-265.

Gas Chromatographic Determination of Trace Amount of Poorly Adsorbed Gases *)

A. A. ZHUKHOVITSKI, V. I. KALMANOVSKI, B. K. KRYLOV, S. N. MOROZOVA, M. L. SAZONOV and N. I. RAKITA

Academy of Sciences of the USSR, Institute of Steel, Moscow

The problem of adsorption enrichment of trace amounts of admixtures sorbed less strongly than the main component can be solved on the basis of a frontal movement of the mixture through the bed of a sorbent either evacuated [1] or filled with a „pushing" gas that is adsorbed less than the other components [2, 3]. The use of an additional gas in the analysis of trace amounts, however, is always undesirable. Moreover it is often difficult or altogether impossible (e. g., in the determination of helium) to choose the proper pushing gas.

This paper describes the development of the frontal enrichment method for trace admixtures involving the use of reduced pressure.

As compared to the first publication [1] in which a simplified solution has been presented, this paper contains a discussion of the possibilities of essentially improving the results of analysis by using high vacuum and by improving the sampling technique.

The frontal method yields a high enrichment since the less strongly sorbed component is considerably concentrated due to differences in the rates of movement along the bed. In addition to the degree of enrichment in a given volume of the sample, the degree of extraction is also of interest.

When the mixture is introduced to the evacuated sorbent, the admixtures can not be completely extracted. The maximum extraction value may be easily estimated from the material balance.

Let the volume of sorbent be V, and the Henry coefficients of the main component and of the admixture be Γ_1 and Γ_2, respectively. Then upon filling the bed with the mixture, the quantity of the main component adsorbed will be $V\Gamma_1$, and that of the admixture,

*) Paper was sent, but not held at symposium.

$V\Gamma_2 N_2$ (N_2 is the molar fraction of the trace component). The volume $V\Gamma_1$, however, contained $V\Gamma_1 N_2$ of the trace component. In this way, the trace component with a volume of $(V\Gamma_1 N_2 - V\Gamma_2 N_2)$ is obtained in a pure state. Hence, the extraction coefficient is defined by the equation

$$K_{ex} = \frac{V_1 \Gamma_1 N_2 - V\Gamma_2 N_2}{V_1 N_2} = 1 - \frac{\Gamma 2}{\Gamma 1},$$

and K_{ex} does not depend either on the length of the bed or on the concentration of the trace component.

From the above it follows, for instance, that with an enrichment the extraction of oxygen present in nitrogen under conditions of the given experiment (molecular sieve 12 X; temperature, 20 °C; $\Gamma_{N_2} = 3.8$; $\Gamma_{H_2} = 2.2$) cannot be higher than 42 %, while the extraction of hydrogen (activated carbon SKT; temperature, 20 °C; $\Gamma_{N_2} = 5.7$; $\Gamma H_2 = 0.82$) cannot exceed 86 %.

An important role in the extraction of poorly sorbed components is played by the degree of pressure reduction. There are two factors impairing the extraction of trace components with insufficient pressure reduction. In the first place, when the mixture under the higher pressure passes through the column in which the mixture under the reduced pressure is contained, the trace component evolved from the first mixture may only increase the pressure of the second mixture to that of the first. Moreover, the second mixture is being continuously compressed, and the enrichment process will cease when the pressure of the two mixtures has been equalized. Subsequent changes in the concentration of the poorly adsorbed components may take place only due to diffusion. Fig. 1 shows a plot of the degree of extraction versus pressure (helium extraction from air).

Let us consider the mechanism of movement of a mixture through an evacuated vessel. If the free path length exceeds the average distances between the grains, the concept of gas dynamic flow is not valid. In this case the gas does not follow the laws of viscous flow, and its discrete molecules move between grains according to Knudsen's diffusion laws. When the pressure is of the order of 10^{-3} atm. however, the free path length is less than the grain size (about 10^{-2} cm) and the movement within the bed is similar to that in frontal analysis. The role of the pushing gas is as it were played by the vacuum zone. At the boundary of this and the adjacent zone additional strong broadening takes place.

The validity of this concept is confirmed by the dependence of the outlet pressure of the evacuated column packed with the sorbent (SKT carbon) on time when the mixture is passed through it (Fig. 2). At it will be seen from the plot when helium (curve 1) and nitrogen (curve 2) are passed through the column the pressure is for some time practically constant, but sharply increases when the zone boundary referred to above is approached. Hence, the physical picture and the theory of the discussed and of the convential version of frontal analysis, closely resemble each other.

The point of distinction between them lies in the fact that the rate of movement in the new version is determined not by the flow rate of the pushing gas or of the mixture to be analyzed but by the pressure gradient, and hence, decreases with time. To illustrate this difference, the simplest case of frontal movement of one substance along the evacuated sorbent in the first column will be considered.

Let the length of the bed saturated with the mixture be y, and the velocity of the gas within it assumed to be constant and equal to α. The pressure in this bed is equal to one atmosphere, and that in the remaining part is zero. From the laws of gas dynamics, $\alpha = \frac{K}{y}$, where K is a constant, and according to the balance requirements at the boundary,

$$\frac{dy}{dt} = \frac{\alpha}{\Gamma}.$$

From these equations it follows that

$$t = \frac{\Gamma}{2K} y^2.$$

The saturation time of an evacuated vessel (column) is thus proportional not to the length of the bed (as it is the case with the conventional versions of frontal analysis), but to its square (as in the diffusion processes).

The above relationship is well confirmed by experiment. Fig. 3 shows the measured values of the time of filling (t) of the bed as a function of the square of the length of the latter (y^2). All the points fall into a straight line passing through the origin of the coordinates, which, in fact, confirms the validity of the assumptions made.

From the results obtained in preliminary for the analysis of trace amounts of helium, neon, and hydrogen in air and in other gaseous mixtures has been designed and constructed [4], the sensitivity of determinations being 10^{-4} to 10^{-5} vol.%. The apparatur is assembled on a panel and enclosed in a metallic housing with double walls and an air gap between them. It is supplied with a forevacuum pump and electric recording potentiometer with a 1.5 mv scale, a 13 volt accumulator and a cylinder with the carrier gas (argon or nitrogen). Detection is made with a katharometer.

A set of enrichment columns of different shapes and sizes enables the volume of samples of the gas to be analyzed to be varied within a wide range, e. g., to enrich admixtures in air, nitrogen or argon, samples with a volume of 50 to 1500 cc may be introduced. Columns of small diameter (0.4 cm) are packed with fine-grained activated carbon (SKT, 0.25 to 0.5 mm). Columns of a diameter exceeding 0.8 cm are packed with this grade of carbon without additional crushing. The volume of the vessel in which the admixtures of the poorly adsorbed components are collected and from which they are sampled to the analytical column is chosen in accordance with the concentration of the admixtures and the dimensions of the analytical column from the set with volumes ranging from 1 to 5 cc. The analytical column, 5.5 m long, 0.4 cm in diameter is also packed with activated carbon (SKT, grain diameter 0.25 to 0.5 mm).

The main features of the apparatus are the sensitivity and precision of determinations, as well as the degree of extraction of admixtures from the gas analyzed. The limit of sensitivity for different components has been determined experimentally. Samples to be analyzed were prepared by mixing a volume of air with pure argon, in which the concentration of He was lower than $5 \cdot 10^{-6}$ vol.%, and the Ne concentration, about $1 \cdot 10^{-5}$vol.%. The composition of the gaseous mixture thus prepared was both calculated and determined by the proposed method.

The data obtained were used to construct a calibration plot, with peak height and concentration as coordinates. As shown by Fig. 4, the peak heights are directly proportional to the values of concentration of He and Ne in the mixture. The lowest values of the concentration of admixtures which could be determined under conditions of this particular experiment were $4 \cdot 10^{-6}$ vol. % for helium, and $1.2 \cdot 10^{-5}$ vol.% for neon.

The sufficiently linear relationship observed in Fig. 4 is evidence of a constant nature of the coefficient of extraction (K_{ex}) of admixtures from the gas analyzed, their concentrations varying within the above range. It is of interest to note that the value of the extraction coefficient remains constant with much higher concentrations of trace components. Table 1 presents data showing that upon tenfold and even hundredfold increases in the concentration of trace components as compared to their concentration in air, the values of the extraction coefficients do not differ more than 17% from the mean.

Table 1. Values of extraction coefficients of helium, neon and hydrogen with enrichment

No.	Helium			Neon			Hydrogen		
	Conc. found	K_{ex}	Deviation %	Conc. found	K_{ex}	Deviation %	Conc. found	K_{ex}	Deviation %
1.	0.00052	0.52	0.0	0.0018	0.55	4.6	0.00007	0.38	0.8
2.	0.0013	0.52	0.0	0.0026	0.50	5.1	0.0008	0.33	12.5
3.	0.021	0.47	9.6	0.0035	0.46	12.6	0.0017	0.34	9.8
4.	0.0030	0.57	9.6	0.0049	0.52	1.3	0.0025	0.38	0.8
5.	0.0055	0.50	3.8	0.0068	0.55	4.6	0.0051	0.39	3.5
6.	0.0105	0.54	3.8	0.0118	0.58	10.1	0.0101	0.44	16.7
Mean:		0.52	4.7		0.53	6.4		0.38	7.3

In the experiments described above, the volume of the enriched gaseous mixture was 650-670 cc. The mean values of the extraction coefficient were 0.52 for helium and neon, and 0.38 for hydrogen. The mean arithmetic relative error in determining the value of K_{ex} is 5 to 7%.

In this way, the technique developed on the basis of chromatography without carrier gas permits to determine helium, neon and hydrogen in gaseous mixtures with a high sensitivity (10^{-5} to 10^{-6} vol. %) by using a conventional katharometer.

The trace components referred to above are of great interest to the geochemist in the prospecting for useful minerals and in the study of problems related to their origin. The proposed technique can also be useful in solving such vital problems as the determination of traces of poorly adsorbed components in inert gases (e.g., hydrogen in argon, oxygen in nitrogen, etc.) or traces of hydrogen, oxygen and nitrogen in high purity ethylene.

References

1. Turkeltaub, N. M., Rjabtchuk, L. H., Morosowa, S. N., and Zhukhovitski, A. A.: Journal analyt. chemie (russ.) 19, 133 (1964).

2. (siehe oben)

3. Zhukhovitski, A. A., Turkeltaub, N. M., Naljasova, L. A., Schljachov, A. F., Naumova, V. V., and Pogrebnaja, T. P.: Savodskaja Laboratorija (russ.) 10, 1162 (1963)

4. Rakita, N . I., Zhukhovitski, A. A., Morsova, S. N., and Rjabtchuk, L. N.: Avtorskoje svitedelstvo No. 192 467. Bulletin Isopretenii No 5 (1967)

APPENDIX I

Summaries of Papers Published in German in the Preprints of the Sixth Symposium on Gas Chromatography, held in Berlin 1968

The English translations of these papers may be ordered from
Unterkommission für Gas-Chromatographie
DDR-705 Leipzig, Permoserstr. 15

Comments on Calculation of the Retention Behaviour of Hydrocarbon Isomers

K. ALTENBURG, Berlin-Adlershof

It is demonstrated that the retention index I of isomeric acyclic alkanes is a linear function of the density d (at 20 °C) and the effective value of radius[1] R (which largely corresponds to the mean moment of inertia of the molecule) - referred to length a_1 of a C-C linkage:

$$I = I_0 + 22{,}307\ M\ d + 136{,}7\ R^2/a_1^2$$

I_0 is an empiric constant depending on the molecular weight, M is the molecular weight. The density can be eliminated by introducing n_3 which is the number of C_4 groups (C atoms which are separated from each other by 3 C-C linkages):
Thus, the formula reads:

$$I = I^* + 23{,}89\ (n_3 - n_3^*) + 196{,}39\ (\overline{R^2}/a_1^2 - \overline{R^{*2}}/a_1^2)$$

(* represents the values of isomeric non-branched compounds).

These equations allow determination of the retention index of the isomeric alkanes C_6 to C_8 with a mean error of 2.9 and 5.6 units, respectively.

It is demonstrated with dimethylcyclohexanes, X-cyclohexylalkanes, and X-phenylalkanes, that these relations are also applicable to other substances. Finally, some rules concerning the dependence of the retention index on the number of C atomes in homologous series, the correlation between boiling temperatures and retention index, and the dependence of retention index of substituted n-alkanes on the position of the substituent are quantitatively interpreted and precised.

Preprints p. 1

---

1) effective value of the radius = term for r. m. s. value (root mean square value)

On Modified Polymers and Their Use as Liquid Phase in Gas Chromatography

K. ASSMANN, G. GEPPERT, H. G. STRUPPE and O. SERFAS, Leipzig

By reaction of the free hydroxyl groups of polyethylene glycol and of polyvinyl alcohol with various reagents products could be obtained that represent polar and comparatively temperatur-stable liquid phases. Two methods are suggested by which, in tests in the gas chromatograph, the maximum working temperature of the liquid phases for isothermal and temperature-programmed working conditions can be estimated. On the basis of the δ-values for the separation of primary alcohols and alkyl benzenes from n-alkanes the selectivity of the modified liquid phases is discussed. Example, show, how the highly polar products are able to repress the disturbing adsorption activity of the solid support. Finally the potential applications of the described liquid phases are discussed.

Preprints p.33

Derivatographic Study of Gas Chromatographic Supports and Partition Liquids

S. BÉKÁSSY, G. LIPTAY and J. TAKÁCS, Budapest

A derivatographic method was developed for the determination of the upper temperature limit of applicability of gas-chromatographic partition liquids. The influence of the experimental conditions (rate of heating, quality of gas surrounding the sample etc.) on the upper temperature limit was studied. The method also made the examination of interactions between support and partition liquid possible, as well as the measurement both of wetting degree and exhaustion of the column packing.

Preprints p. 73

Gas Chromatographic Studies of Surface Adsorbent Molecular Sieves on Solid Support

V. G. BEREZKIN and N. S. NIKITINA, Moscow

At a given selectivity in regions of higher velocity of carrier-gas, resolution is limited by the final rate of mass transfer; i. e., adsorption and desorption take place instantaneously. There fore, the use of new sorbents, which are inert supports, with powdered active sorbent in their pores, are of special interest in high-speed gas chromatography.

It ist shown that when coarse-particle surface adsorbent is used, an efficient column can be obtained with low pressure drop to gas carrier flow; consequently such sorbents could be used effectively in rapid analysis.

Preprints p. 89

Gas Chromatographic Investigation of Vapor Phases

H. BINDER, Graz

A method for vapor phase analysis, using the first section of a chromatographic column for the concentration of trace impurities in gases, is described and compared to the well known injection with gas-tight syringes. The disturbing influence of variation of the diffusion rate in vapor phases which are in equilibrium with liquid mixtures is shown. After the elimination of this error-cource the vapor-liquid equilibrium of the system ethanol-benzene is investigated.

Preprints p. 99

A New Coating Procedure for Glass Capillary Columns

J. BOUCHE and M. VERZELE, Ghent

This paper describes a new method to coat glass (or metal) capillaries producing very high plate number per unit length. In principle the column is filled with a solution of stationary phase (5 to 10 mg/ml), the column is sealed at one end and the solvent evaporated through the other end under reduced pressure.

The main drawback to this procedure is its duration. Problems can be encountered with the stability over longer periods of some stationary phases in solution. However, excellent results were obtained, even with stationary phases which are reputedly difficult to use in capillary columns (SE-30).

Preprints p.603

Molecular-Statistical Calculation of Retention Volumes on Zeolites

P. BRÄUER and A. A. LOPATKIN, Jena and Moscow

A general equation of the isotherm for adsorption on zeolite was derived by the method of statistical mechanics. The HENRY's constant was calculated in classical approximation on the assumption of pairwise interactions and with a simple potential equation. The retention volumes determined from this constant are presented as a function of temperature in Fig. 1 and are compared with experimental data.

Preprints p.129

Preparative Gas Chromatography - Investigation on the Operation of Large Diameter Columns

M. B. DIXMIER, B. ROZ, J. TABERÉAU and G. GUIOCHON, Orsay, Chaton and Paris

Experiments designed after a theoretical study have allowed optimization several factors which have a large effect on the output of a preparative chromatograph, whilst at the same time conzidering the quality and purity of the products prepared.

The choice of the support, the ratio of stationary phase and the procedure adopted for conditionning the columns allow to obtain good load diagrams for 40 mm diameter columns. Furthermore, optimization of the column temperature, together with careful calculation and design of the injector and of the connections between the columns, made it possible to obtain peaks of almost Gaussian shape at a higher loadings and permitted a shorter retention time for a given resolution. All these results contribute to a high level of output.

The overall output efficiency of the apparatus was improved by the inclusion of a new temperature-gradient trap.

Since gas chromatography has become practicable for production purposes, many authors have studied the functional characteristics of large-diameter columns, and from the beginning the importance of loadability and efficiency has been stressed.

However, most of the research has been either on a purely theoretical plane or very empirical and only a few workers - such as Bayer et al., in Germany and ROSE et al. in the United States - have considered the design and operational aspects of a column intended for the production of pure chemical substances. In this case the design and working parameters of the columns must be optimized to obtain the highest possible rate of production, whilst insuring the best standards of purity combined with low production costs.

The work described in this paper supplements an earlier publication which stressed the importance of column productivity as the major factor to optimize, and expounded a number of theoretical concepts which were derived from this new optimization principle. The present purpose is to provide the answers to some problems encountered in the experimental stage, which had not previously been covered by detailed research.

It is obviously of interest to ascertain the conditions necessary for constructing an efficient column, and to know why some columns perform satisfactorily whereas others give only poor results.

It is also extremely important to learn how to recover, with a good yield, the substances separated in the column. From a theoretical point of view, this might appear to be only a minor difficulty associated preparative chromatography: in many instances, the low recovery yield is due to mistformations consecutive to quenching of the diluted vapor, which are vitually impossible to trap. Practically, however, this is an essential problem because the trapping yield has a direct influence on the overall productivity of the equipment, but it has been the subject of only very limited investigation.

Preprints p. 261

A New Monitoring System for Liquid Chromatography and Some Applications

M. T. DIXON, Cambridge

A detection system, for use with liquid chromatographic columns, which employs a moving stainless steel wire and a gas chromatographic detector, is described. The dependence of the response and sensitivity of the system are discussed with relation to the wire speed, and applications using liquid/solid, liquid/liquid and gel permeation techniques are given to show the versatility of the detection system.

Preprints p. 137

Kinetics of Reactions of the Type $A_{(gas)} + B_{(liqu.)} \longrightarrow C_{(gas)} + D_{(liqu.)}$ by Means of Reaction Gas Chromatography

C.-E. DÖRING, W. PEHLE and G. SCHMID, Leuna

The kinetic investigation of reactions of the type $A_{(gas)} + B_{(liqu.)} \longrightarrow C_{(gas)} + D_{(liqu.)}$ requires the continuous maintenance of the distribution equilibrium of the gaseous substances between the liquid and the gas phase. The reaction $2\,NO + (C_4H_9O)_3\,P \longrightarrow N_2O +$

$(C_4H_9O)_3PO$ serves as an example by means of which the authors report on experiments to study such a reaction with the help of reaction gas chromatography. The kinetic law is set up and the activation parameters are determined.

Preprints p. 143

Solution of Automatic Analysis of Alcohol-Acetaldehyde Mixture in Butadiene Production and Experience with Application of CH-04 Process Chromatograph

J. DVOŘAK, Prague

In the production of synthetic caoutchouc there was the requirement to solve and realize the automatic, analytical method of alcohol-acetaldehyde mixture in the contact-production of butadiene. The mixture consists of the following components: 0,7 % ether, 1,9 % acetaldehyde, 2 % methanol, 82-85 % ethanol, 0,6 % propanol, approx. 10 % water, 0,2 % butanol and further components, which are no more important for the production. An experimental test of the methods has been carried out in cooperation with the research workers of the factory and it was proposed to utilize the automatic process chromatograph type CH-04 for the process analysis.

Preprints p. 155

Chromatography of Polymer Solutions on Macromolecular Sieves

YU. A. ELTEKOV and E. K. BOGACHOVA

A chromatographic investigation has been carried out of non-fractinated polystyrene, having an average molecular weight of 300,000, by means of the frontal method. Macromolecular sieves having pore dimensions 280, 550 and 800 Å were used as adsorbents. The results show good prospects for the chromatographic method of fractionalization of polymers using a series of macromolecular sieves.

Fractions of polystyrene having molecular weights ranging from 1,200,000 up to 130,000 have been isolated from the original polymer.

Preprints p. 163

Results of the Determination of Normal Paraffins by Segregation Gas Chromatography

G. EPPERT, I. SCHINKE and S. RENNHAK

Experimental possibilities and results of a new method for the determination of normal paraffins by segregation gas chromatography on Molecular Sievė 5 A are described. The principle of this method, which has been reported earlier, is based on the use of the molecular sieve effect at higher temperatures in the gas phase. Both non-normal paraffins and normal paraffins can be determined with very short analysis times and without regeneration of the molecular sieves. Contrary to the methods for the determination of normal paraffins previously known, the new method is excellently suitable for automatic process control.

Preprints p. 169

Polyglycerine and its Esterification Products as Liquid Phases in Gas Chromatography

F. FALK and P. G. DIETRICH, Berlin

The synthesis and application of polyglycerine as a liquid phase in gas chromatography is described. It is excellently suited for the analysis of strongly interacting compounds such as alcohols, diols and polyethylene glycol. Liquid phases are obtained - by means of cyanethylation - which are strong π-electron acceptors and thus separate compounds having equal chain lengths from compounds with one or more double bonds. Such phases are suited both for alcohols and diols as well as for less polar compounds such as Ketones, esters, etc. By mixing the aforementioned phases or by partial esterification of polyglycerine with acrylonitrile, phases are obtained which have shown good results in separating of medium and long-chained aliphatic compounds.

Preprints p. 197

Group Identification of Organic Substances by Gas Chromatography and Catalytic Dehydrogenation

J. FRANC and F. MIKEŠ, Pardubice-Rybitví

The authors have elaborated a procedure for group identification of organic substances by gas chromatography, in which the substances leaving the chromatographic column are first passed over the first detector and then over a dehydrogenation catalyst. The liberated hydrogen is separated from the other products of dehydrogenation in the second chromatographic column and recorded by the other detector.

Preprints p. 209

Gas Chromatography of Tropan-Alcaloids

Part II: Study of Thermo-catalytical Decomposition Problems of Atropine and Scopolamine under GLC Conditions

H. FRAUENDORF, Ingelheim/Rhein

Under GLC determination conditions atropine and scopolamine often show extremly high decomposition rates. This decomposition was determined as thermo-catalytical causing dehydratation of the alcaloids to the corresponding apoalcaloids, apoatropine and aposcopolamine. Optimum conditions have been worked out to prevent this decomposition and to allow GLC determination even of very small amounts of this alcaloids as obtained from drugs.

Preprints p. 215

Comparative Investigation of Supports for Gas-Liquid Chromatography

T. B. GAVRILOVA and N. I. BRYSGALOV, Moscow

A comparative investigation of inorganic supports of liquid phases for gas-liquid-chromatography has been carried out, using the mercury-porosimetric, electron-microscopic and chromatographic methods. Supports obtained on the basis of diatomaceous earths - Chromosorb W and Support ADS-possess the optimum structural characteristics: pore volume of about 2 cm^3/g; pore size of about 5 μm, and the most homogeneous pore size distribution. To obtain satisfactory separation of complex mixtures in gas-liquid chromatography, supports should be used which have all these optimum structural characteristics simultaneously, coupled with a sufficiently low specificity of the surface.

Preprints p. 223

Report on Molecular-Sieve-Gas-Chromatography
Separation of a Complex Gas Mixture under Optimal Conditions

G. GNAUCK and E. TREUTLER, Berlin

The paper deals with studies of the separation of a gas mixture consisting of argon, nitrogen, krypton, methane and carbon monoxide with various molecular sieve types. The optimal conditions are dependent primarily on the thermal pretreatment, i. e., on a certain degree of hydration of the molecular sieve. The separation of nitrogen and krypton with well activated zeolite, which was possible only a column temperature of more than 50 °C, was of special interest. The results are discussed. The reason is to be sought in the interaction between the adsorbate and the adsorbent and in a pre-adsorption of polar water molecules.

Preprints p. 231

Properties of Adsorption Layers in Gas Chromatography

I. HALÁSZ and F. HOLDINGHAUSEN

Problems in the preparation of gas adsorption layers are discussed. Even if an active solid is coated with a multilayer of liquid film, the separation properties of the adsorbed liquid can be considerably different from that of the pure liquid. The stationary phases used in GAC are characterized by means of the „BROWN-triangle", which may also be useful for the determination of poisoning of catalysts. The advantages and disadvantages of GAC are discussed especially for the analytical application of porous layer beads and open tubes.

Preprints p. 311

Studies on the quantitative analyses in programmed flow gas-chromatography

M. JEDRYCHOWSKA, P. SZYNAGEL and S. GLINSKA, Krakow

Application of a programmed flow technique to the quantitative gas chromatography analyses was studied. Dependence of the change of carrier gas flow rate on the peak area for hot wire detectors was investigated.
It has been stated that these changes were identical for various hot wire detectors, were similar for n-paraffin and aromatic compounds and depended on the column temperature.
An equation for the peak area correction has been given in order to enable quantitative estimation of the chromatographic curves obtained with gas flow programming technique.
(English translation is not available.)

Preprints p. 343

Relation of Gas Chromatographic Behaviour and Chemical Structure - Mechanism of the Separation of Benzene and its Methyl Derivates

T. L. KWA and C. BOELHOUWER, Amsterdam

There is a growing interest to correlate chemical structure and behaviour in gas-liquid chromatography. In this study an approach is proposed that uses simple and straightforward mathematical expressions of quantum theory and statistical mechanics. Our results show agreement with those derived from other techniques, viz. gas heat capacity and IR intensity measurements.

Preprints, p. 365

Data on Double Programmed Gas Chromatography

L. MAZOR, J. BALLA, S. BÉKÁSSY and J. TAKÁCS, Budapest

Some theoretical and practical problems of gas chromatography could be clarified by double programming, that is, a temperature program followed by a carrier program. Especially good results were obtained with the analytical method, when the column was operated at the upper temperature limit of the partition liquid.

Preprints p. 379

Chromatography on Aerosilogel

YU. S. NIKITIN, Moscow

A method is described for the preparation from aerosil of aerosilogel a homogeneous, wide-pore silica of high chemically purity. The feasibility of its application as an adsorbent in gas chromatography is demonstrated for the separation of hydrocarbons and highly polar substances. Even such compounds as ethers, alcohols and amines, which are strongly and specifically adsorbed on silica surface, produce symmetrical peaks on aerosilogel.

Preprints p. 391

Selected Applications for Process Chromatographs

M. OREANS, Karlsruhe

The sampling problem is of equal importance for the process chromatograph as for the classic gas analysing techniques, such as chose employing infrared, paramagnetic or thermal conductivity methods. For this report, we have selected some applications of process chromatographs in which, apart from the analytical questions, the sample preparation requires special solutions.

Preprints p, 399

Reversion Gas Chromatography

H. OSTER and R. KAISER, Ludwigshafen

Based on theoretical considerations, the correlation between isothermal chromatography and thermochromatography is explained. The distribution coefficient of the moving furnace is calculated and the most favorable operating parameters are discussed. With respect to the apparatus, the method is essentially improved by an axially symmetrical furnace and by blocking cold fields. The sensitivity of the method under experimental conditions is said to be $5 \cdot 10^{-3}$ ppb.

Preprints p. 419

Experiments with a Selective Adsorption Detector

H. PAUSCHMANN, Tübingen

For a considerable time, adsorption detectors with thermo couples have been used in gas chromatography [1, 2, 3]. In the following paper, a thermistor equipped adsorption detector as used in liquid chromatography (4) is examined for its properties upon application to GLC systems.

Preprints p. 429

A Frequency Response Analysis Method to Determine the Mean and Variance of Residence Time Distributions

A. PETHÖ, Budapest

The possibility of the application of the frequency response analysis, well known from process dynamics and control, is outlined for determining the mean and variance of residence time distributions in a continuous-flow system.

Preprints p. 439

Calculations of Retention Volumes on the Basis of Molecular-Statistical Theory of Adsorption

D. P. POSHKUS, Vilnius

The results of the statistical calculation of the gas-solid chromatographic retention volumes for methane, ethane, propane, n-butane, n-pentane and benzene on the basal face of graphite at zero surface coverages are presented. The semi-empirical potential functions of the interaction between the CH_3 and CH_2 groups of the n-alkane molecules and the graphite surface are obtained.

Preprints p. 443

On Sorption Decrease in Gas Chromatographic Trace Analysis of Sulphur Organic Compounds in Separation Columns

M. REINHARDT, R. OTTO and H. KOCH, Bitterfeld and Leipzig

A special method for the treatment of separation columns for gas chromatographic trace analysis of sulphur organic compounds is described. With columns thus treated and an argon ionization detector as little as 0,1 g of sulphur organic compound in 100 m^3 heating gas can be quantitatively determined without enrichment.

Preprints p. 453

Problems of Combining IR-Spectroscopy and Gas Chromatography

O. SERFAS, G. GEPPERT and W. ZIRKLER, Berlin and Leipzig

The problems of combining IR-spectroscopy and gas chromatography for the indentification of unknown compounds are summarized and discussed from the standpoint of the IR-spectroscopist. A simple method combining IR-spectrometers and gas chromatographs is described.

Preprints p. 489

Gas Chromatographic Separation of Structural and Stereo-Isomers on Graphitized Carbon Black

K. D. SHCHERBAKOVA, Moscow

In this paper it is shown that the sequence of emergence of substances from the surface of graphitized carbon black principally depends on the energy of non-specific interaction, which is connected with the orientation of the molecules on the plane surface. The gas chromatographic behavior on the surface of graphitized carbon black of some of the structural isomers of octane, of the olefins pentene-2 to octene-2 and of hydrocarbons with five and six-membered rings is discussed. For these examples, it is shown that stereo-isomers separate well on graphitized carbon black when the differences in the geometrical structures lead to a different number of points of contact of their molecules with the surface.

Preprints p. 533

Gas Chromatography of Glycol Ethers

M. ŠINGLIAR and J. DYKYI, Nováky

The resolution of ethylene glycol/monoethylether ethylene glycol and diethylene glycol/ monoethylether ethylene glycol is very important, because ethylene glycol is always present in glycol ethers mixture. The authors, proved several stationary phases with the purpose to reach the best resolution all components in the mixture of ethers. With the aid of standards is sufficient to measure the increments of elution indices for etoxy and CH_2 groups and on the basis these results to calculate the elution times other ethers.

Preprints p. 503

The Voltampere Characteristics of Several Typical Versions of the Thermionic Detector

V. SVOJANOVSKY, M. DRESSLER and J. JANÁK, Brno

It has been shown by means of voltampere characteristics that the two systems of combined detectors interact with changes in the level and character of the flame ionization signal. Such changes are due to electrode polarities and to detector design and material. The response of the thermionic system in a combined detector is a function of the distance between the two detector systems and, at the same ionization efficiency, it is always lower than the responce of a single flame thermionic detector (TID). The voltampere charcteristics of single flame TIDs are influenced, above all, by the temperature of the fused alkali salt coating.

Preprints p. 545

Surface Treatment of Glass Capillaries for Gas Chromatography

K. TESAŘÍK and M. NOVOTNY, Brno

A new treatment of glass capillaries has been tested, involving their etching with a gaseous agent. Both efficiency and suitability have been determined for treated capillaries of less than 0.1 mm internal diameter. The performance of etched capillary columns coated with polar stationary liquids has been investigated.

Preprints p. 575

Nitrilesilicone as a Stationary Phase in Gas Chromatography; Especially for Fatty Derivatives

V. VASILESCU, Rodleben

The need for highly polar stationary phases more stable than polyesters at higher temperatures led to the examination of the nitrilesilicone oil OE 4178 from the „Institut für Silikon- und Fluorcarbonchemie, Radebeul bei Dresden". This material was employed for the gaschromatographical analysis of C_8-C_{22} fatty acid derivatives, dodecanol isomers, naphthalene hydrogenation products, etc.
Providing suitable conditioning is carried out, this product can be used up to 250 °C for the separation of saturated and unsaturated components of equal molecular length, such as stearic, oleic, linoleic and linoleic acid methyl esters or of the corresponding fatty alcohol acetates.
The possibility of separating saturated and unsaturated components with free hydroxyl groups, such as those present in the rapeseed oil ocenol (polyunsaturated C_{10}-C_{22} fatty alcohol) and in the castor oil fatty acid methyl ester, is to be pointed out.
Included in this paper are chromatograms of the separated components and a discussion of the apparatus employed.

Preprints p. 585

The Evaluation of Measurement Data in Gas Chromatography

O. WEBER, Frankfurt (Main)

Automatic integrators at present are not generally satisfactory in their method of operation, above all when used on work other than routine analyses under optimal conditions. The main disadvantage is that measuring values cannot be stored individually in order to make subsequent decisions of their use.
This is possible only in digital computers with corresponding storage capacities, where costs are so high that their use is meaningful only where many data inputs are connected. Large initial investments may be difficult to justify unless it is clearly appreciated whether and how a computer can effect economies under varying conditions in everyday laboratory work.

Preprints p. 605

Radio Gas Chromatography with Labeled Hydrocarbons

W. WEILAND, R. OTTO and H. KOCH, Leipzig

A heated internal gas flow counter of Geiger-Müller-type is applied to radio gas chromatography of ^{14}C- or T-labeled hydrocarbons. The dependency of counting rate upon the operating conditions of the counter is investigated. Results from analysis of a test mixture and of radioactive components of a reforming product are given.

Preprints p. 611

Separation of a Mixture of Methyl-, Ethyl-, and Vinylchlorsilanes by Gas Chromatography

M. WURST, Kolin

The retention data of 12 methyl-, ethyl- and vinylchlorosilanes are reported. These compounds have very near or equal boiling points. The compounds was separated on 4 types of silicone stationary phases with different composition and polarity. In the present paper the application of these results to the analysis of the reaction products from the production of methylchlorosilanes, methylvinylchlorosilanes, vinylchlorosilanes and distallation residue from the production of methylchlorosilanes is illustrated.

Preprints p. 643

APPENDIX II

Figures

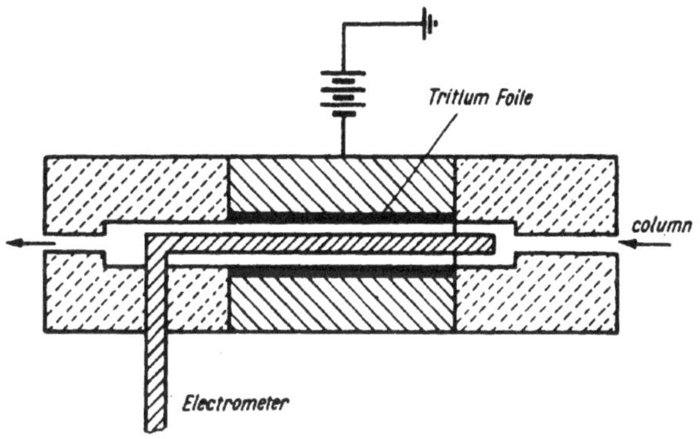

Fig. 1. Cross-section detector by LOVELOCK, SHOEMAKE and ZLATKIS [4]

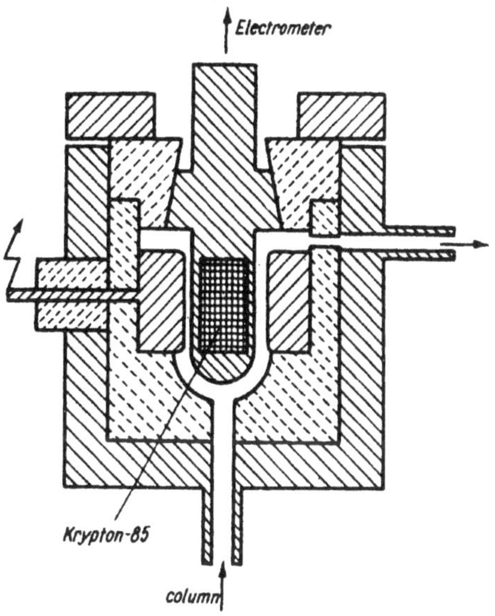

Fig. 2. Cross-section detector also usable as argon- or helium detector [22]

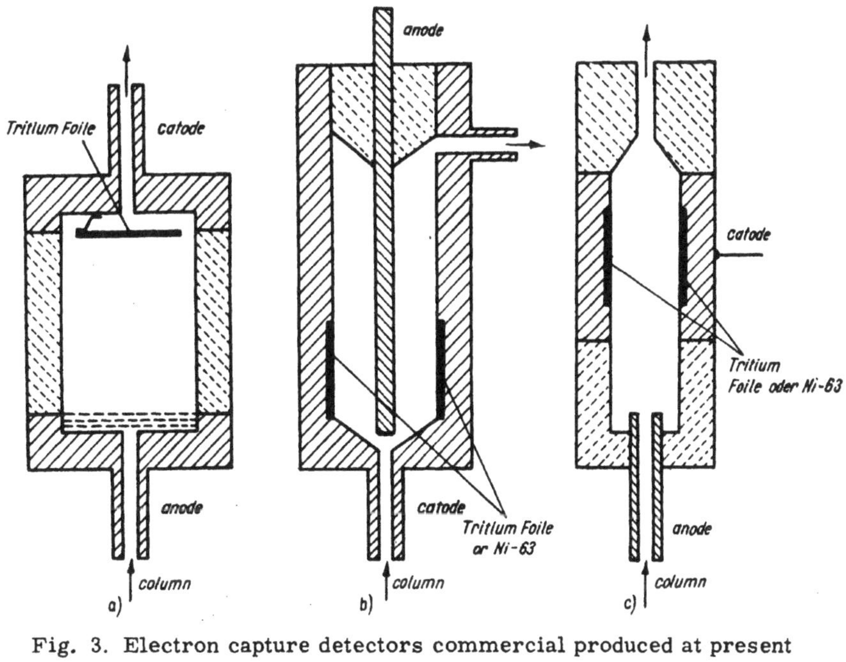

Fig. 3. Electron capture detectors commercial produced at present

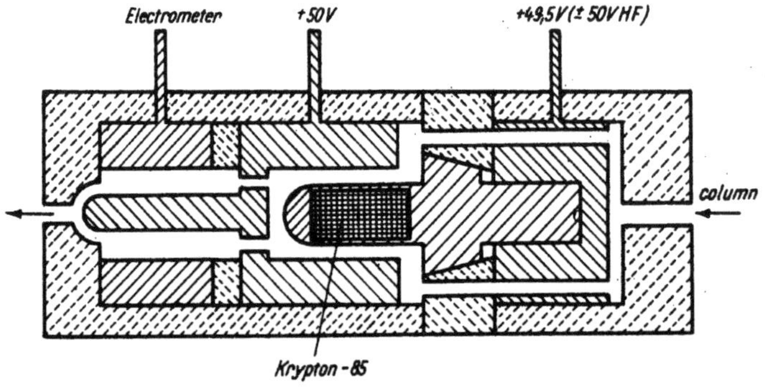

Fig. 4. Electron capture detector for immediate detecting negative ions

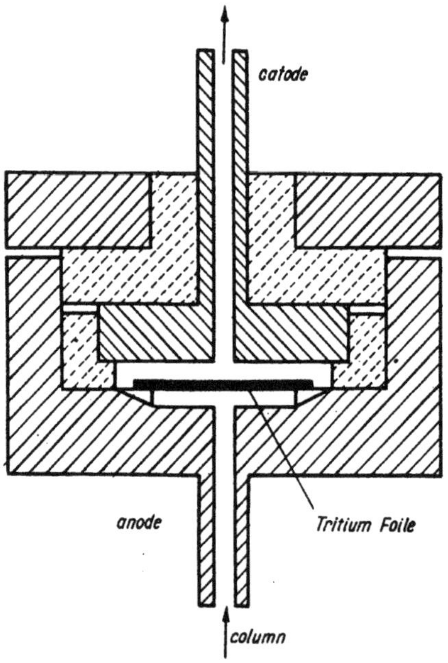

Fig. 5. Helium detector by HARTMANN and DIMICK [19]

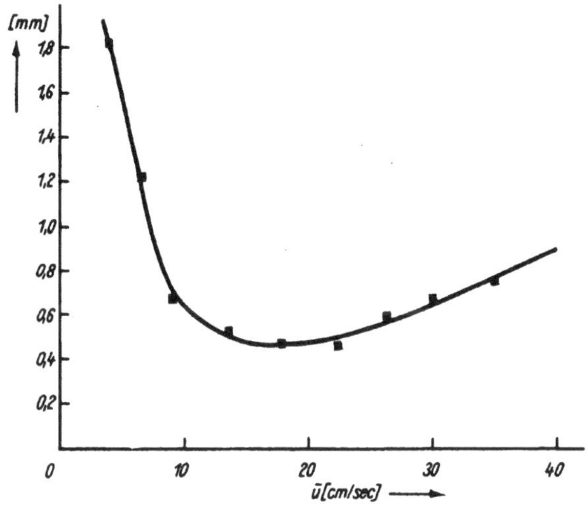

Composition of the porous layer:
97.57% b. w. highly dispersed ZrO_2
2.44% b. w. triethylene glycol
sieve fraction: 0.09-0.1 mm
carrier gas: hydrogen
2 = 1.3 μg
FID
K = 1.6×10^{-7} cm^2
t_c = 100 °C
L = 2 m
i.d. = 2 mm

Fig. 1. h vs. ū curve for n-heptane (k′ = 2.9) on a PLB column [11]

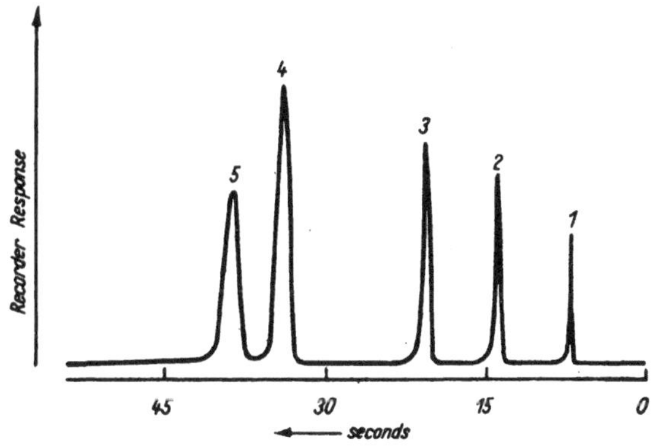

Composition of the porous layer:
90.9% highly dispersed β-ferric oxide
9.1% triethylene glycol
sieve fraction: 0.09-0.1 mm
carrier gas: hydrogen
p = 4.0 at
ū = 29 cm/sec
s = 0.5 μg
FID
t_c = 100 °C
L = 2 m
i.d. = 2 mm

(1) methane
(2) benzene
(3) toluene
(4) m- and p-xylene
(5) o-xylene

Fig. 2. Separation of aromatic hydrocarbons on porous layer beads [11]

Fig. 3. Quick quantitative analysis with PC column [15]

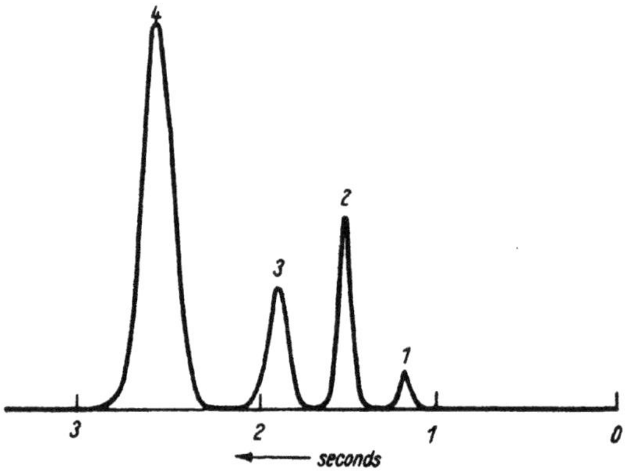

alumina (130 m^2/g) activated at 500 °C
sieve fraction: 0.10 - 0.15 mm
carrier gas: hydrogen
Δp = 4 at
$\bar{u}$ = 80 cm/sec
s = 0.1 μg
FID

N/t [sec^{-1}] values for:
ethane 92 (k' = 0.36)
ethylene 112 (k' = 0.70)
propane 250 (k' = 1.31)

(1) methane
(2) ethane
(3) ethylene
(4) propane
t_c = 70 °C
L = 1 m
i.d. = 0.3 mm

Fig. 4. Separation of C_1 - C_5 hydrocarbons on alumina with PC column [24]

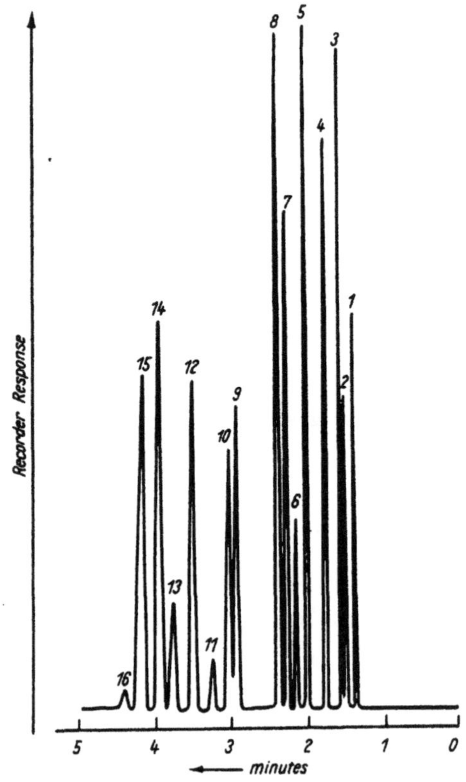

(1) methane
(2) ethane
(3) ethylene
(4) propane
(5) propene
(6) acetylene
(7) isobutane
(8) n-butane
(9) n-butene
trans-butene-2
(10) isobutene
(11) cis-butene-2
(12) isopentane
(13) n-pentane
(14) butadiene-1.2
propyne
(15) butadiene-1.3
(16) 3-me-butene-1

K = 4.4 x 10^{-7} cm^2
t_c = 132 °C
L = 8 m
i. d. = 0.25 mm
sieve fravtion =
0.09-0.10 mm
carrier gas:
nitrogen moistened
over $Na_2SO_4 \cdot 1\ OH_2O$
at room temperatur
Δp = 4 at
$\bar{u}$ = 9.6 cm/sec
s ~ 1 μg

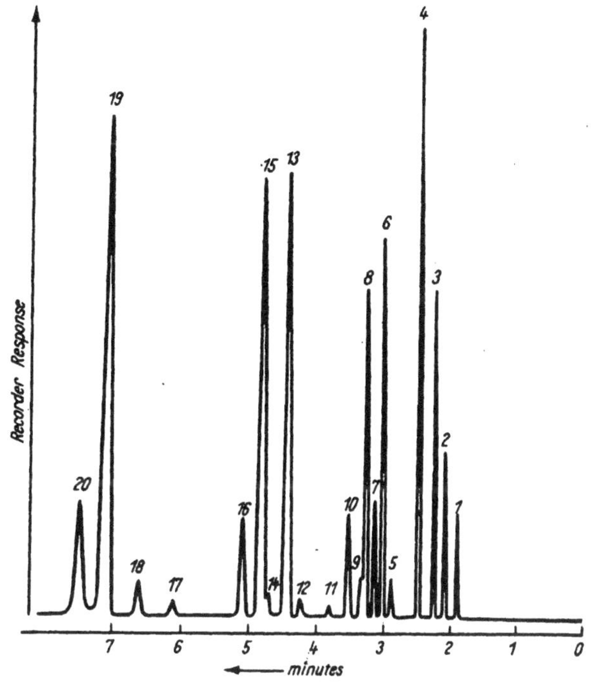

Fig. 5. Separation of C_1 - C_5 hydrocarbons on alumina coated with NaCl in PC column [24]

(1) methane
(2) ethane
(3) ethylene
(4) propane
(5) cyclo propane
(6) propene
(7) isobutane
(8) n-butane
(9) acetylene
(10) propadiene
(11) unknown
(12) neopentane
(13) n-butene trans-butene-2
(14) isopentane
(15) isobutene
(16) cis-butene-2
(17) butadiene-1, 2
(18) 3-me-butene-1
(19) butadiene-1, 3
(20) propyne

$K = 5 \times 10^{-7}$ cm^2
t_c = 124 °C
L = 10 m
i. d. = 0.25 mm
sieve fraction: 0.09 - 0.10 mm

carrier gas: dry nitrogen
Δp = 4 at
$\bar{u}$ = 8.6 cm/sec
s ~ 1 µg

Fig. 6. h vs. u_o curve for propene (k′ = 1.46) in an aerogel column [26]

column packing: highly dispersed silica (Aerosil 2491/380), 7,5 mg/meter column length, corresponding 2.85 m^2 of surface/meter

$K = 3 \times 10^{-8}\ cm^2$

$t_c = 25\ ^oC$

i.d. = 0.4 mm

carrier gas: hydrogen

s = 0.2 µg

FID

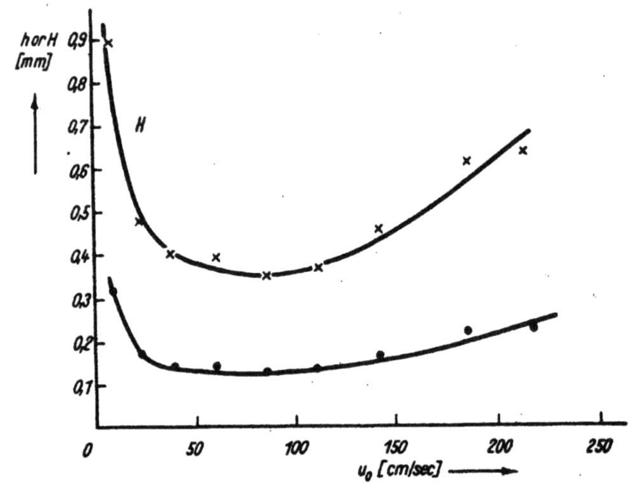

Fig. 7. H vs ū curves in PLOT on graphited carbon black (recalculated from [7])

xxx n-heptane

ooo 2, 2, 3-trimethylbutane

PLOT

K = $9 \times 10^{-6}\ cm^2$

t_c = 175 °C

L = 30 m

i. d. of the uncoated tube: 0.5 mm

particle size: <1 micrometer

k′ for n-heptane: 4.07

k′ for 2, 2, 3-trimethylbutane: 1.42

phase ratio: 63

carrier gas: hydrogen

s~ 1 µg

FID

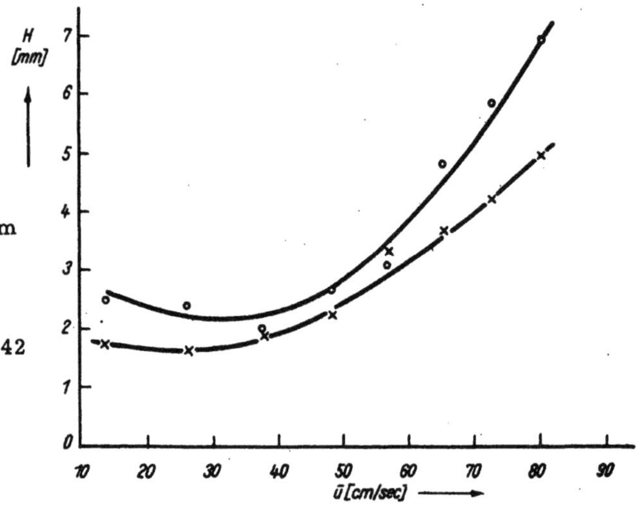

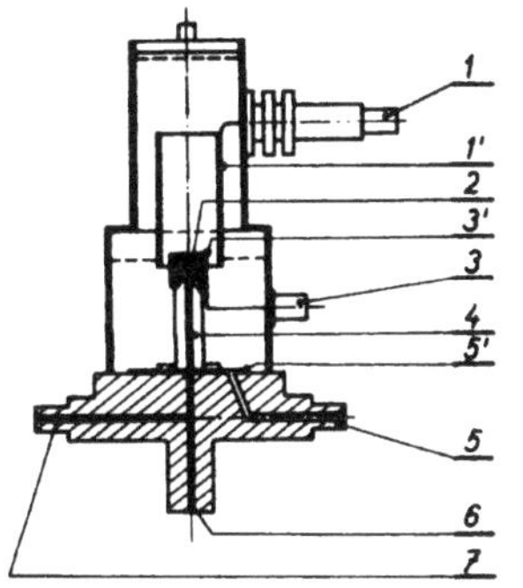

Fig. 1. Single flame thermionic detector
1 - collector connector; 1′ - collector electrode; 2 - pressed alkali metal salt tip; 3 - lower electrode connector; 3′ - lower electrode; 4 - quartz glass jet; 5 - air inlet; 5′ - air flow rectifier mesh; 6 - carrier gas inlet; 7 - hydrogen inlet

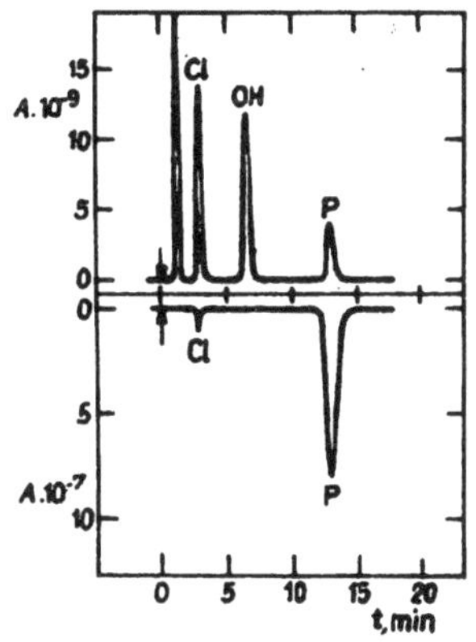

Fig. 2. Chromatogram of a model mixture
OH - n-hexyl alcohol;
Cl - monochlorobenzene;
P - 0, 0-diisopropylmethyl phosphonate
(the upper record, flame ionization detection; the lower one, thermionic detection)

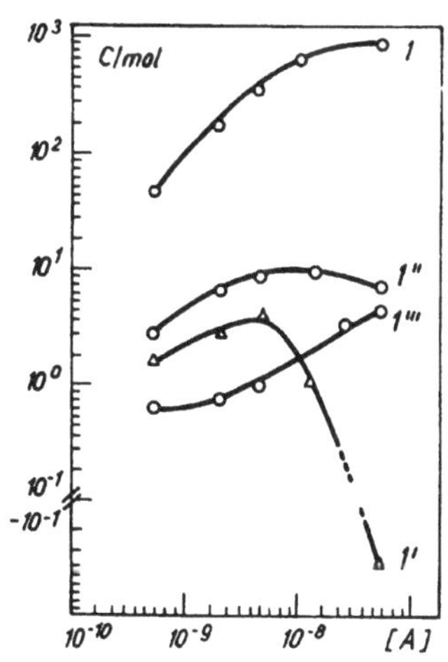

Fig. 3. The plot of ionization efficiency against beck ground current at various carrier gas flow rates (carrier gas, nitrogen; air flow, 660 ml/min; indices 1-4 carrier gas flow rates 45, 54, 60 and 70, 8 ml/min., resp.)

1, 2, 3, 4 — 0, 0-diisopropylmethyl phosphonate; diameter of jet tip bore 0, 6 mm; Na_2SO_4;

1′, 2′, 3′, 4′ — pyridine; diameter 0, 6 mm; K_2SO_4;

1′′, 2′′, 3′′, 4′′ — pyridine; diameter 2, 5 mm; K_2SO_4;

1′′′, 2′′′, 3′′′, 4′′′ — chlorobenzene; diameter 0. 6 mm; Na_2SO_4.

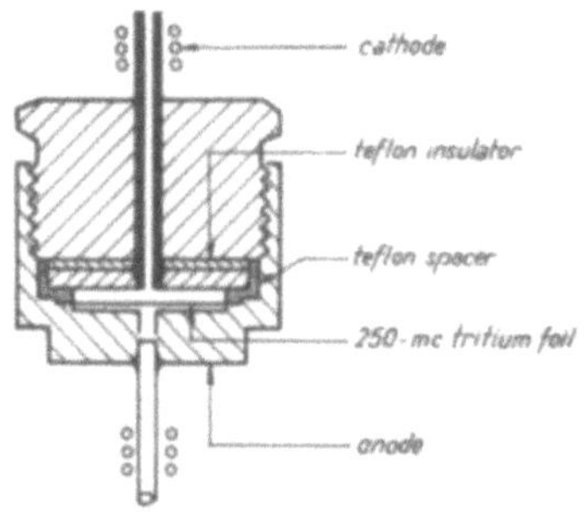

Fig. 4. Helium detector

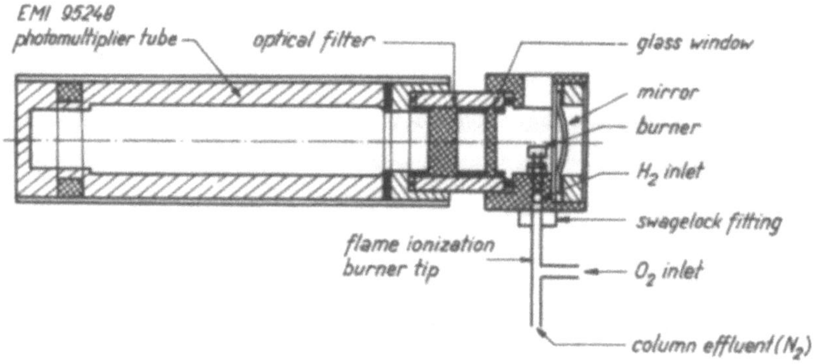

Fig. 5. Flame photometric detector

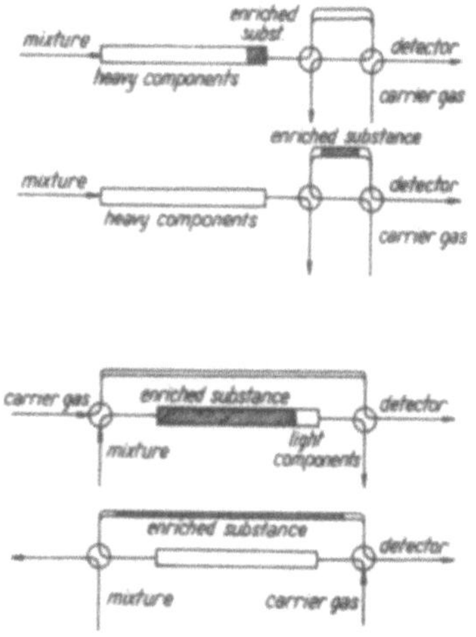

Fig. 6. Concentration trace substance by frontal chromatographic enrichment

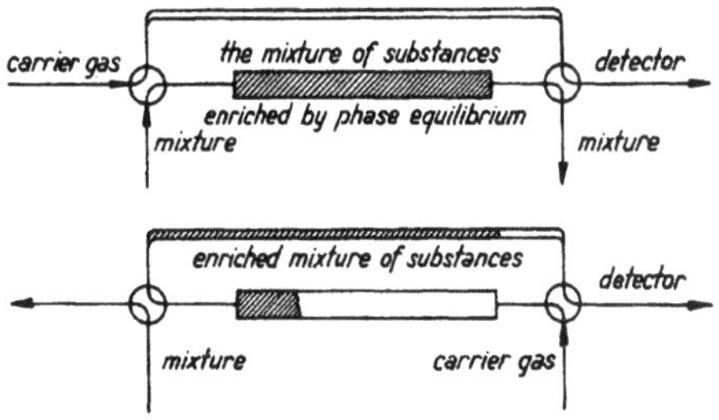

Fig. 7. Concentration of trace substance by frontal chromatographic equilibration

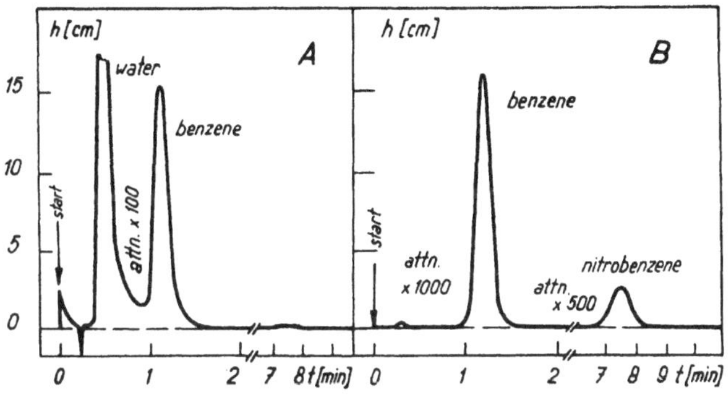

Fig. 8. Determination of nitrobenzene in the presence of benzene and water (nitrobenzene, 400 mg/m^3, water vapor, 2500 mg/m^3).

a) direct injection of 5 ml sample; SE-30 on Celite 545;

b) concentration tube; SE-30 on Celite 545

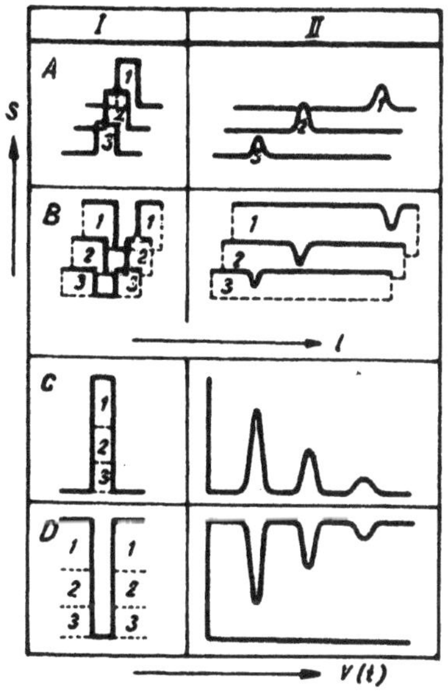

Fig. 9. Diagrammatic illustration of the vacancy chromatography

I - sample

II - schematic travelling of the substances

A, C-injection of the substance and substance mixture into the carrier gas

B, D- injection of the pure inert gas into the streaming mixture through the column

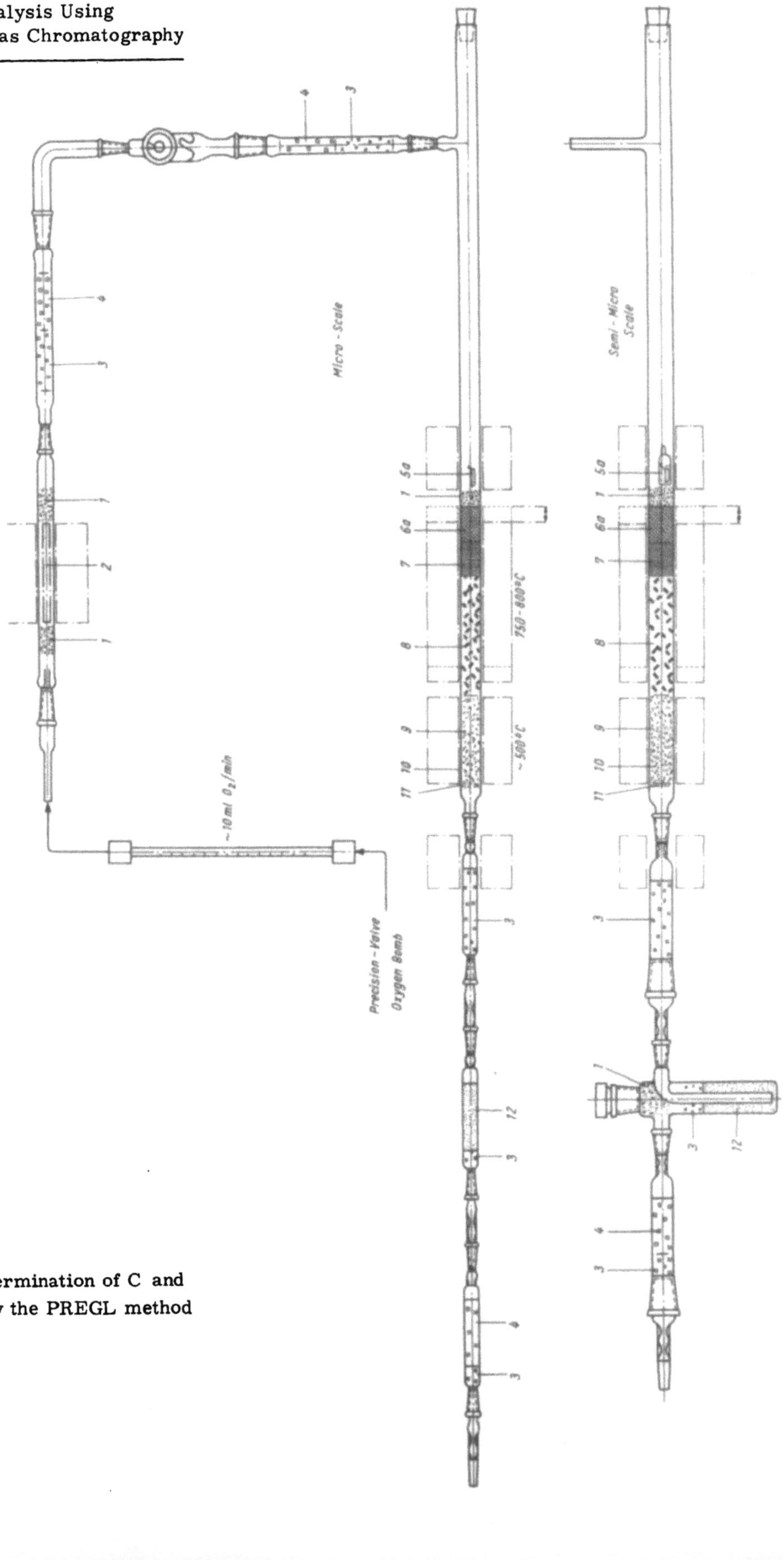

Fig. 1. Determination of C and H by the PREGL method

to fig. 1.

1 - quartz wool,

2 - catalyst for cleaning the oxygen (platinum-, quartz-wool or silver permanganate or cobalt oxide),

3 - $MgClO_4$,

4 - Ascarite,

5 - plainum boat, 5a - quartz cylinder with platinum boat,

6 - roll of platinum wire gauze,

6a - roll of wire gauze of cupric oxide (CuO),

7 - cerium dioxide or magnesium oxide (granulated),

8 - oxydation catalyst (CuO or Co_3O_4),

9 - silver catalyst (silver pumice stone or silver wool an electrolytic silver)

10 - silver wire,

11 - gold wool,

12 - adsorption catalyst for nitrogen oxide (active manganese ocide MnO)

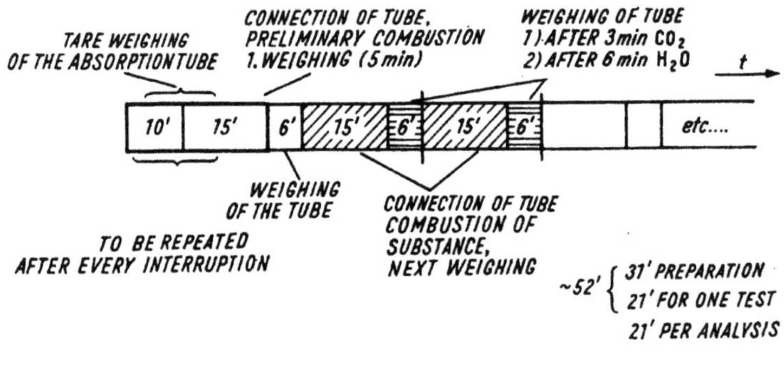

Fig. 2. Timetable for PREGL method

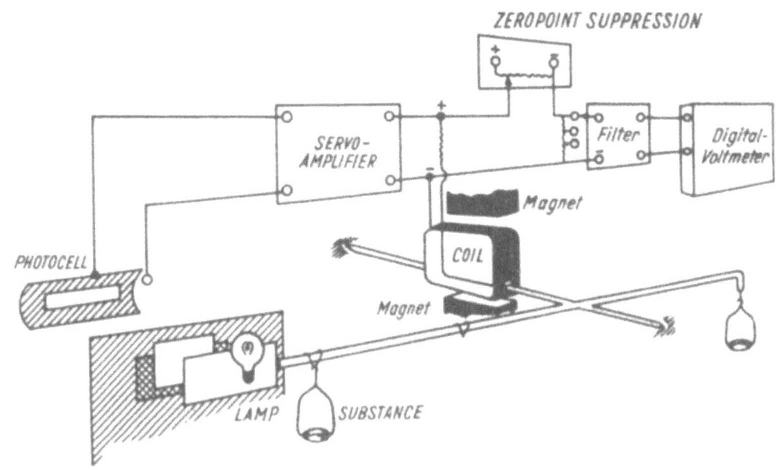

Fig. 3. CAHN electrobalance (principle)

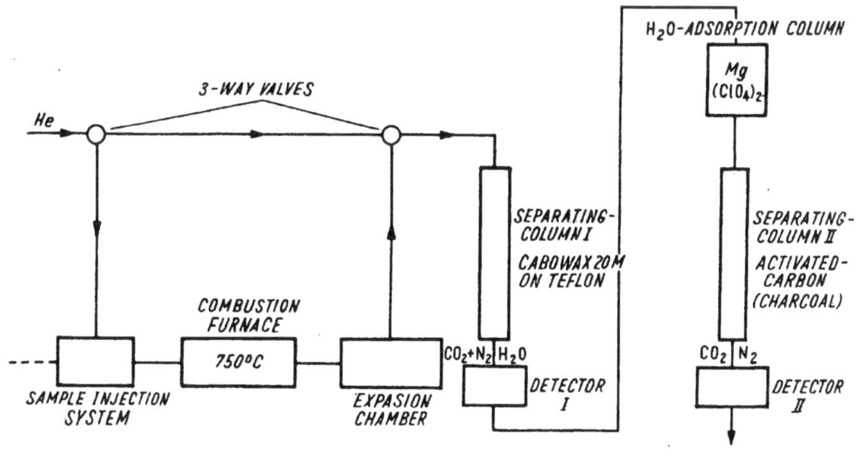

Fig. 4. CHN Analyzer F & M Scientific Div., model 180

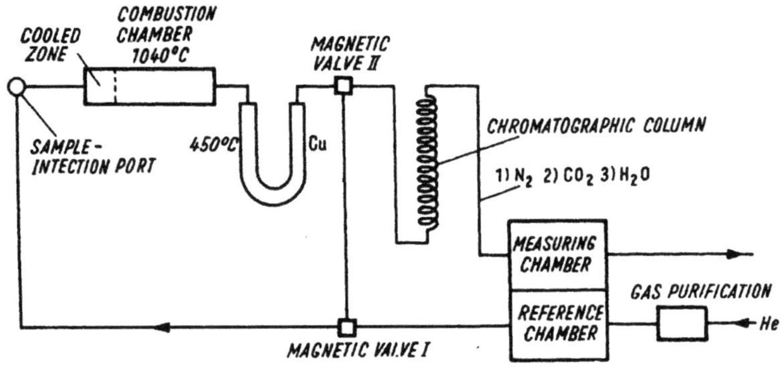

Fig. 5. CHN Analyzer HEWLETT-PACKARD/F & M, model 185

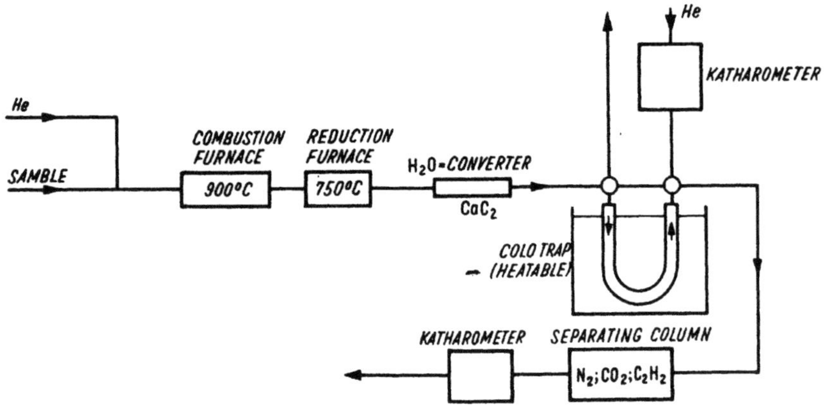

Fig. 6. FISHER CHN Analyzer

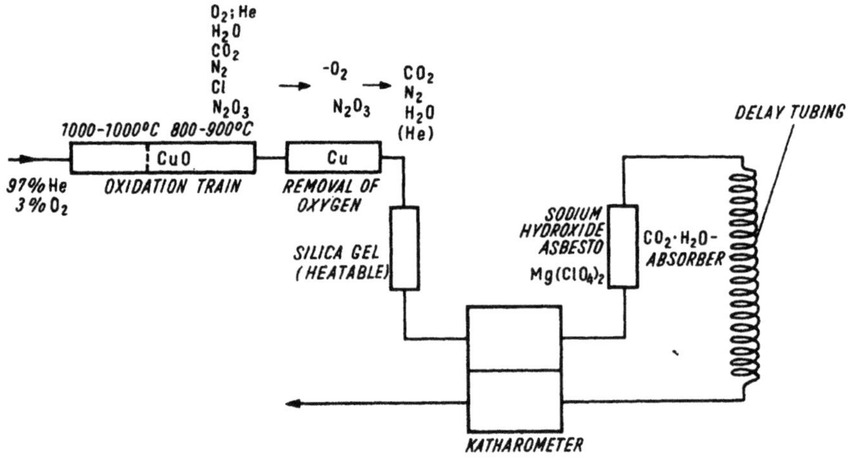

Fig. 7a. CHN Analyzer, type WALISH (Technicon/WEITKAMP-KORTE)

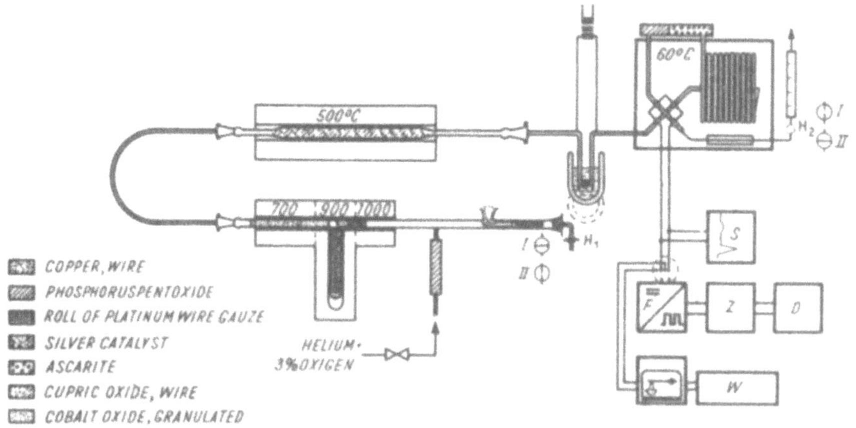

Fig. 7b. CHN-O ultra (principle)

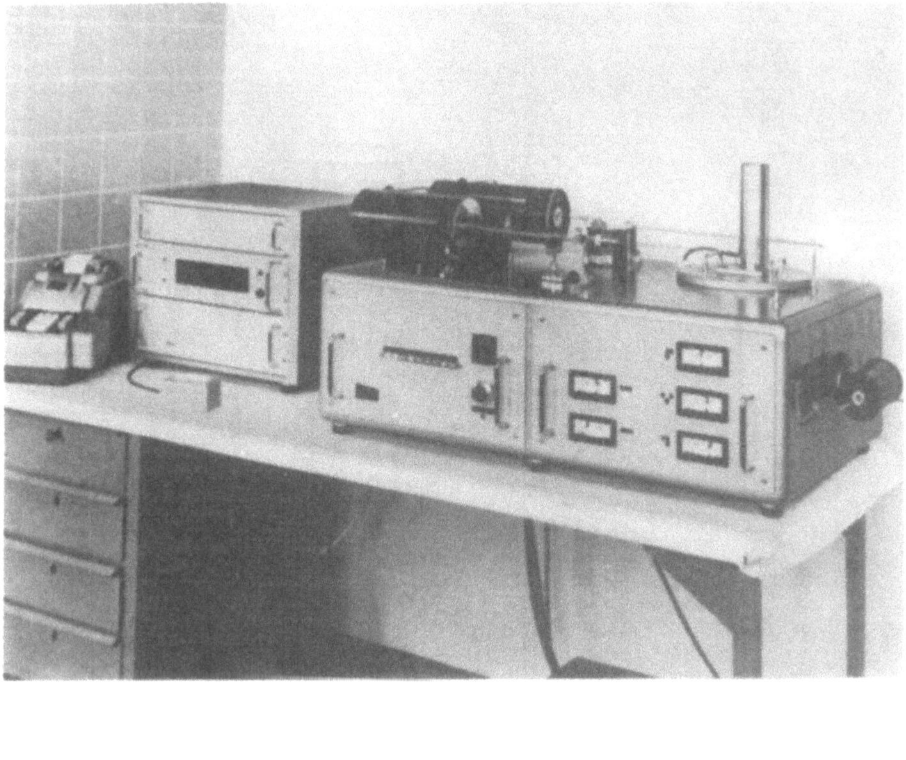

Fig. 8. CHN-O ultra (general view)

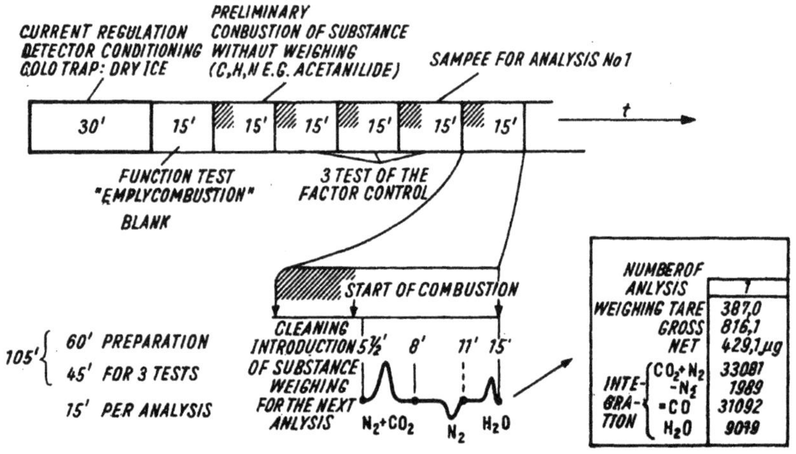

Fig. 9. „GC" CHN-determination

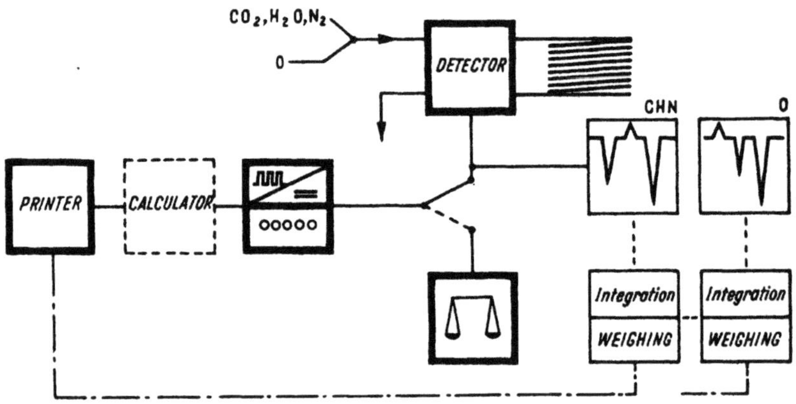

Fig. 10. Circuit of balance and integration (CHN-O ultra)

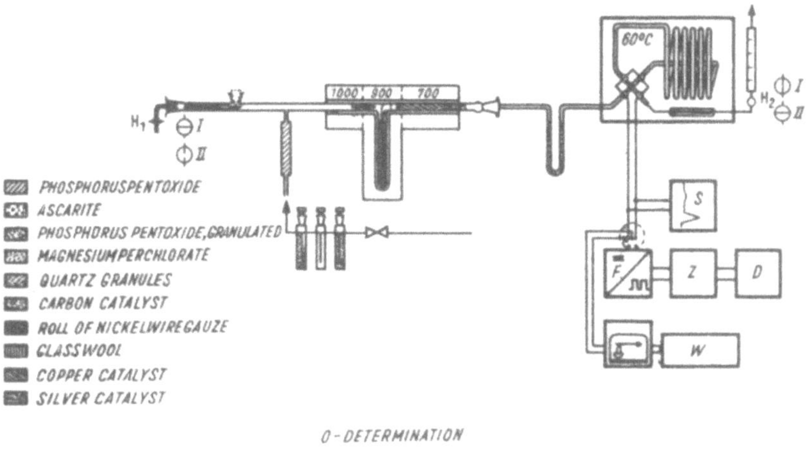

Fig. 11. The determination of oxygen (CHN-O ultra)

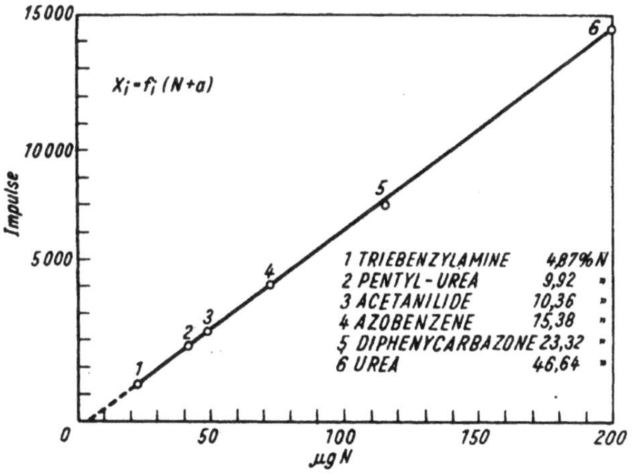

Fig. 12. Calibration curve for nitrogen (CHN-O ultra)

substance	amount weighed g	n	$\%^{C}$calc.	$\%^{C}$found (average)	d_c	$\%^{N}$calc.	$\%^{N}$found (average)	d_N	$\%^{H}$calc.	$\%^{H}$found (average)	d_H	
I Acetanilide	300-600	13	71,07	f_C=4,808 gC/ 1000 Impulse		10,36	f_N=13,037 gN/ 1000 Impulse		6,72	f_H=1,423 gH/		S_C = 0,34 S_N = 0,47 S_H = 0,13
II Phenacetin	300-600	2	67,02	67,00	-0,02	7,82	7,16	-0,66	7,31	7,34	+0,03	S_C = 0,42
Cholesterol		2	83,87	84,48	+0,61	-	-	-	11,99	11,81	-0,18	S_N = 0,41
Rastinone		2	53,31	52,97	-0,34	10,36	10,52	+0,16	6,78	6,78	+0,07	S_H = 0,26
Azobenzene		2	79,09	79,02	-0,07	15,38	15,04	-0,34	5,53	5,60	+0,07	
Pentylurea		2	55,29	55,26	-0,03	9,92	9,52	-0,40	6,43	6,55	+0,12	
Urea		4	19,99	20,12	+0,13	46,64	46,76	+0,12	6,71	6,57	-0,14	
Acetyl-salicylic Acid		2	60,00	60,78	+0,78	-	-	-	4,48	4,61	+0,13	
Diphenyl Carbazone		2	64,00	65,37	+0,39	23,32	22,59	-0,73	5,04	5,72	+0,68	
Azobenzene		5	79,09	79,06	-0,03	15,38	15,38	0,00	5,53	5,52	-0,01	
Polyacrylnitrile		2	67,09	68,50	+0,59	26,39	26,55	+0,16	5,70	5,72	+0,02	
2-Thio-urea		2	15,78	16,08	+0,30	36,78	37,26	+0,48	5,30	5,34	+0,04	
Chloralmonofluoro-acetamide		2	21,40	22,02	+0,62	6,24	6,21	-0,03	2,25	2,58	+0,33	

n = number of combustions; d = average deviation; s = standard deviation

Fig. 13. Experimental results of test substances (CHN-O ultra)

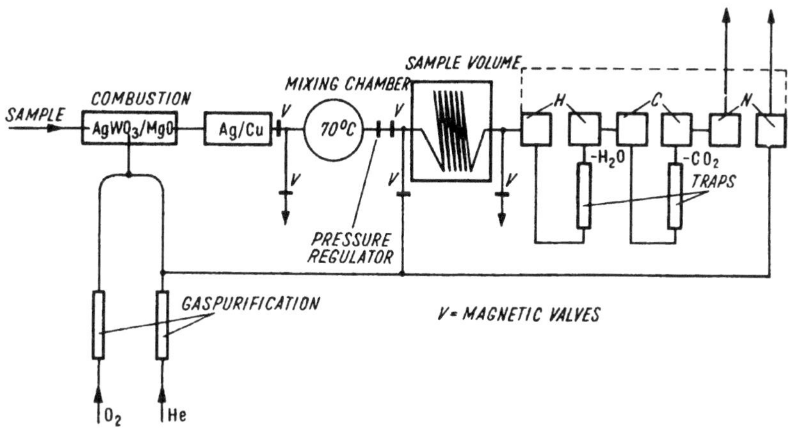

Fig. 14. CHN-Analyzer PERKIN-ELMER, model 240

Fig. 15. CHN-Analyzer PERKIN-ELMER, model 240

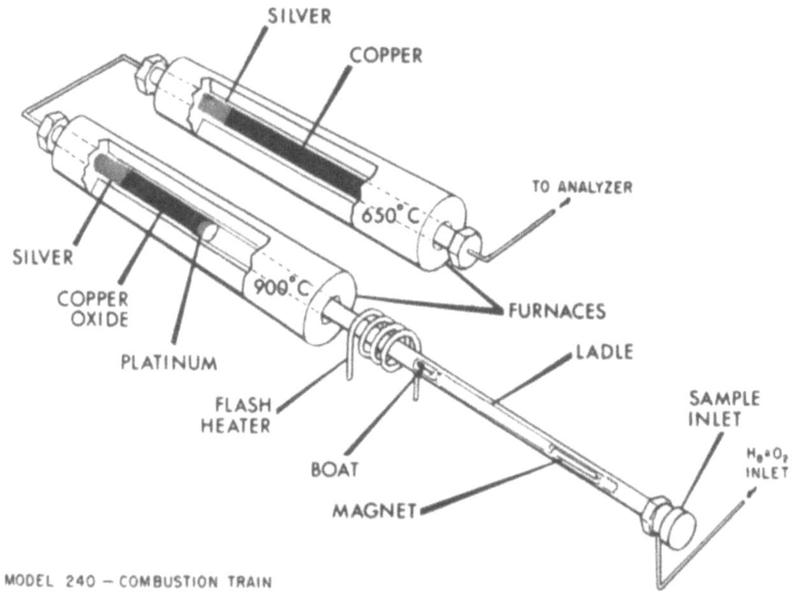

Fig. 16. Schematic view of the combustion and reduction train. Both furnace heating elements are contained in one housing.
(PERKIN-ELMER model 240)

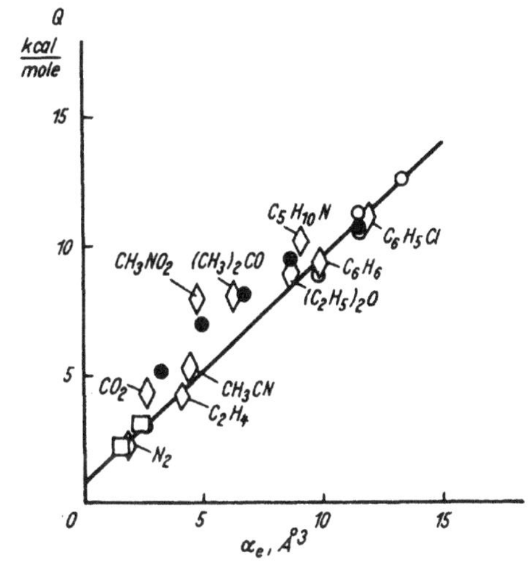

Fig. 1. Dependence of heat of adsorption on graphitized thermal carbon black of group A (n-alkanes, noble gases), group B (indicated) and group D (n-alcohols) molecules on electron polarizability of molecules

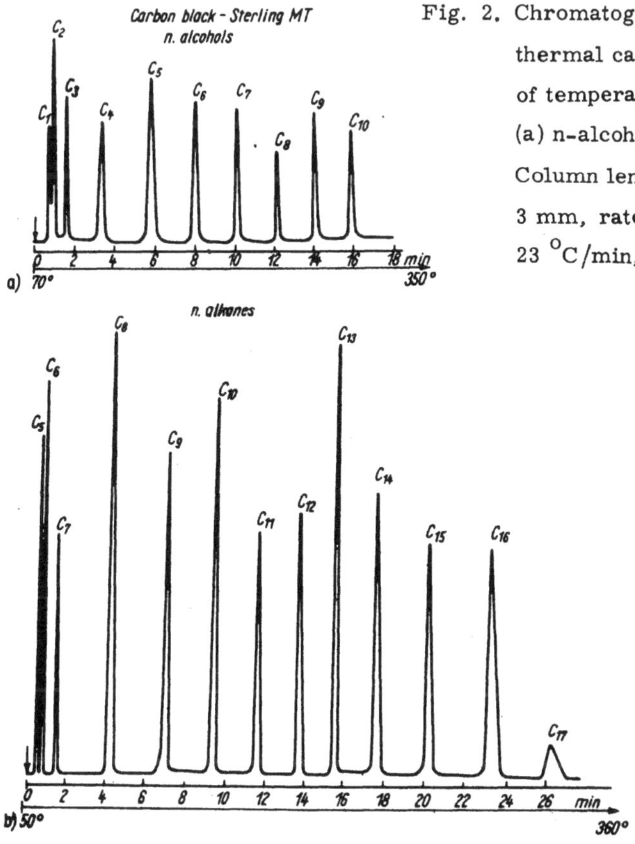

Fig. 2. Chromatograms on graphitized thermal carbon black under conditions of temperature programming: (a) n-alcohols, (b) n-alkanes. Column length 1 m, diameter 3 mm, rate of temperature rise 23 °C/min, flame-ionisation detector

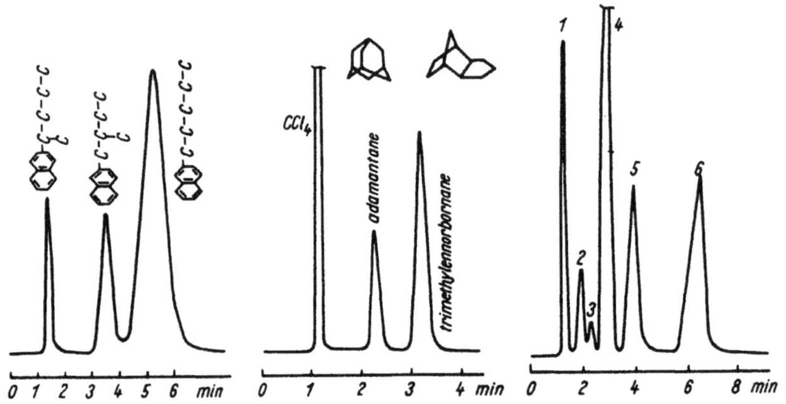

Fig. 3. Chromatograms on graphitized thermal carbon black with specific surface area of 8.1 m^2/g: (a) amylnaphthalenes, (b) adamantane and trimethylene-norbornane in CCl_4 solution and (c) cis- and trans-isomers of silicon - and germanium - element-organic compounds of $(CH_3)_3ECH{=}CHE(CH_3)_3$ type

1 - trans $Me_3SiCH = CHSiMe_3$
2 - cis ——— " ———
3 - trans $Me_3GeCH = CHSiMe_3$
4 - cis ——— " ———
5 - trans $Me_3GeCH = CHGeMe_3$
6 - cis ——— " ———

Fig. 4. Potential function of van-der-Waals interaction between two carbon atoms $\varphi_{C \ldots C}$

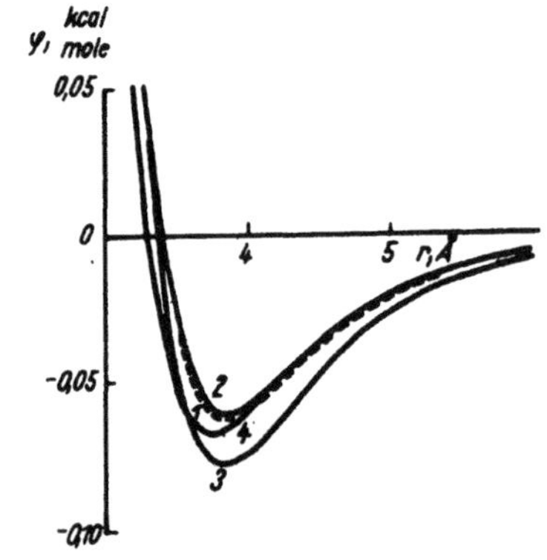

1 - for adsorption of hydrocarbons of graphite

2 - for interaction of graphite layers

3 - for interaction in hydrocarbon crystals

4 - for interaction of methane molecules in real gas

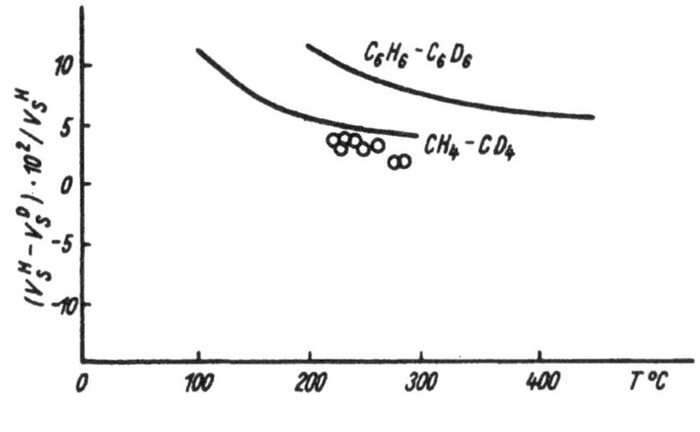

Fig. 5. Calculated (curves) and experimental (points) values of relative difference of retention volumes on graphite (curves) and on graphitized thermal carbon black (points) for CH_4, CD_4 and C_6H_6, C_6D_6

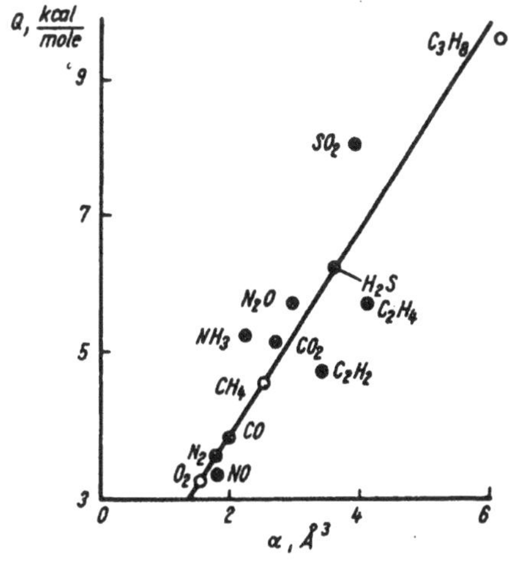

Fig. 6. Dependence of heats of adsorption on Saran carbon at low surface coverages on electron polarizability of group A (n-alkanes, open circles), group B (circles with crosses) and group D (black cizcle) molecules

Fig. 7. Chromatograms on Saran carbon:

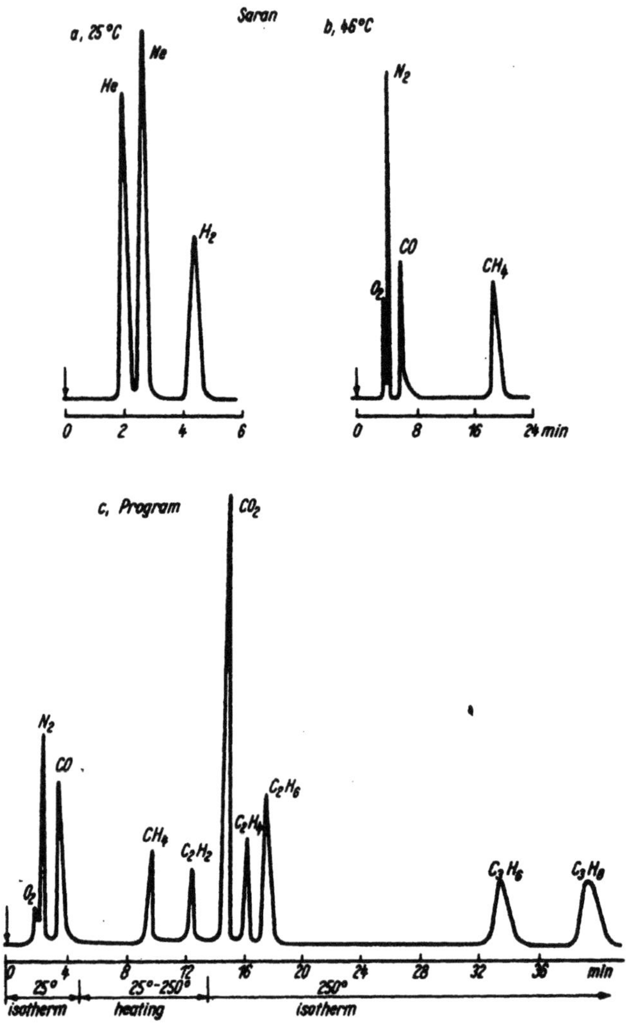

(a) He, Ne and H_2 at 25 °C (column length 300 cm, diameter 4 mm),

(b) O_2, N_2 CO and CH_4 at 46 °C, and

(c) O_2, N_2 and CO at 25 °C, CH_4 and C_2H_2 on heating of column to 250 °C at the rate of 30 °C/min and CO_2, C_2H_4, C_2H_6 and C_3H_8 at 250 °C

Fig. 8.

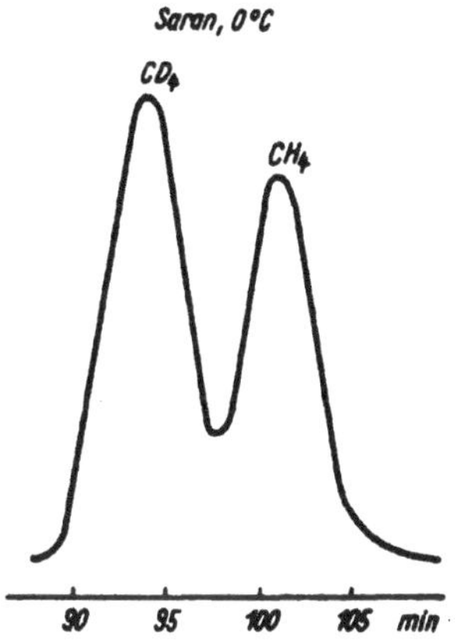

Chromatograms of mixture of CD_4 and CH_4 on Saran carbon at 0 °C. Column length 5 m, diameter 2 mm, flow rate of helium carrier gas 43 cm^3/min, flame- ionization detector

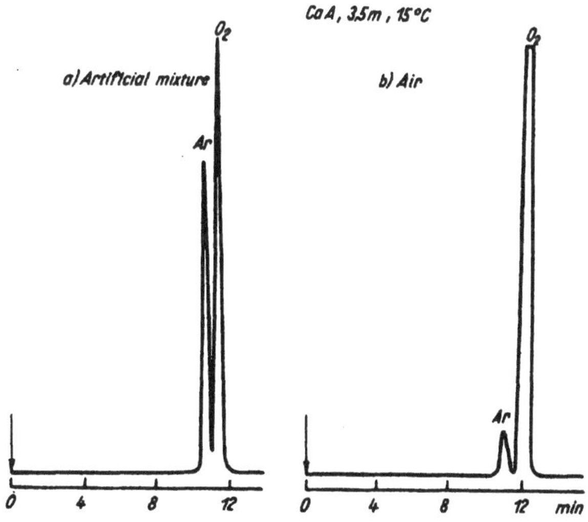

Fig. 9. Chromatogram of artificial and atmospheric mixture of argon and oxygen on column with granulated zeolite CaA (length 3.5 m) heat-treated at 400 °C

Fig. 10.

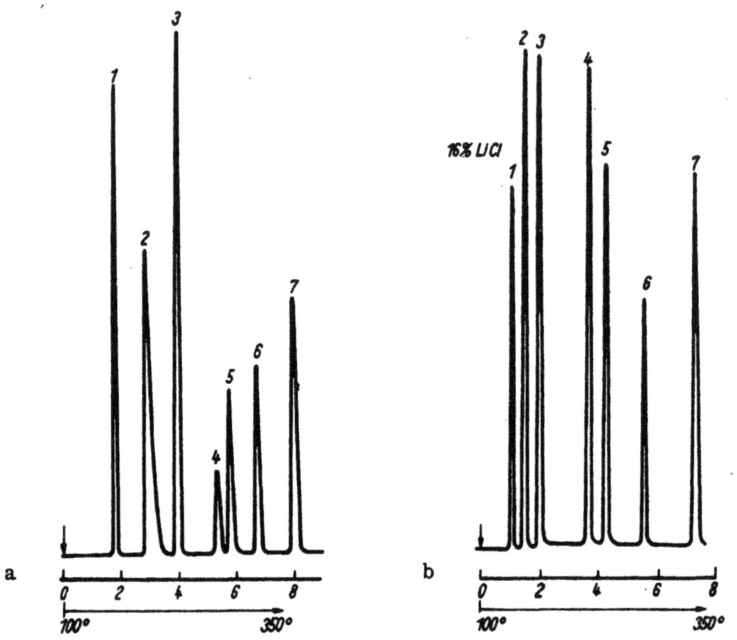

Chromatogram of mixture of

(1) cyclohexane,

(2) benzene,

(3) decalin,

(4) mesitylene,

(5) naphthalene,

(6) diphenyl,

(7) phenanthrene

on

(a) macroporous silica gel,

(b) same silica gel with 16 % LiCl and

(c) same silica gel with 19% CsCl. Column length 200 cm and diameter 4 mm. Rate of temperature rise in column 35°/min

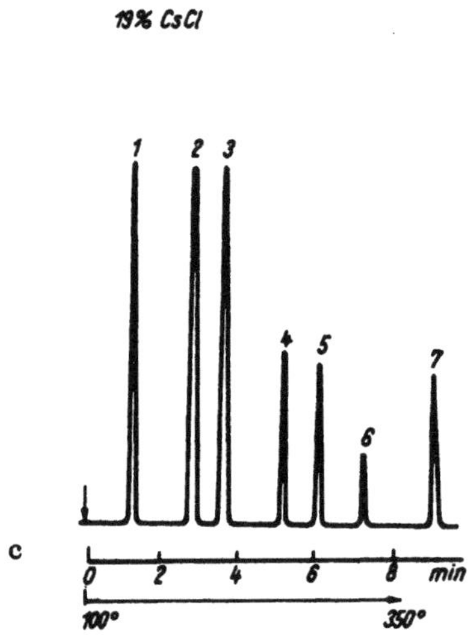

Fig. 11. Chromatograms of unsaturated hydrocarbons C_4: (1) butene-1, (2) 2-methylpropene, (3) cis-butene, (4) butadiene, and (5) trans-butene on (a) graphitized thermal carbon black, and (b) silica gel with deposited KOH

a

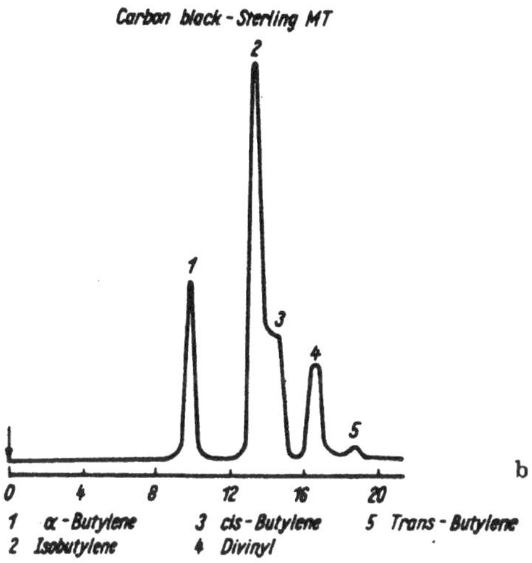

b

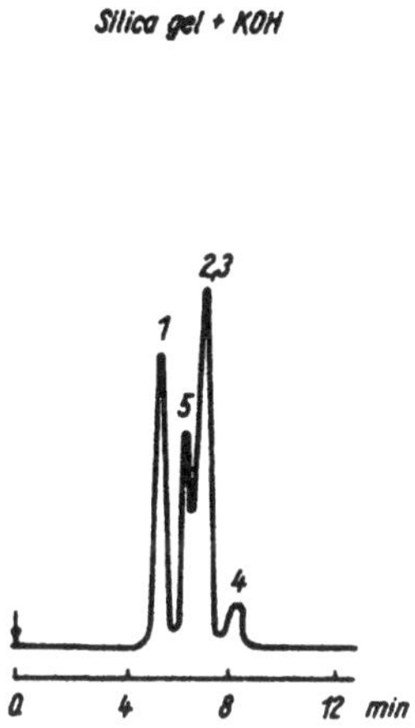

Fig. 12. Chromatograms on aerosilogel of group B molecules of different geometric and electronic structure

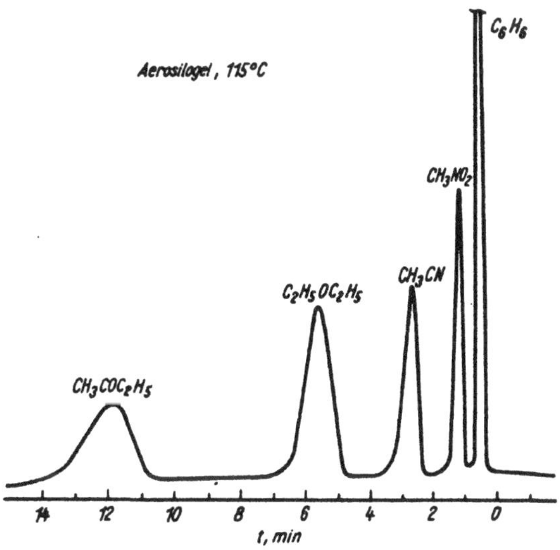

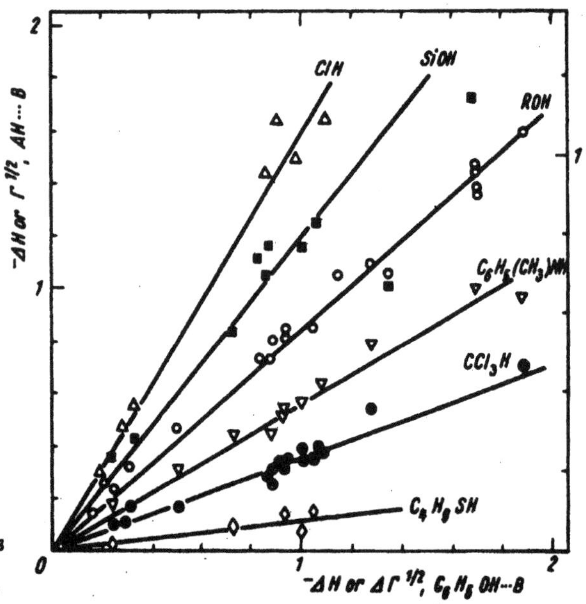

Fig. 13. Mutual dependences of values of $-\Delta H$ and $(\Delta\Gamma)^{1/2}$ for hydrogen bonds A-H... B of different acids A-H (indicated near curves) with the same set of bases B. x-axis: values of $-\Delta H$ or $(\Delta\Gamma)^{1/2}$ for H-bonds $C_6H_5OH...B$. y-axis: values of $-\Delta H$ or $(\Delta\Gamma)^{1/2}$ for H bonds in solution ClH... B (1), in adsorption SiOH... B (2), in solutions ROH... B (3), $C_2H_5CH_3H...B$ (4), $CCL_3H...B$ and $CCL_3D...B$ (5), $(CH_3)_3CSH...B$ (6) - ΔH- black points, $(\Delta\Gamma)^{1/2}$ - white points. For acids (2) and (5) scale is on the left, for acids (1), (3) and (4) on the right. Unit: values of $-\Delta H$ or $(\Delta\Gamma)^{1/2}$ of H-complex $C_6H_5OH...O(C_2H_5)_2$

Fig. 14. Dependence of ion current of flame-ionization detector on column temperature. Amount of deposited monoethanol amine (triangles) and decyl alcohol (circles) corresponds to monolayer, as well as 22% liquid monoethanolamine and 25% decyl alcohol. Dashed lines show boiling points of monoethanolamine (170 °C) and decyl alcohol (231 °C)

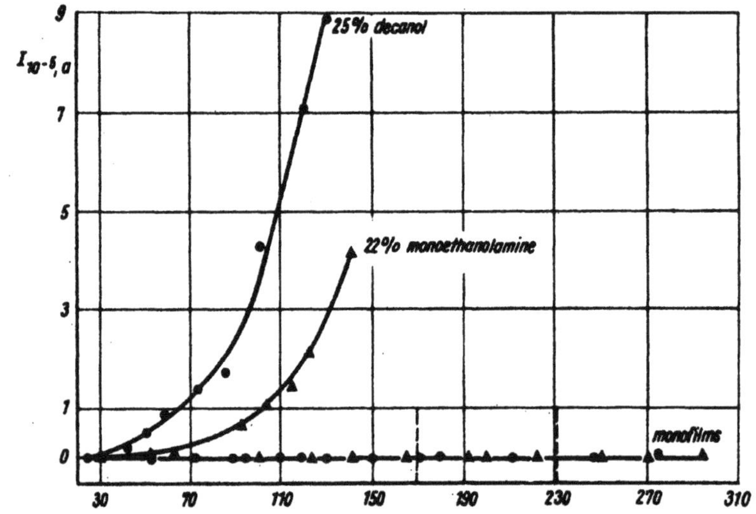

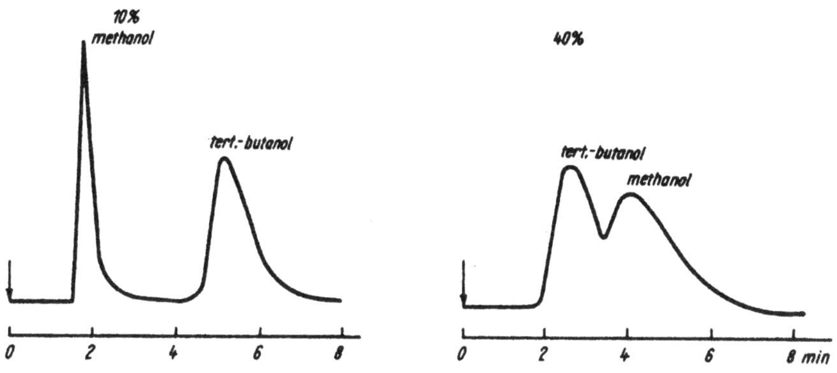

Fig. 15. Chromatogram of methyl and tertiary butyl alcohols in adsorption on glycerol monolayer (10% glycerol on macroporous silica gel with specific surface area of 60 m^2/g) and in solution (40% glycerol) at 103 °C

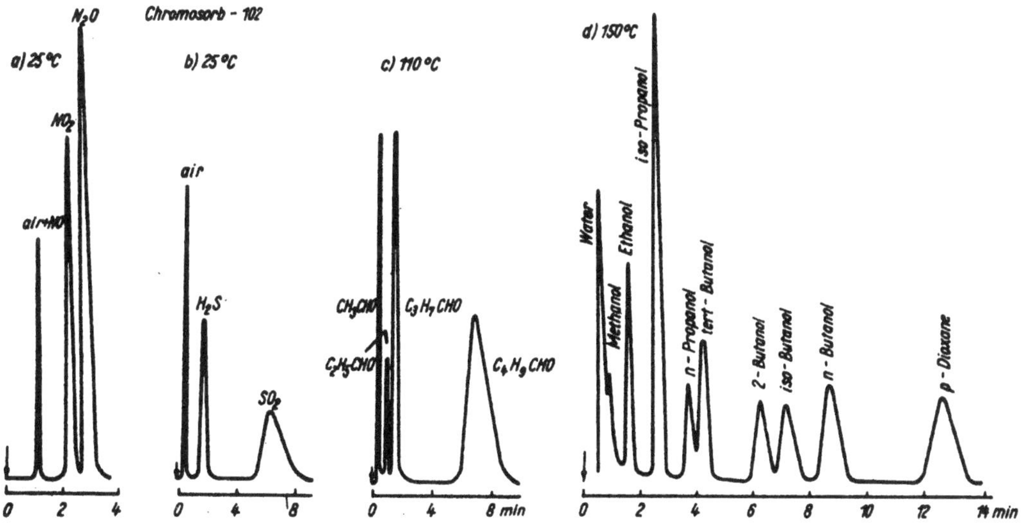

Fig. 16. Chromatograms on Chromosorb 102: (a) of nitrogen oxides, (b) of H_2S and SO_2, (c) of aldehydes, and (d) of water, alcohols C_1-C_4, and 1.4-dioxane

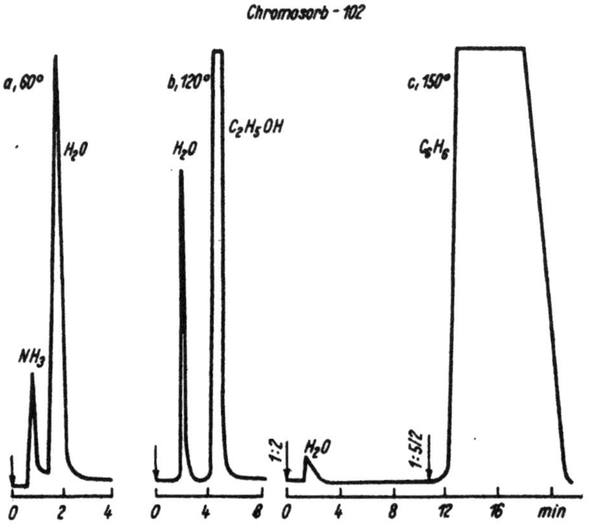

Fig. 17. Chromatograms on Chromosorb 102: (a) of ammonia and water, (b) of water impurity in ethanol, and (c) of water impurity in benzene

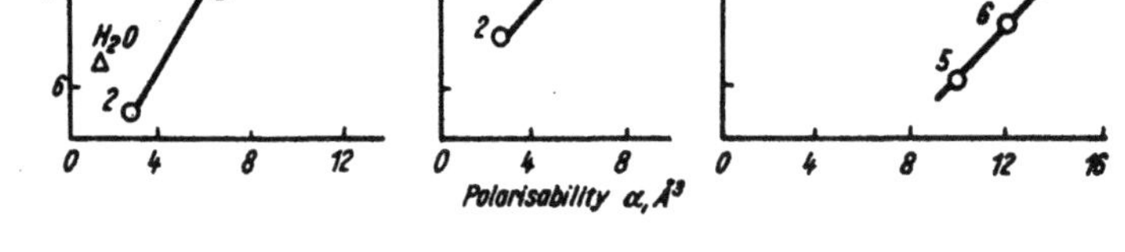

Fig. 18. Dependences of heats of adsorption at low surface coverages (a) on Chromosorb 102, (b) on polyphenyl-disiloxane, and (c) on polyacrylonitrile on electron polarizability of group A (n-alkanes), group B (indicated), and group D (water, n-alcohols) molecules

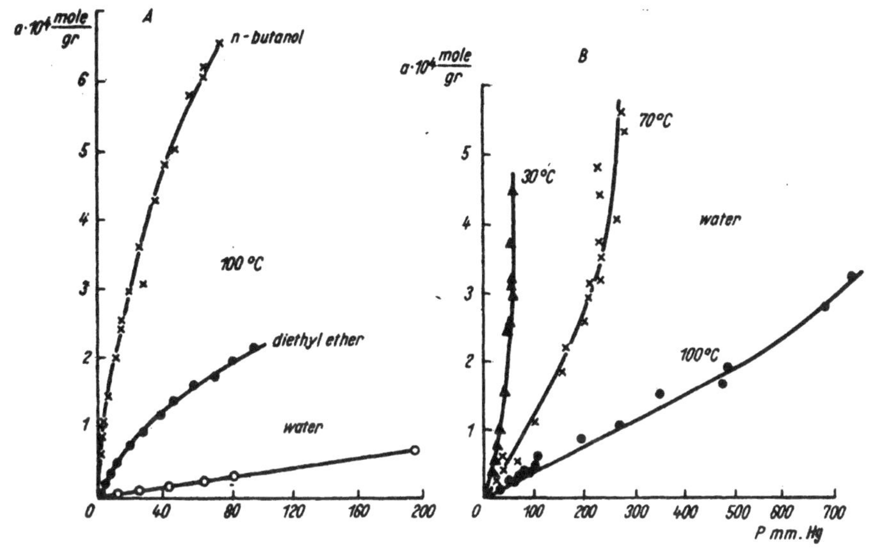

Fig. 19. Adsorption isotherms on Chromosorb 102 of (a) water, diethyl ether, and n-butanol at 100 °C, and (b) water at different temperatures, obtained from chromatograms

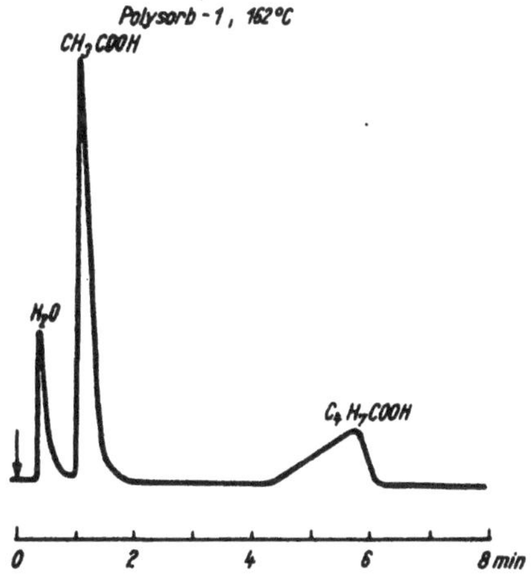

Fig. 20. Chromatogram of mixture of water, acetic and n-butyric acids on Polysorb-1 at 162 °C, sample size 0.3 ml, column length 1.6 m, diameter, 18 mm

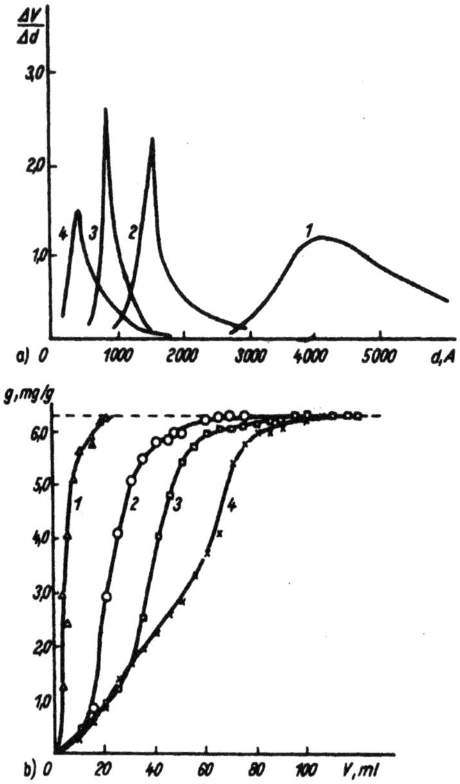

Fig. 21. (a) Curves of pore volume distribution of four macroporous silica gels according to their effective diameters. (b) Frontal chromatograms of polystyrene solution of mean molecular weight of about 300, 000 on columns with same silica gels

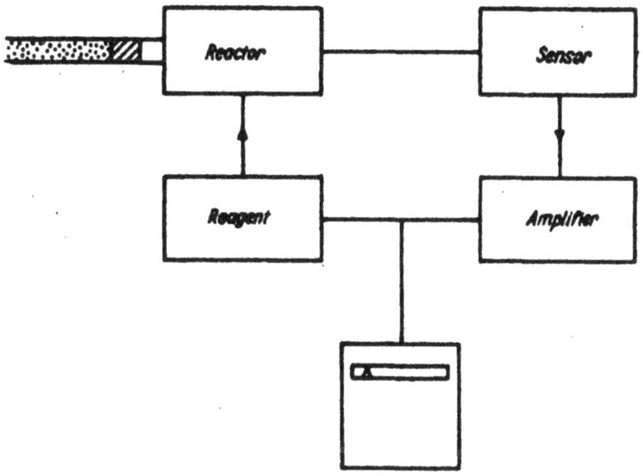

Fig. 1.

Block Schematic Diagram of a Servo-Detector for Gas Chromatography

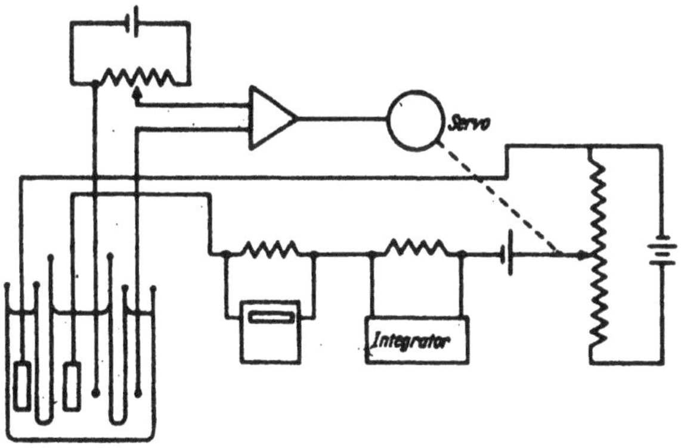

Fig. 2.

Schematic diagram of a servo-coulometer [24]. For gas-chromatographic purposes, the effluent enters in the main limb of the titration vessel, which contains generator and sensor electrodes. Reference electrodes are in side limbs

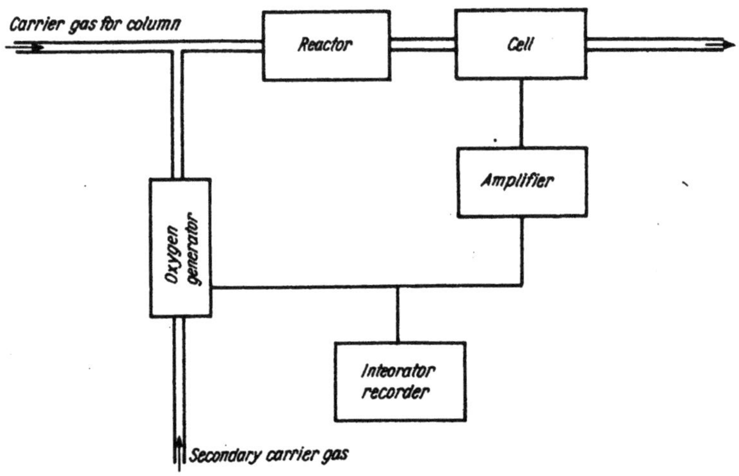

Fig. 3. Schematic diagram of Reaction Coulometer working in the combustion mode [31]

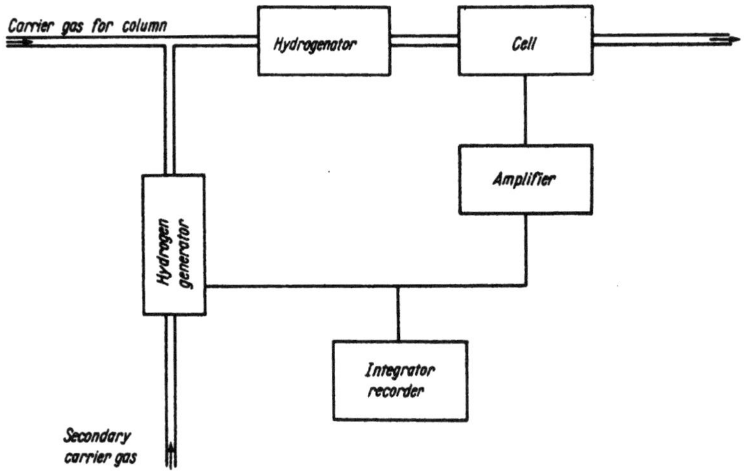

Fig. 4. Hydrogenation Coulometer [32]

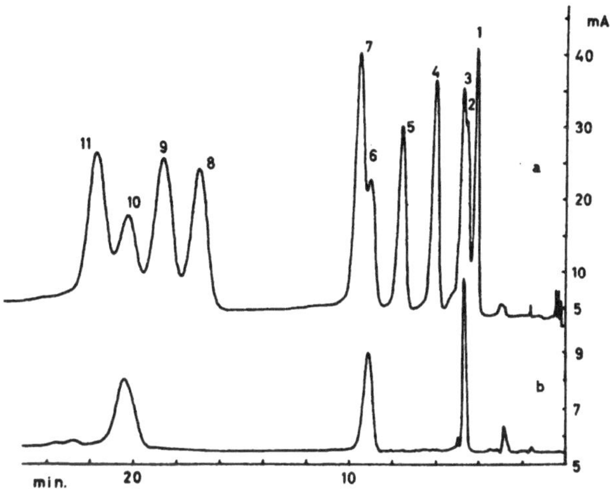

Fig. 5. Comparison of chromatograms of a mixture of alkanes and alkenes, first on a combustion coulometer, and second on a hydrogenation coulometer [32]

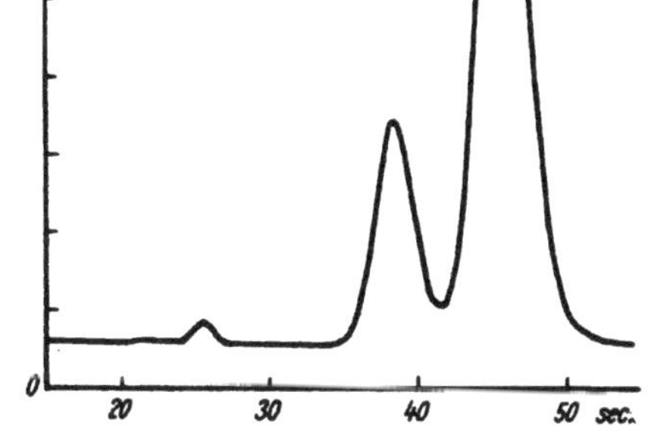

Fig. 6. To illustrate rapid response of a servo-detector

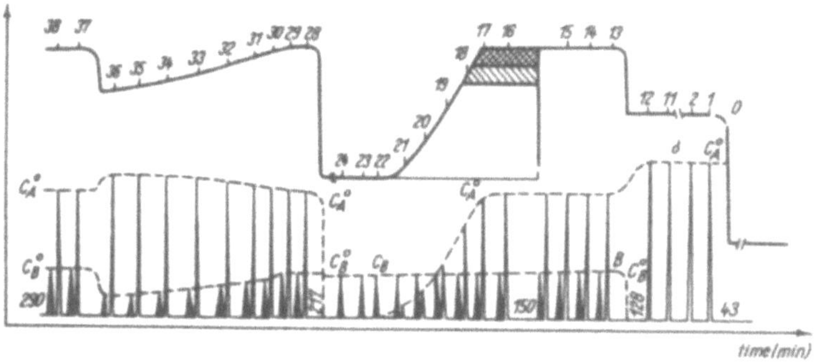

Fig. 1. The principle of chromatographic measurement of the individual adsorption isotherms of the gases from their mixtures.

abscissa - time

ordinate - concentration

↕ - moment of inlet of gas mixture

↓ - moment of entrance of vacancy

Fig. 2.

Photography of the radio-chromatograph

Fig. 3.

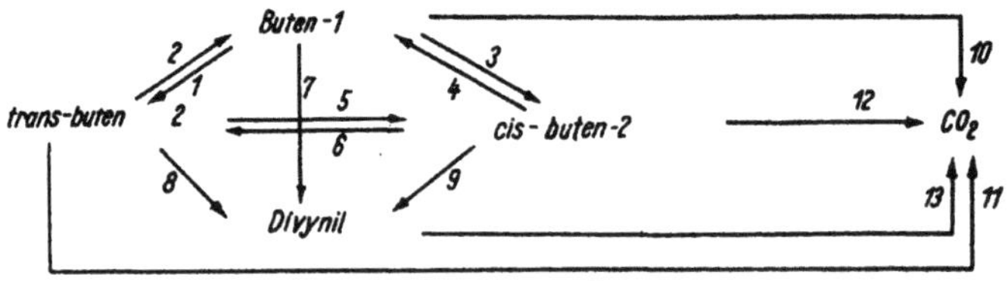

Scheme of the main reaction during a catalytic oxidative dehydrogenation of butylenes

Fig. 4.

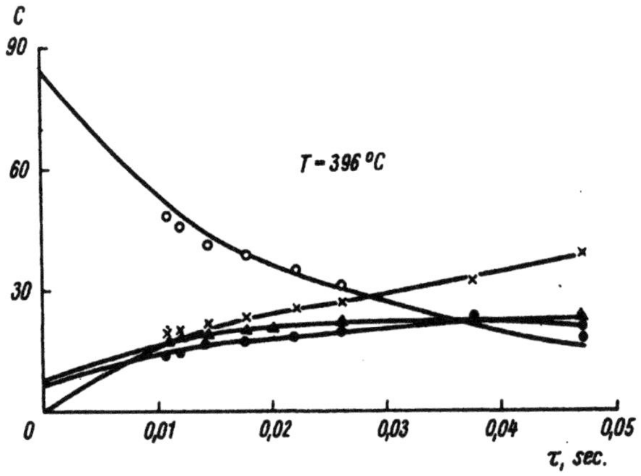

Calculated kinetic curves for the oxidative dehydrogenation and for the isomerization of n-butylene isomers. The points indicate experimental data, measured at 396 °C, of

1) butene-1;
2) trans-butene-2;
3) cis-butene-2;
4) divinyl.

Fig. 5.

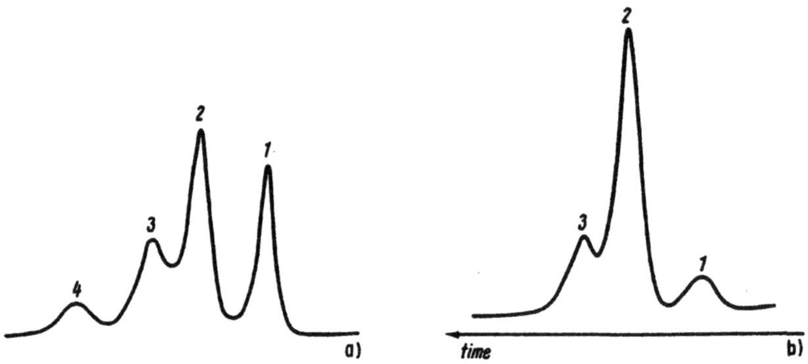

Gas chromatograms of the products of oxidative dehydrogenation of a binary mixture of non-labelled butene-1 and labelled cis-butene-2 at 310 °C. The peaks are marked by the same numbers as the curves in fig. 4

a) direct detection of the concentrations
b) detection according to C^{14} radioactivity

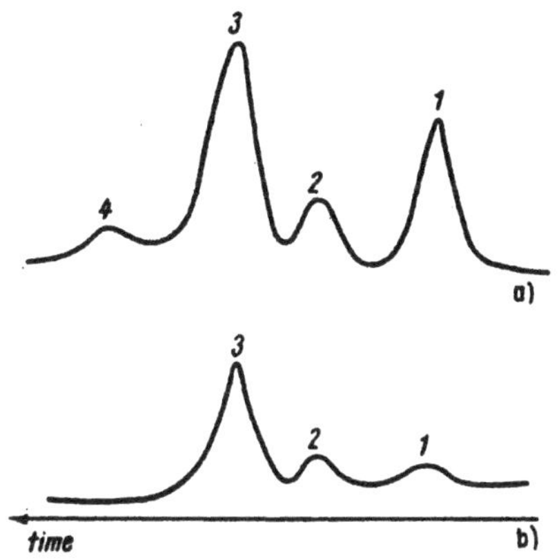

Fig. 6.

Thes gas chromatograms of the same products as in fig. 5. but for the oxidation of a binary gas mixture of butene-1 and labelled butene-2 at 310 °C.

a) direct detection of the concentrations
b) detection according to C^{14}.

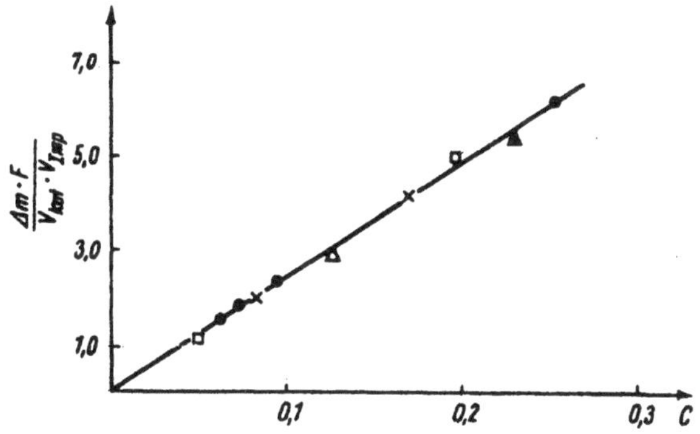

Fig. 7.

Dependence of the reaction velocity on the concentration

a) series 1
b) series 2
c) series 3
d) series 4

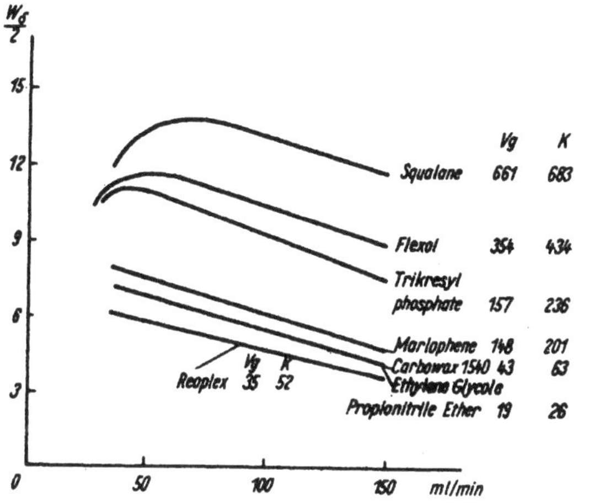

Fig. 1. Dependence of the separating power on the coefficient of distribution

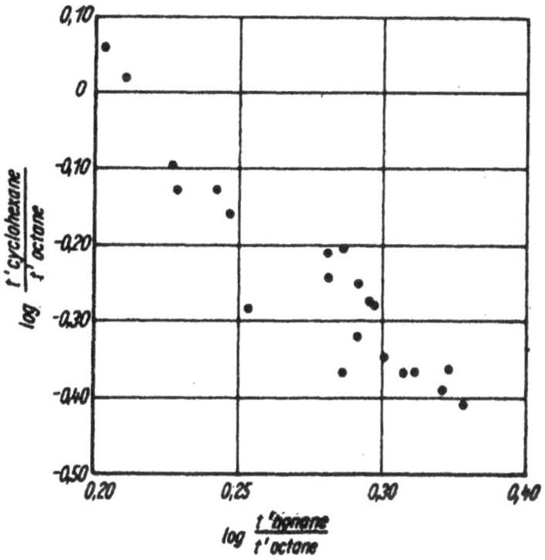

Fig. 2. Retention ratios of cyclohexane/n-octane and of n-nonane/n-octane for various separating liquids

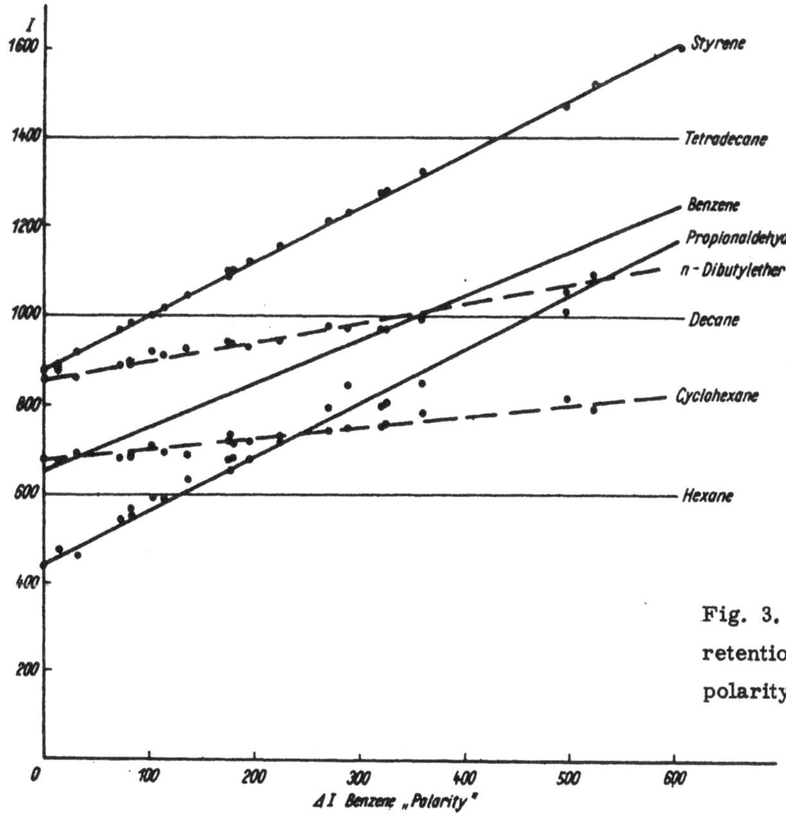

Fig. 3. Dependence of the retention index on the polarity from ref. [9]

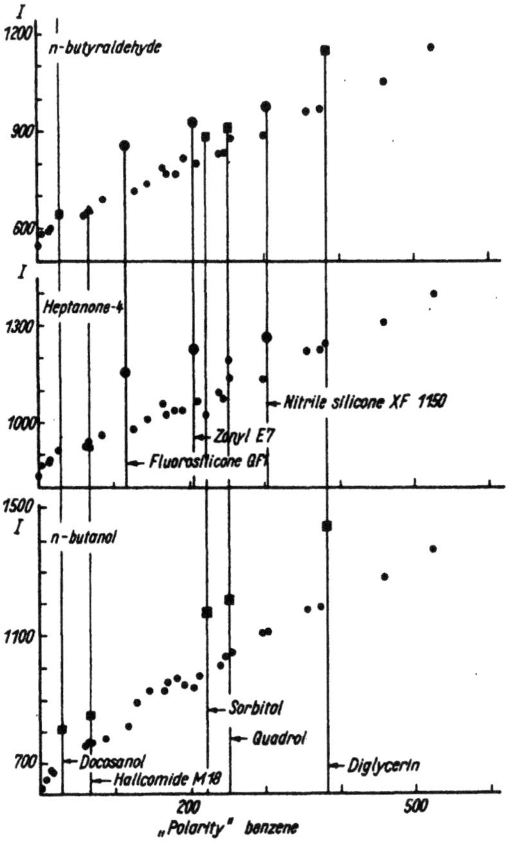

Fig. 4. Specific selectivities of different separating liquids

H. ROTZSCHE

On Column Support Materials

Fig. 1. Effect of the column support material on the peak shape of benzotrichloride

(a) Methyl silicone on Sterchamol

(b) Methyl silicone on Heydeflon (PTFE)

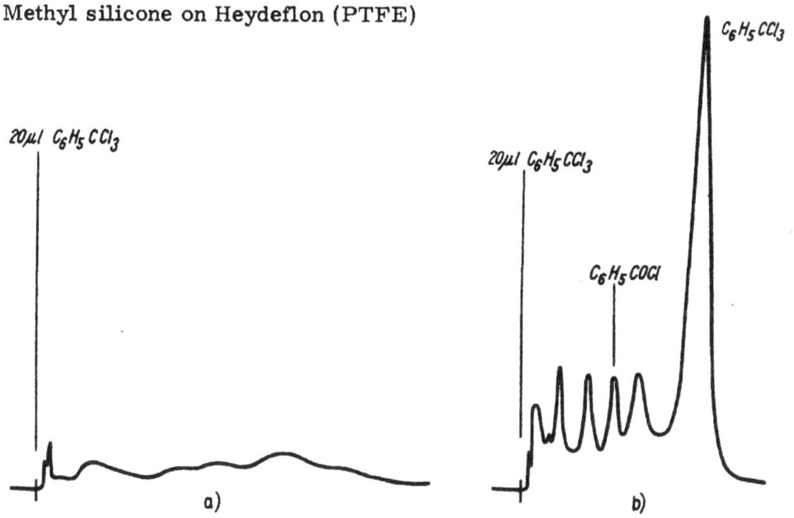

Fig. 2. Effect of the column support material on the peak shape of phenyltrichlorosilane

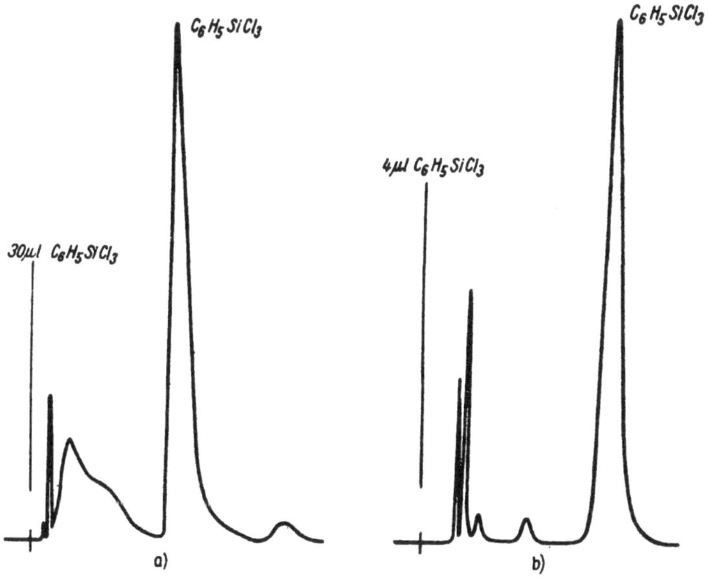

(a) Methyl silicone on Sterchamol

(b) Methyl silicone on Heydeflon (PTFE)

Fig. 3. Effect of the column support material on the peak shape of triethylaminosilane

(a) Methylphenyl silicone oil on Rysorb BLK

(b) Methylphenyl silicone oil on Heydeflon (PTFE)

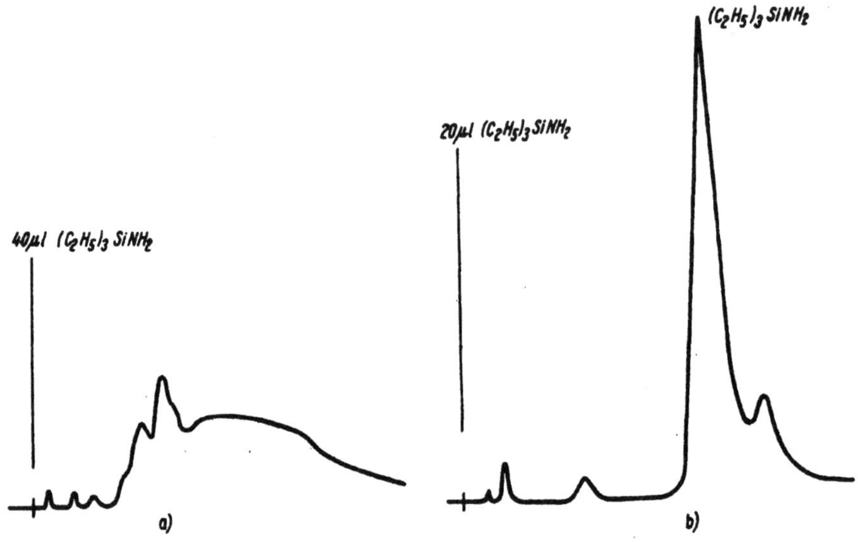

Fig. 4. Effect of the column support material on the peak shape of pyridine

(a) Methyl silicone rubber on Rysorb

(b) Methyl silicone oil on Heydeflon (PTFE)

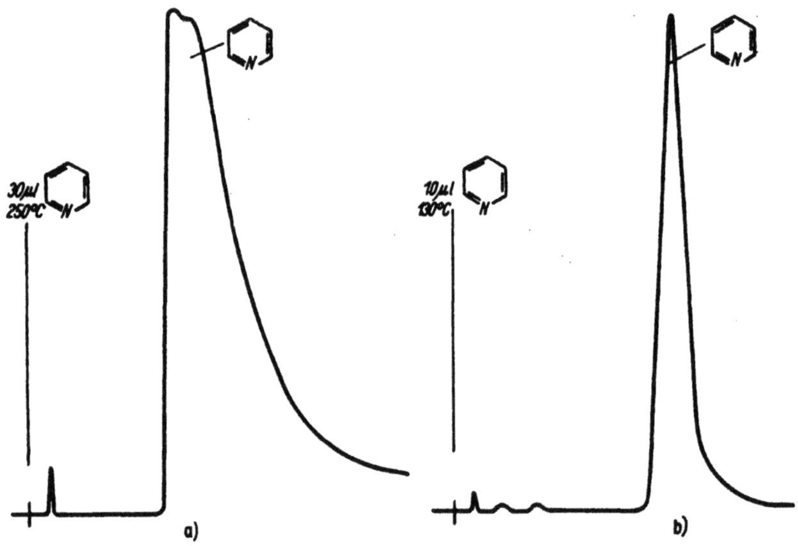

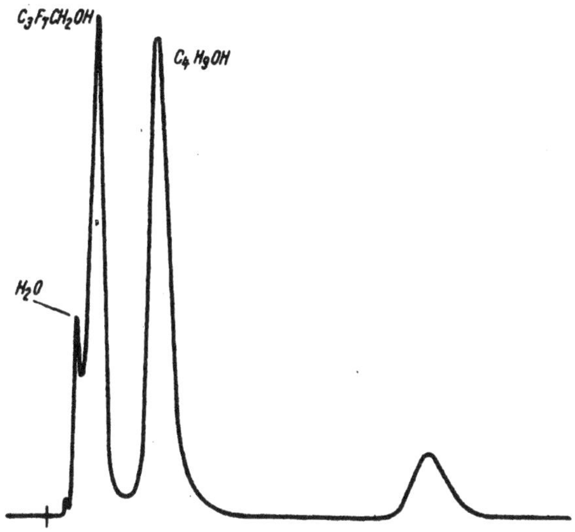

Fig. 5. Separation of polar substances with hydrogen bond formation by means of methylphenyl silicone oil on PTFE supports

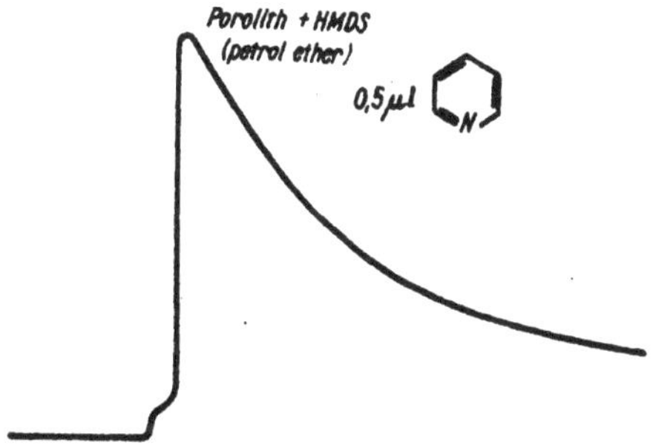

Fig. 6. Influence of the solvent on the deactivation of solid supports by means of hexamethyl disilazane Porolith with HMDS deactivated in petroleum ether

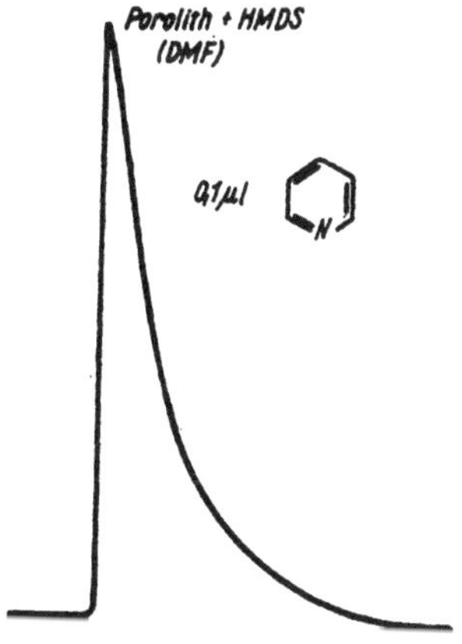

Fig. 7. Influence of the solvent on the deactivation of solid supports by means of hexamethyl disilazane Porolith with IIMDS deactivated in dimethylformamide

In both cases (Fig. 6, Fig. 7) pyridine was introduced in the GC system for the first time. Hence previous saturation of active centres with pyridine was impossible

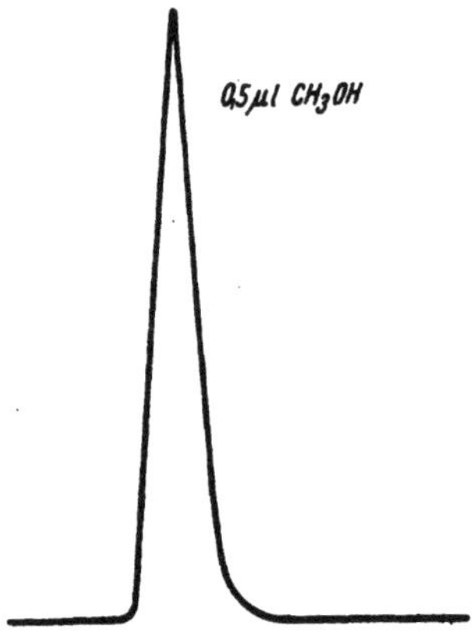

Fig. 8. Utilization of methyl silicone acid as column support material.
Column charged with 5 per cent methylphenyl silicone oil OE 4008 on methyl silicone acid, the Si-OH groups of which have been silylated by means of HMDS (IX)

Fig. 9 and 10. Peak shape of methanol, pyridine, and acetic acid on chemically modified Porolith

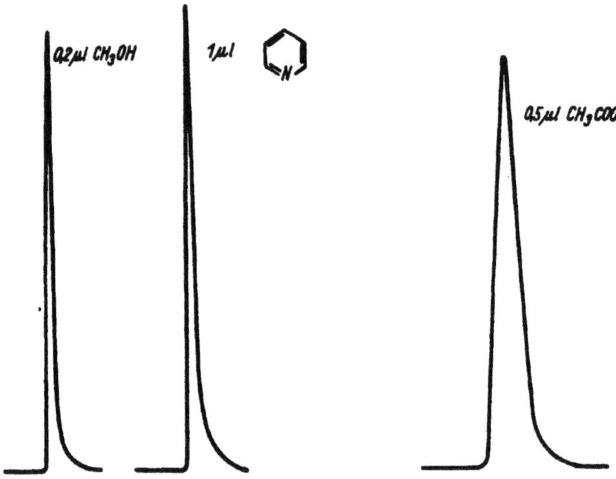

Eventually existing active centres have not been saturated prior to treatment; all three compounds contacted the stationary phase for the first time

$$CH_3SiCl_3 + \left[(CH_3)Si\right]_2 NH$$

Stationary phase: 10 per cent Squalan on Porolith which has been deactivated with

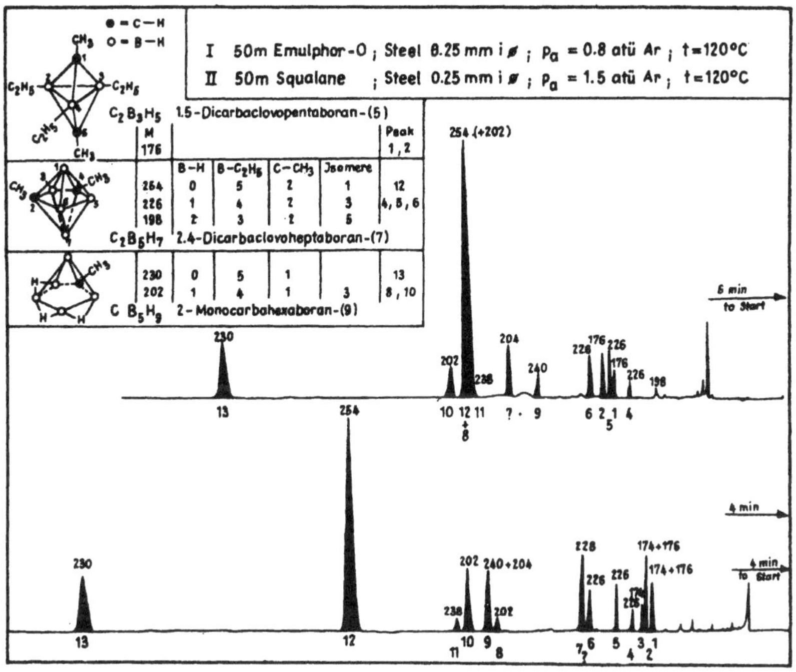

Fig. 1. Mixture of carboranes in two capillary columns of different polarity (squalane, Emulphor O)

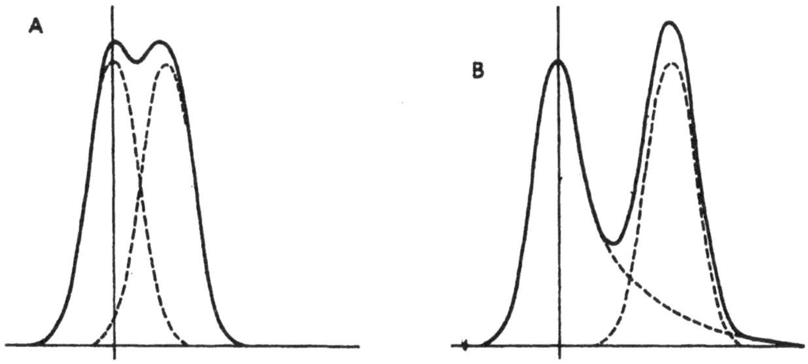

Fig. 2. A) Superposition of two gaussian shaped peaks with a certain partial resolution

B) Superposition of one gaussian peak on a peak with tailing. Separation of both peaks is twice as in A)

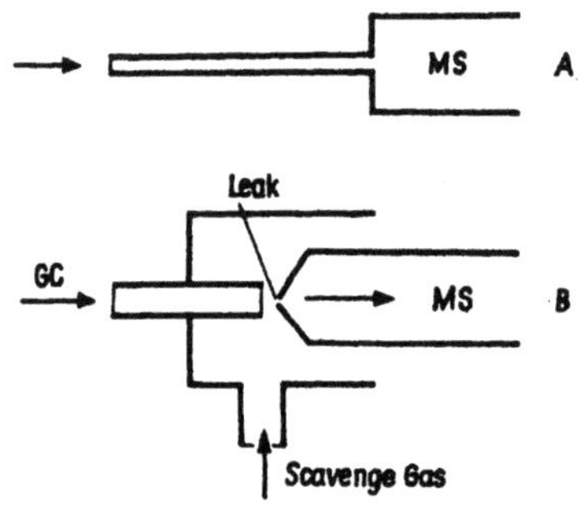

Fig. 3. Introduction of eluates into mass spectrometers

A) direct

B) by split

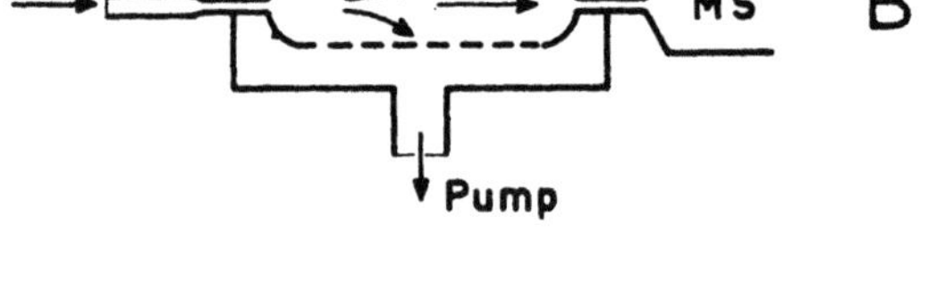

Fig. 4. Principle of WATSON-BIEMANN separator (A) and RYHAGE-STENGHAGEN separator (B)

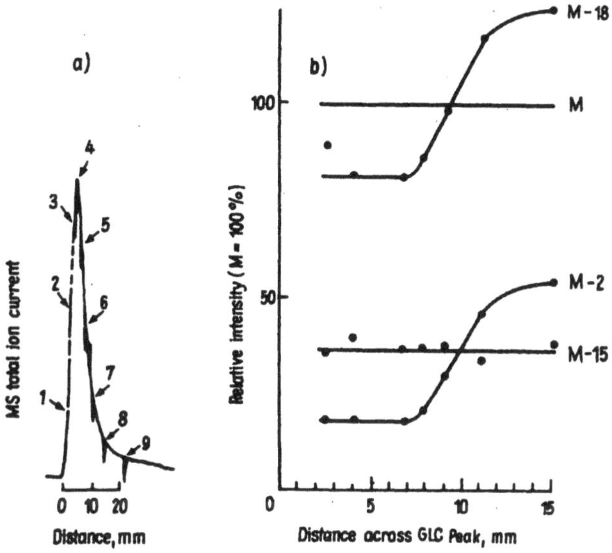

Fig. 5. a) Gas chromatographic peak of cholesterol behind WATSON-BIEMANN separator. Numbers indicate positions where MS were taken

b) Intensities of ions M-2, M-15 and M-18 relative to M taken from the spectra measured across the peak of fig. a). Increasing intensities of the ion M-2 and M-18 behind spectrum 7 indicates decomposition

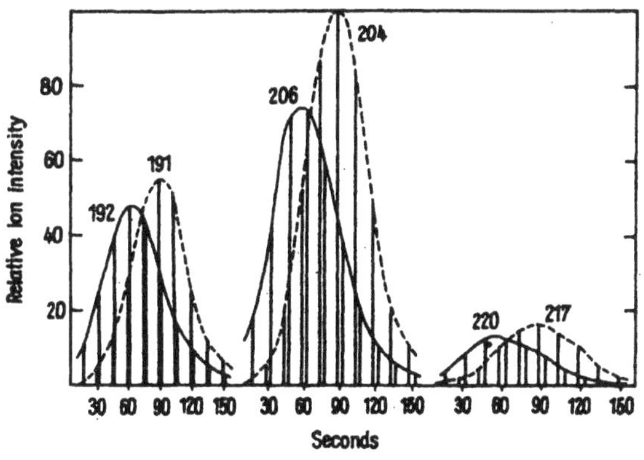

Fig. 6. Changes in fragment ion intensities during elution of glucose and glucose-d_7 mixture from 3% SE-30 column

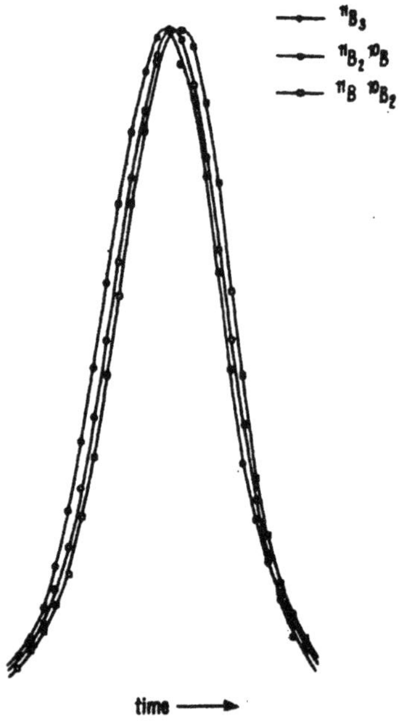

Fig. 7. Elution of three isotope carboran species containing

$^{11}B_3$, $^{11}B_2\,^{10}B_1$, $^{11}B_1\,^{10}B_2$

evaluated from „isotopic scan"

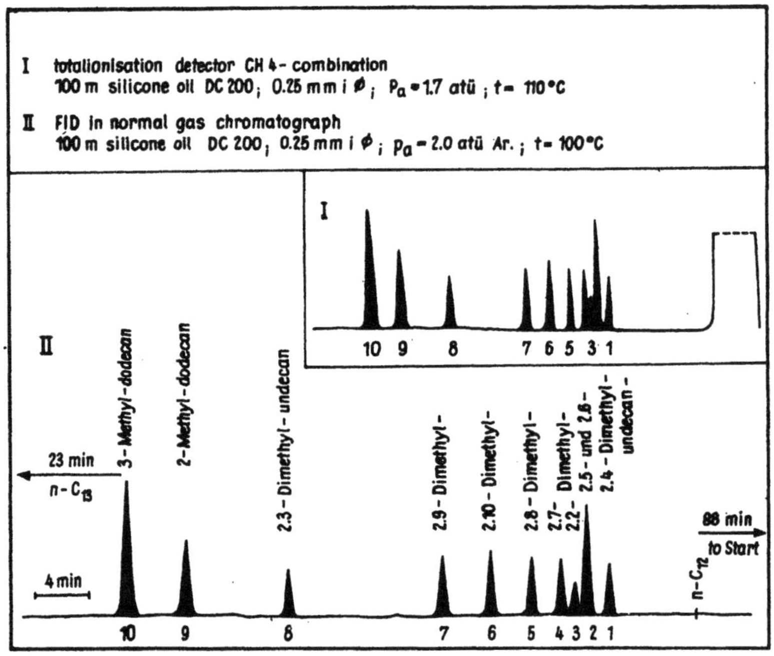

Fig. 8. Chromatograms of methylene-insertion-reaction products from 2-methylundecane measured with total ionization detector within the combination (I) and with the flame ionization detector in a normal gas chromatograph (II)

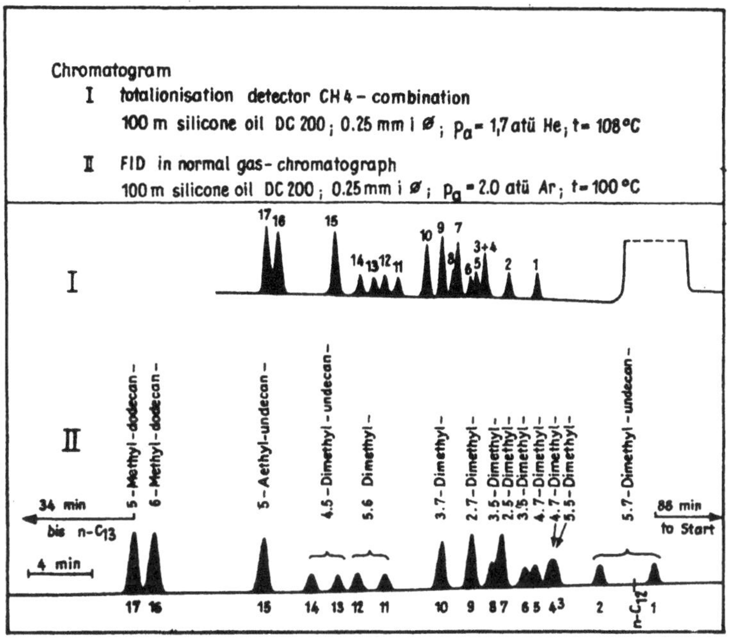

Fig. 9. Chromatograms of methylene-insertion-reaction products from 5-methylundecan measured with total ionization detector within the combination (I) and with the flame ionization detector in a normal gas chromatograph (II).

Fig. 10. and 11. Comparison of calculated and measured I^S values of dimethylundecanes.

Comparison of calculated and measured I^S values of dimetylundecanes

	I^S meas.	i	I^S cal.	D
2-Me-Undecane	1164,0	-36,0		
3-	1169,6	-30,4		
4-	1158,6	-41,4		
5-	1153,6	-46,4		
6-	1151,8	-48,2		
2,2-DM-Undecane	1215,2		1228,0	-12,8
3,3-	1233,1		1239,2	- 6,1
4,4-	1214,6		1217,2	- 2,6
5,5-	1205,4		1207,2	- 1,8
6,6-	1200,0		1203,6	- 3,6
2,3-DM-Undecane	1251,4		1233,6	+17,8
3,4-	1247,0		1228,2	+19,4
	1248,2			
4,5-	1230,4		1212,2	+19,8
	1233,6			
5,6-	1223,4		1205,4	+19,9
	1227,2			
2,4-DM-Undecane	1208,2		1222,6	-14,4
3,5-	1207,2		1223,2	-14,2
	1210,7			
4,6-	1193,0		1210,4	-14,4
	1199,0			
5,7-	1190,4		1207,2	-13,2
	1197,6			

Comparison of calculated and measured I^S values of dimetylundecanes

	I^S meas.	i	I^S cal.	D
2-Me-Undecane	1164,0	-36,0		
3-	1169,6	-30,4		
4-	1158,6	-41,4		
5-	1153,6	-46,4		
6-	1151,8	-48,2		
2,5-DM-Undecane	1210,4		1217,6	- 7,2
2,6-	1210,4		1215,8	- 5,4
2,7-	1215,8		1217,6	- 1,8
2,8-	1221,2		1222,6	- 1,4
2,9-	1232,6		1233,6	- 1,0
2,10-	1227,3		1228,0	- 0,7
3,6-DM-Undecane	1213,5		1221,4	- 7,1
	1215,1			
3,7-	1217,9		1223,2	- 5,3
3,8-	1226,8		1228,2	- 1,4
3,9-	1238,1		1239,2	- 1,1
4,7-DM-Undecane	1204,6		1212,2	- 6,6
	1206,6			
4,8-	1212,4		1217,2	- 4,8

values at 100 °C measured

i = branding increments

D = differences of calculated an measured values

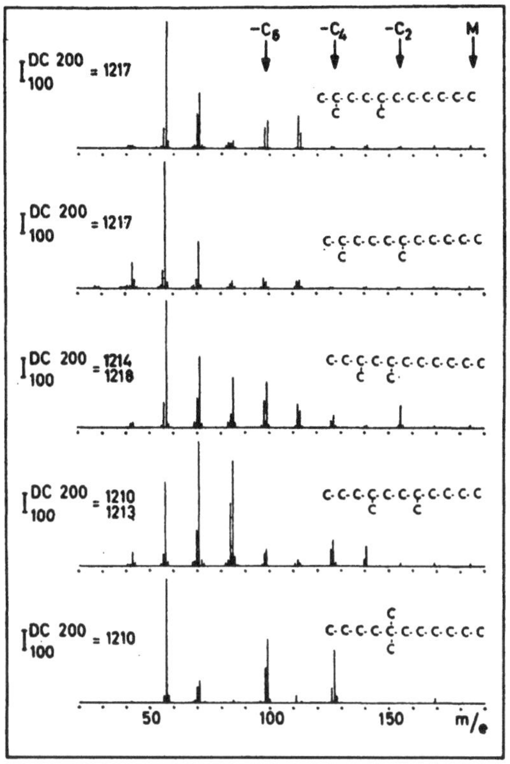

Fig. 12. Mass spectra of dimethylundecanes with similar retention indices

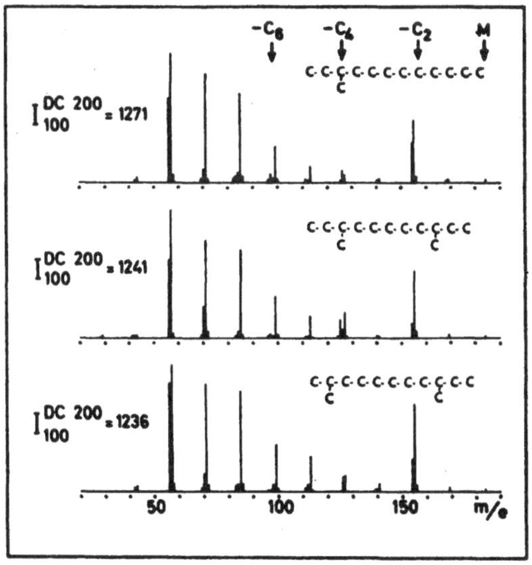

Fig. 13. Mass spectra of three tridecanes with similar retention indices

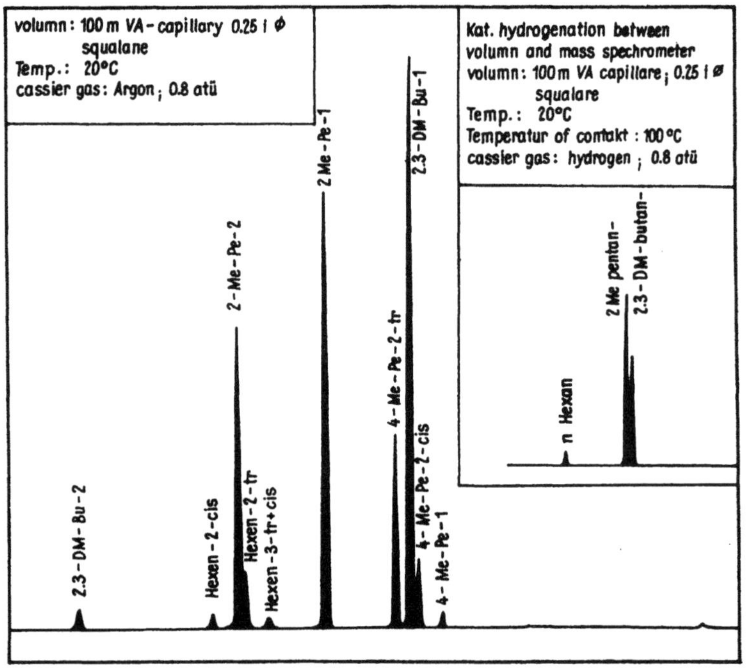

Fig. 14. Chromatogram of an olefin mixture, original on left side, and hydrogenation before the column on the right

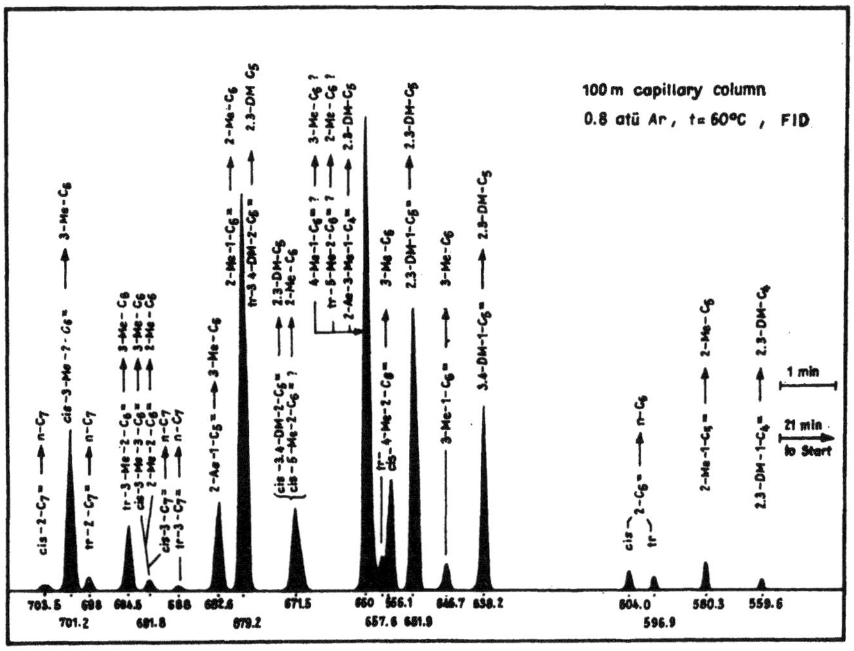

Fig. 15. Olefins from codimerization of propene and 2-butene. The arrows indicate the related saturated hydrocarbons identified by their mass spectra resulting from hydrogenation between column and mass spectrometer

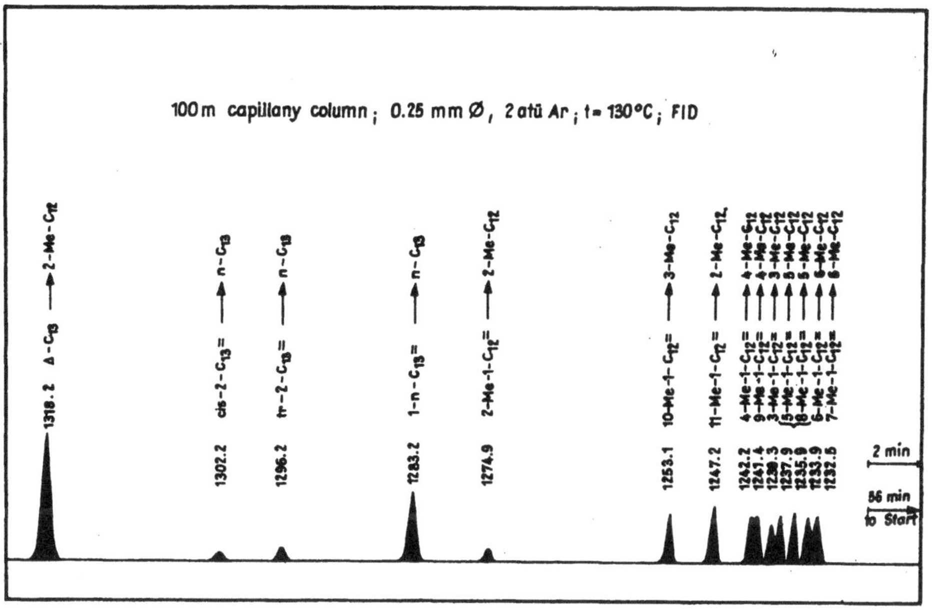

Fig. 16. Olefins from methylene-insertion-reaction with 1-dodecene. The arrows indicate the related saturated hydrocarbons identified by their mass spectra resulting from hydrogenation between column and mass spectrometer

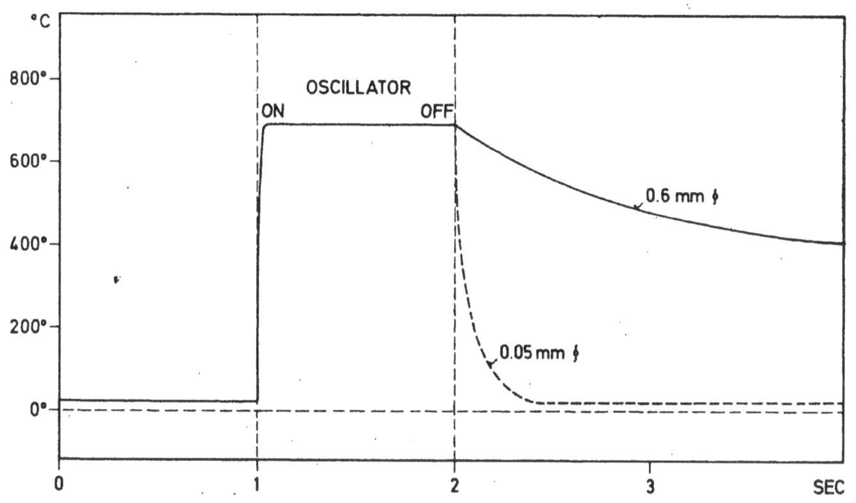

Fig. 1. Temperature Time Profiles of a Cylindric Ferromagnetic Conductor (Oscillator: 0.45 MHz, 1.4 KW)

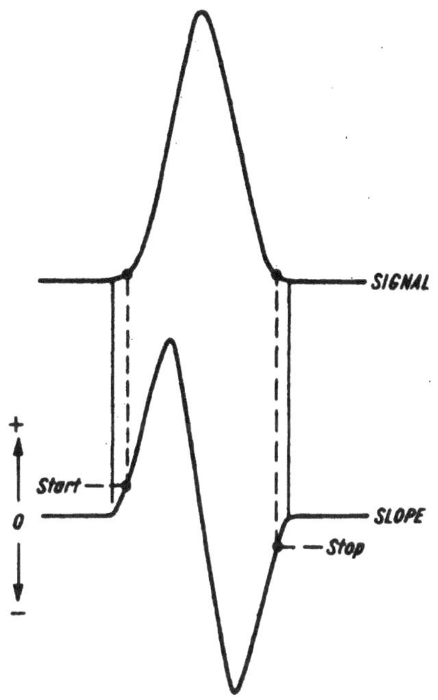

Fig. 1. Determination of the peak start via pass-over of an ordinate-threshold or via pass-over of a slope-threshold [1]

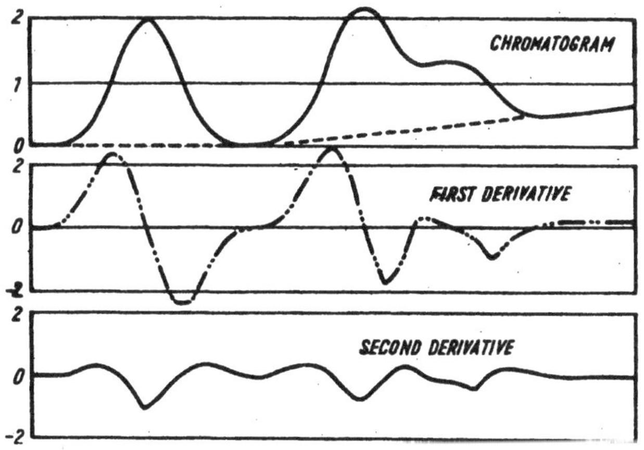

Fig. 2. Figure of baseline trend and peak shoulders on the curves of the first and second derivatives of the chromatogram [2]

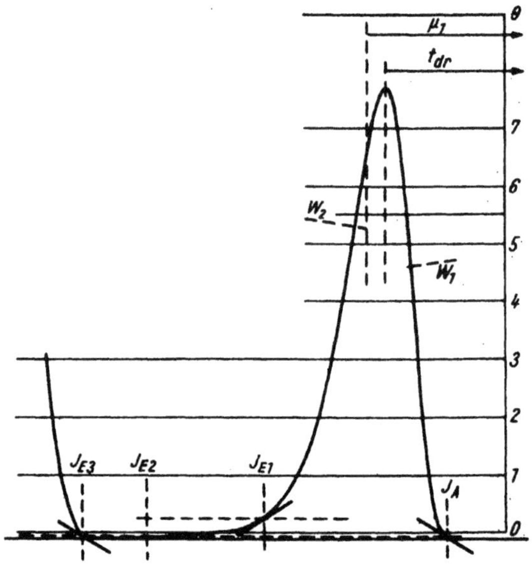

Fig. 3. Typical asymmetric peak of a chromatogram

u_1^o: First moment (location of the centre of gravity)

t_{dr}: Total retention time (determined by the peak maximum)

W_1, W_2: Location of the inflexion points

J_A: Start of integration determined by a given slope-threshold

J_{E1}: Possible end of integration in the case of the same slope-threshold as at J_A but with a negative sign

J_{E2}: Possible end of integration if the curve dropped below the height of the integration start J_A

J_{E3}: Possible end of integration of the same slope as at J_A is reached again (integration start for the next peak)

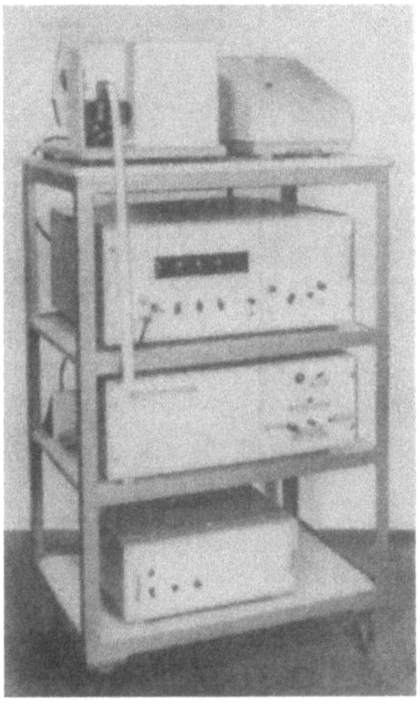

Fig. 4. Device for data-processing, consisting of an analogue-digital converter, tape printer, tape punching, device and converters belonging to

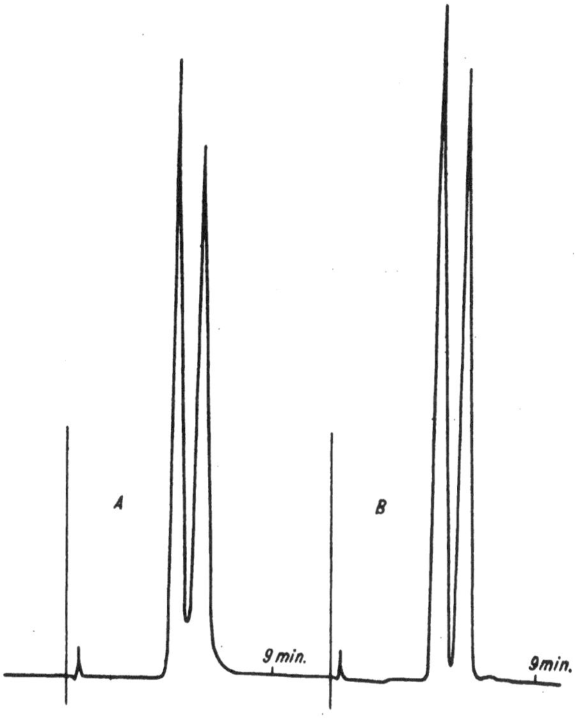

Fig. 1. Injection of 15 μl decalin on Aerograph 90 P - 3 m x 4 mm column
A - in injector
B - on column

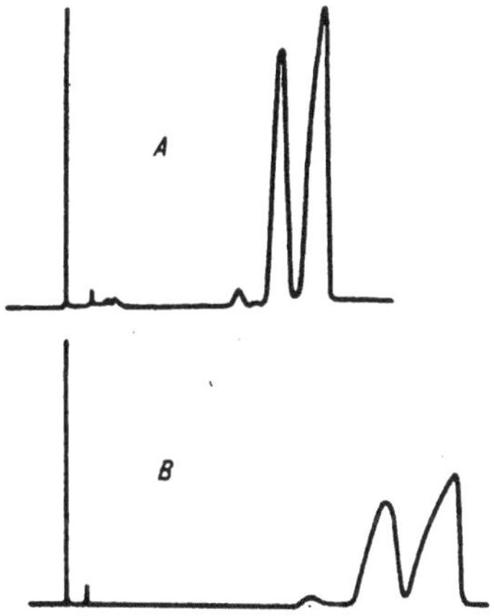

Fig. 2. Separation of 4 ml 3,4- dimethylcyclohexanone cis and trans isomers on column 1 of Table V.
700 ml H_2/min at 150°.
Bridge current 150 mA.
Recovery was expectedly better with a than with B

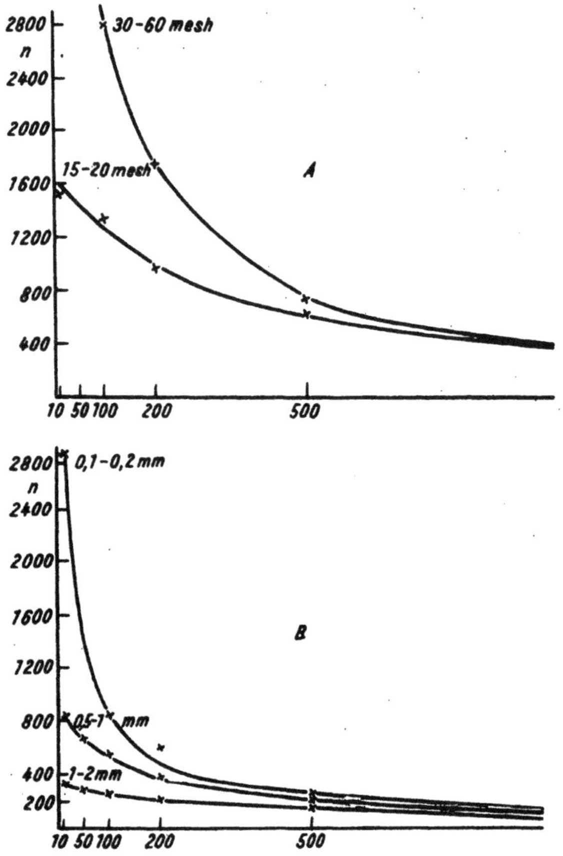

Fig. 3. Shows that N' values for large samples are relatively independent of support material grain size. Upper diagram A shows this for Chromosorb W 30-60 mesh and a self made white support 15-20 mesh with trans decalin as sample at 180° for a 20 m x 9 mm coiled glass column.
Lower diagram B shows this for Sterchamol in the given particle dimensions with iso-octane as sample at 100° and in a 6 m x 9 mm coiled glass column. Gas rate in all cases 200 ml H_2/min

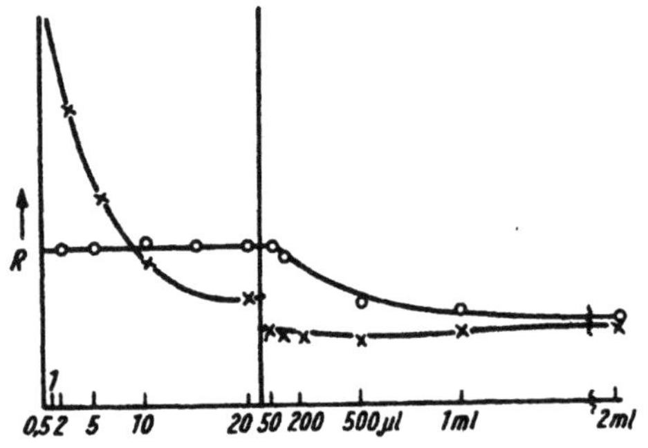

Fig. 4 Response of katharometer detection in hydrogen (o - normal peaks) and nitrogen (x - inverted peaks) as explained in text

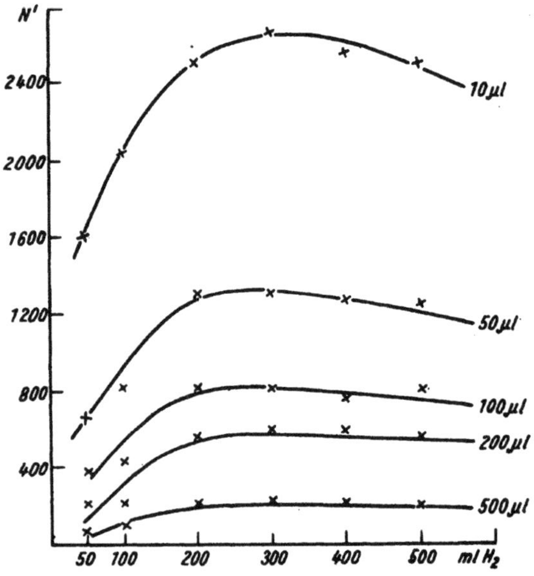

Fig. 5. N' values versus gas rate with hydrogen on a 6 m x 9 mm coiled glass column filled with Chromosorb W 30 - 60 mesh, 5 g SE - 30/100 ml support

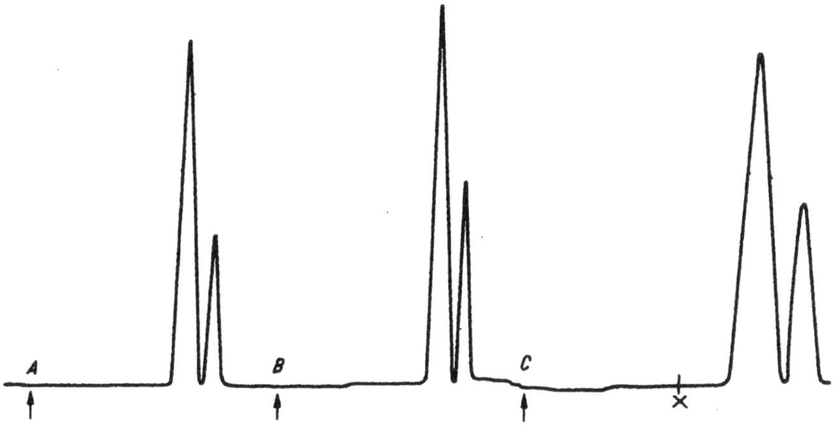

Fig. 6. Chromatograms of decalin on Autoprep 700. Sample load 200 µl; bridge current 150 mA. Attenuator 32 for A and B, 64 for C. Other conditions: (A) isotherm at 170° - 200 ml/min H_2; (B) programmed from 150 to 190° at peak elution; 200 ml/min H_2; (C) programmed from 150 to 200° at peak elution. At point X the gas rate was changed from 200 to 25 ml/min H_2

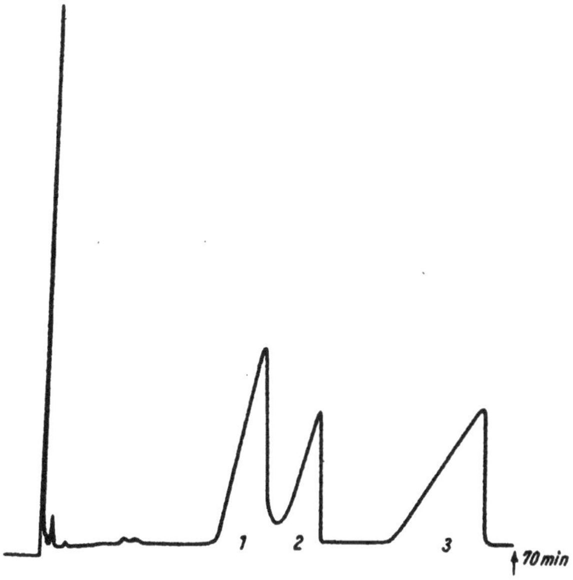

Fig. 7. 150 µl sample solution on coiled glass column of 3 m x 9 mm, filled with Chromosorb W 30-60, coated with Versamid 5 g/100 ml support. 150 ml H_2/min at 230 °C. Autoprep 700 bridge current 150 mA - Attenuation x 4. Collection on silica gel

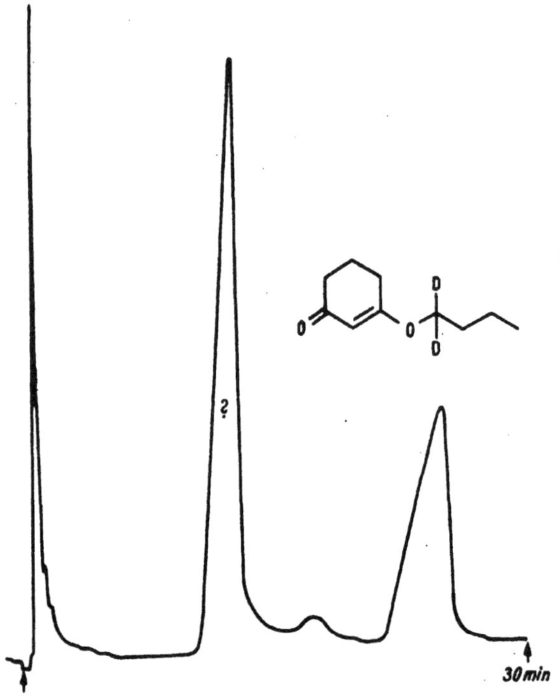

Fig. 8. Separation of 30 μl sample on 2 m x 4 mm column, Chromosorb 60-80 coated with 5 g/100 ml PTMO at 200° with 25 ml H_2/min. Varian Aerograph A 90 P. Bridge current 150 mA. Collection in simple capillary gave a sizable drop of the desired substance

Fig. 9. Apparatus for preparative scale gas chromatography on long columns (see text). Only principles are shown - scale is not respected.

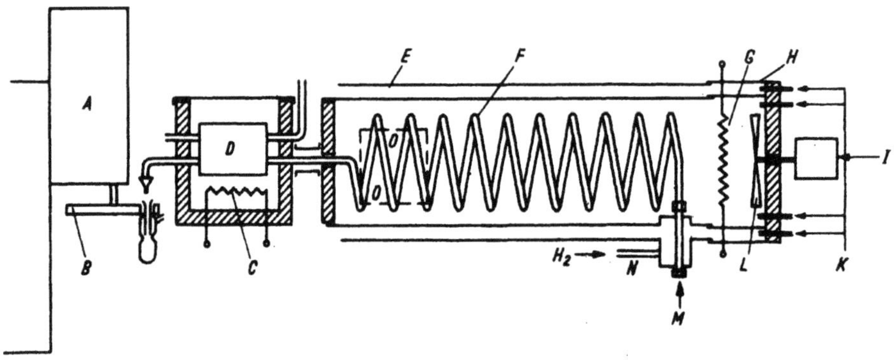

A: Autoprep 700 - B: collector table - C: detector oven heater - D: katharometer detector oven - E: double mantle column oven - F: coiled column - actually the coils are as close as possible to one other - G: column oven calrod heater - H: detachable heater fan motor block - I.: fan motor - K: inlets for air under pressure - L: fan - M: injector - N: valve to close off gas flow - O: trap door

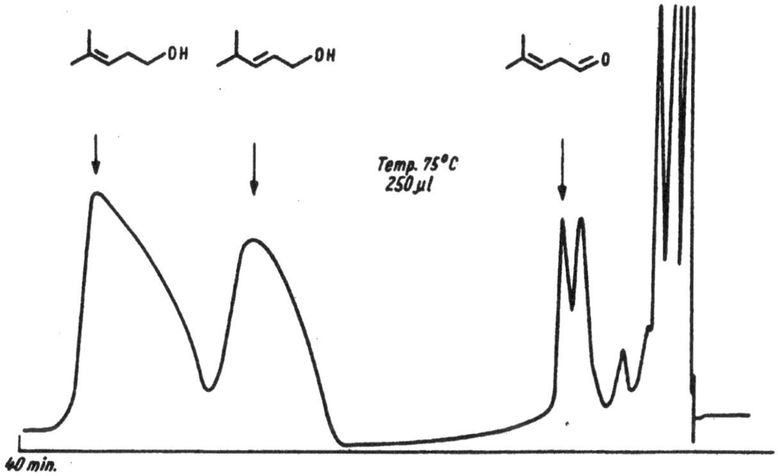

Fig. 10. 250 µl of the reaction mixture produced by $LiAlH_4$ reduction of N-isohexenoala-ziridine. 3 m x 6 mm column - Chromosorb W 60-80 mesh with P.T.M.O. (5 g/100 ml support) in Aerograph 90 P instrument. Working t° - 75°. Structure shown on the graph. Collection of aldehyde and alcohols in cooled U tubes

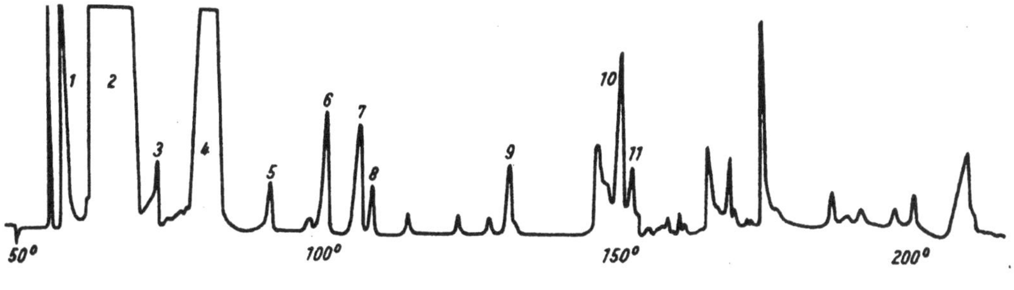

Fig. 11. 4 ml of volatile hops aroma components on 20 m x 9 mm coiled glass column - Chromosorb W 30-60 mesh coated with 5 g/100 ml of P.T.M.O. 250 ml H_2/min. Programmed at 1°/min from 50° to 200°, isothermal afterwards. Modified Autoprep 700. Peak 2 is ether solvent, peak 10 is myrcene. Most other peaks were collected in sufficient amount to be identified by normal NMR and other spectrometric techniques

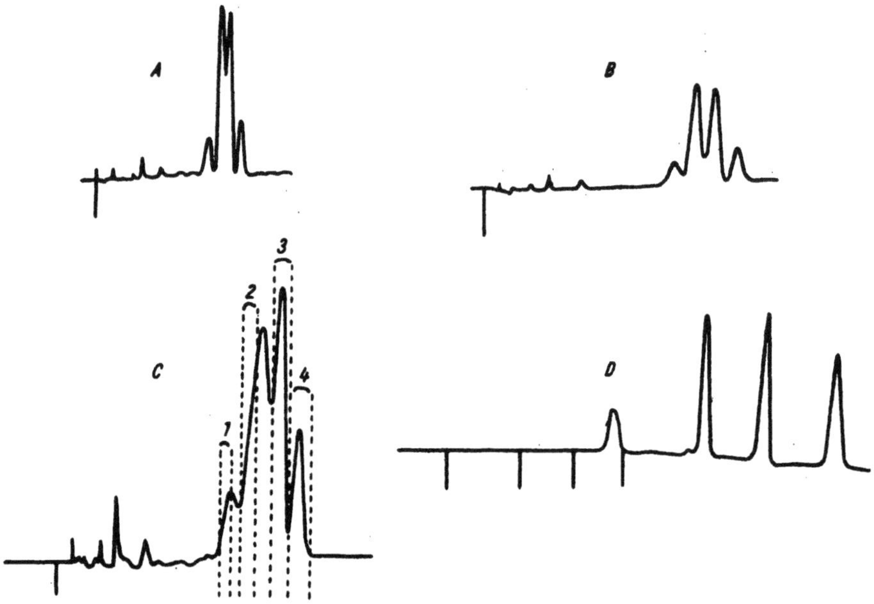

Fig. 12. Separation of methylpentene dimers on a 21 m x (1.5 x 25) coiled aluminium column - Carbowax 20 M on Chromosorb A 20 - 30 mesh. 700 ml H_2/min. See text.

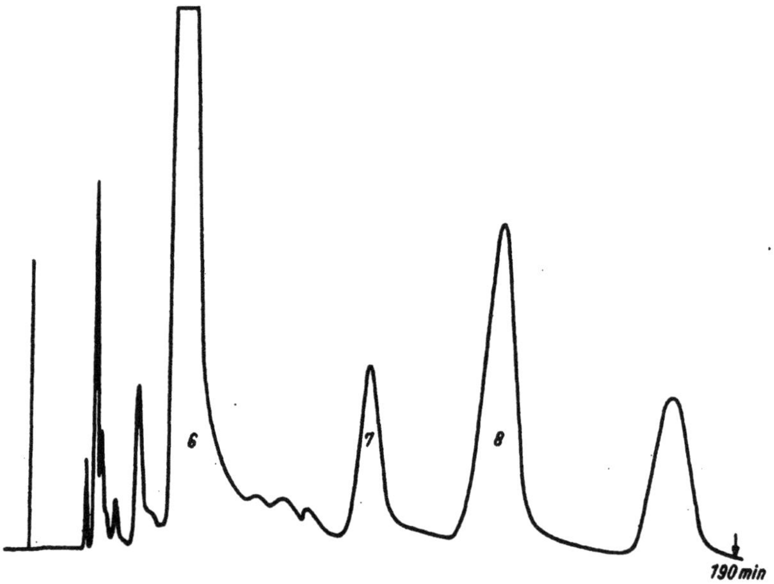

Fig. 13. 2 ml reaction mixture - isothermal at 150°. 600 ml H_2/min. Coiled glass column 20 m x 1.2-2.4 cm filled with 0.8 mm glass beads coated with 2 g Carbowax 20 M/100 ml support

Fig. 1. Separation of various hydrocarbons: methane (1); n-pentane (2) n-hexane (3); cyclohexane (4); benzene (5); cis-2 hexene (6); trans-2 hexene (7); on two different supports - 2 m column length

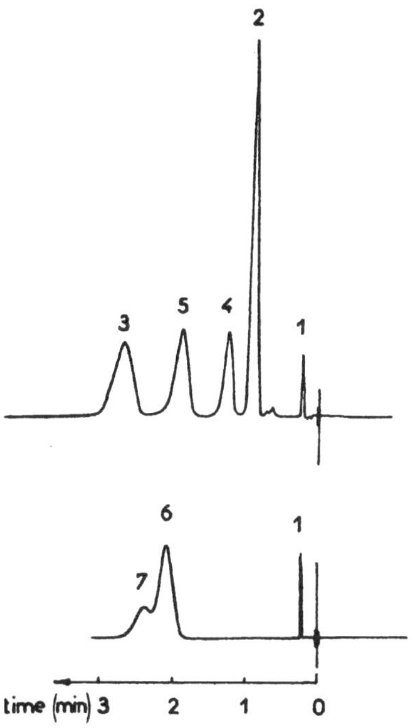

a) 5 % metal-free phthalocyanine on STERLING MT 52.5 °C
Pe = 3.5 kg/cm^2. u_s = 39 cm/sec.

b) 5 % Cobalt phthalocyanine on STERLING MT. 76.5 °C
Pe = 3 kg/cm^2. u_s = 19.5 cm/sec.

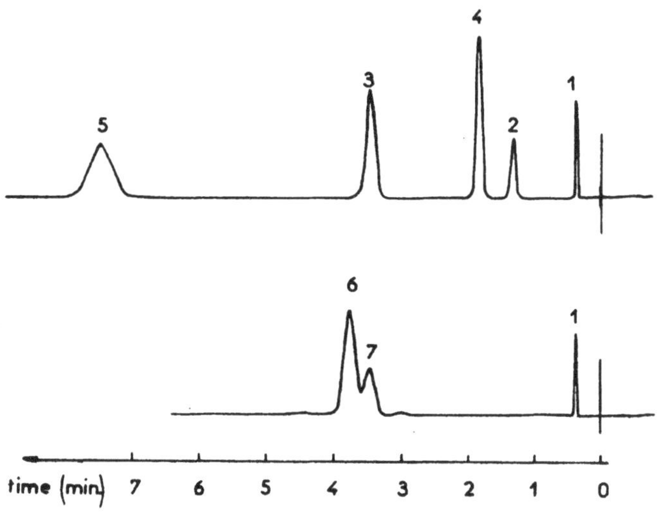

Fig. 2. Separation of a mixture of aromatic hydrocarbons/ benzene (1); toluene (2); ethylbenzene (3); isopropylbenzene (4); meta-xylene (5); para-xylene (6); ortho-xylene (7); n-propylbenzene (8) on various phthalocyanines 4 m column length

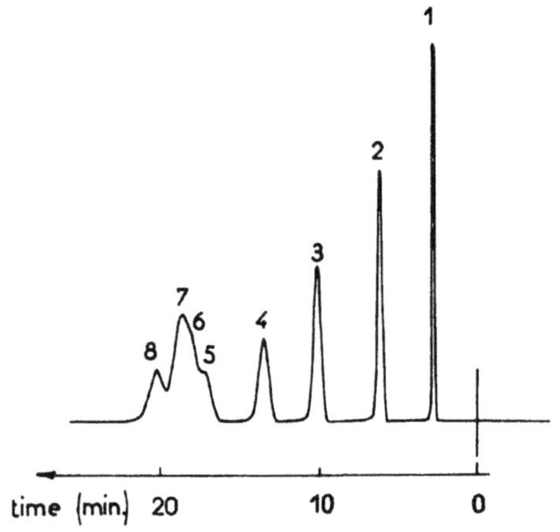

a) 1 % metal-free phthalocyanine on STERLING MT. 113 °C
Pe = 4 kg/cm^2.
u_s = 14. 5 cm/sec.

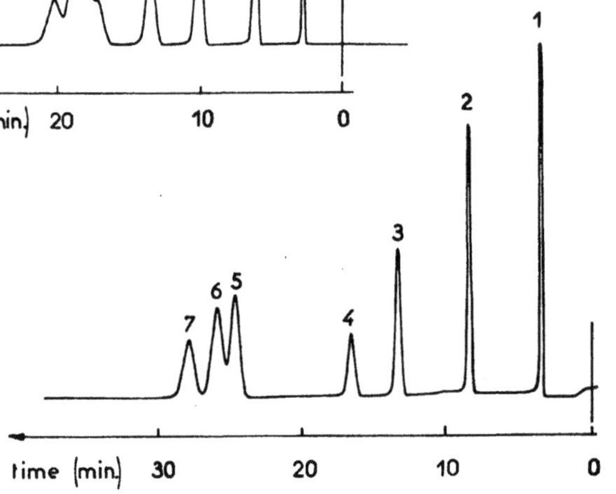

b) 5 % Copper phthalocyanine on STERLING MT 135 °C.
Pe = 6 kg/cm^2.
u_s = 17. 7 cm/sec.

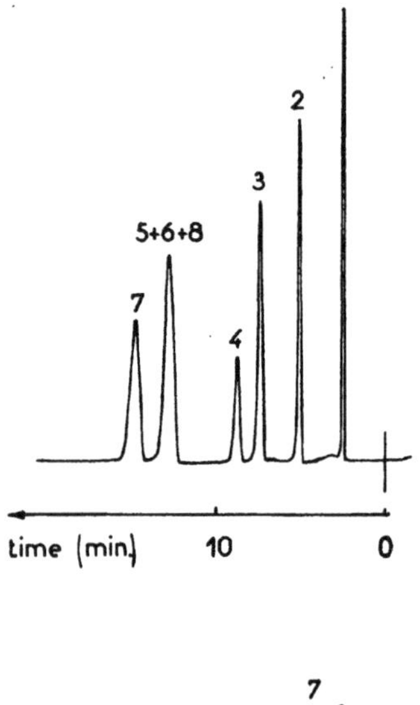

c) 5 % Cobalt phthalocyanine on STERLING MT. 163 °C.
Pe = 3 kg/cm^2.
u_s = 10. 6 cm/sec.

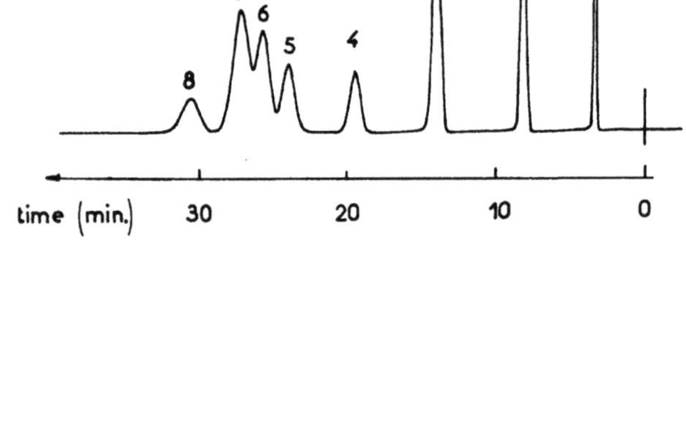

1, 5 % Cobalt phthalocyanine on STERLING MT. 157 °C.
Pe = 4 kg/cm^2.
u_s = 12. 2 cm/sec.

Fig. 3. Separation of a mixture of nitrogen compounds with: 5% Cobalt phthalocyanine on STERLING MT. 197 °C. Pe = 6 kg/cm^2 u_s = 17.6 cm/sec. 4 m column length.

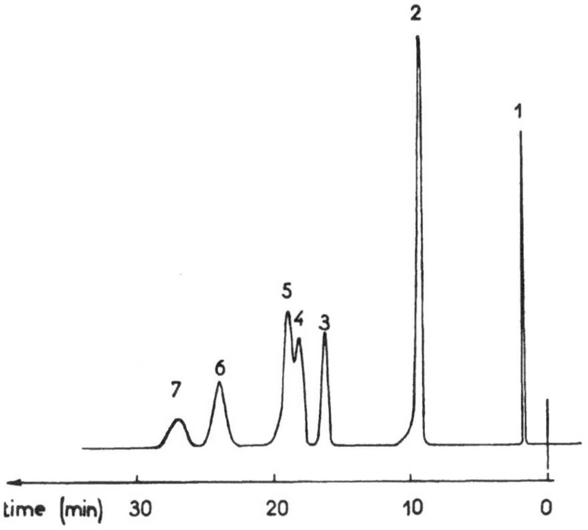

a) methane (1); aniline (2); ortho toluidine (3); N, N-dimethylaniline (4); N-methylaniline (5); para toluidine (6); meta toluidine (7)

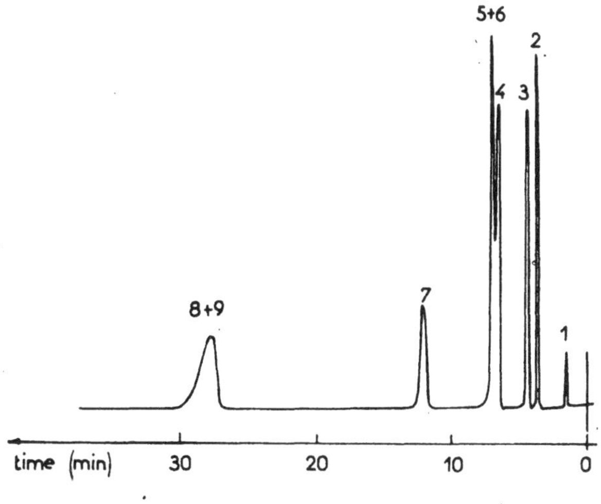

b) methane (1); 2-picoline (2); 2,6-lutidine (3); 2,3-lutidine (4); 2,4-lutidine (5); 2,5-lutidine (6); pyridine (7); 3-picoline (8); 4-picoline (9)

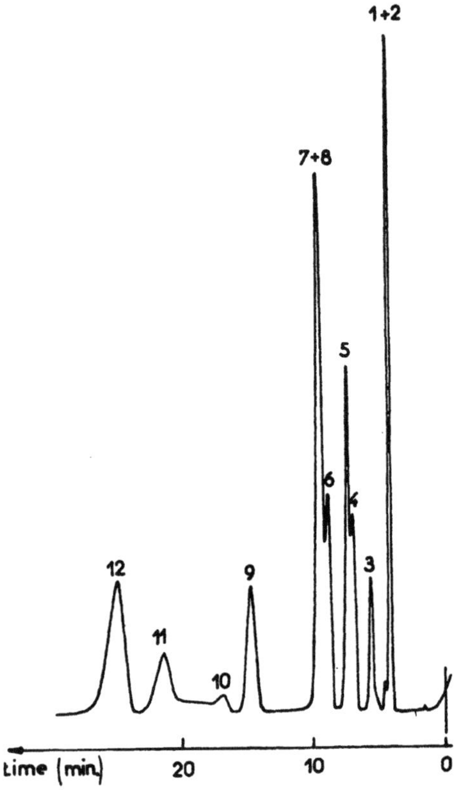

Fig. 4. Separation of a mixture of terpenes:

a) α-pinene (1); camphene (2); sabinene (3); Δ^3-carene (4); fenchone (5); camphor (6); p-cymene (7); limonene (8); n-decane (9); isomenthone (10); menthone (11); carvone (12)

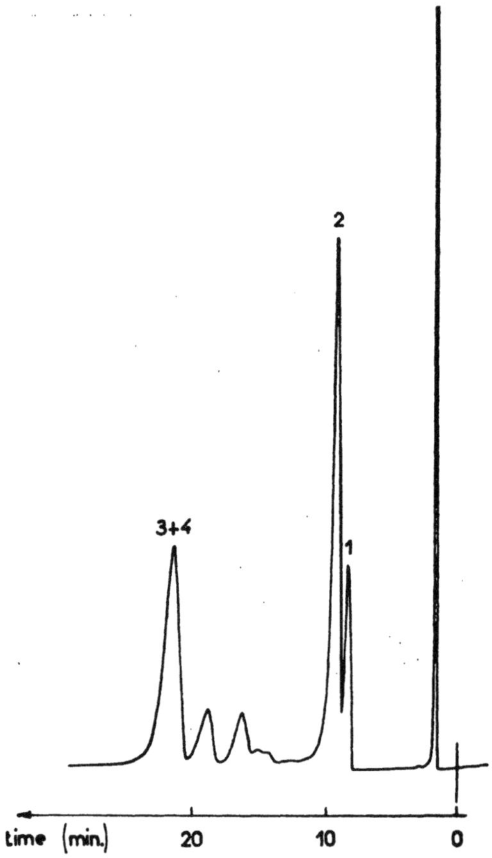

b) α-borneol (1); β-borneol (2); α-terpineol (3); menthol (4).
1 % metal-free phthalocyanine on STERLING MT, 157 °C
Pe = 4 kg/cm^2. u_s = 13.3 cm/sec.
4 m column length

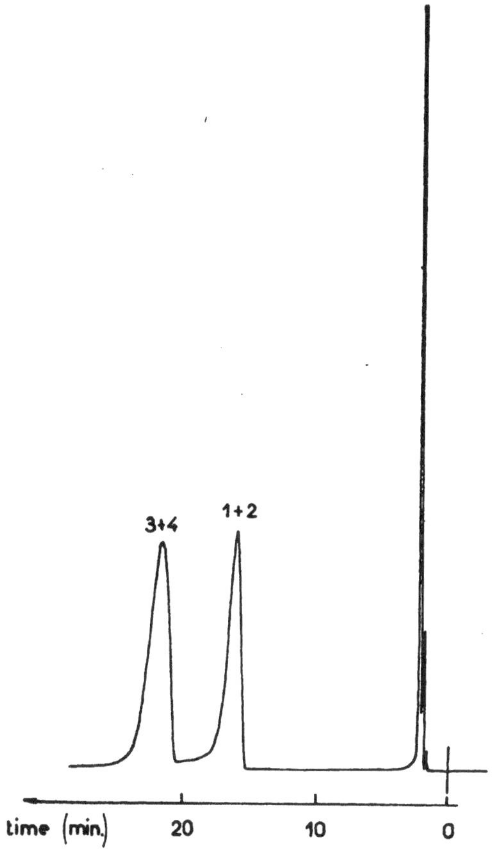

Fig. 5. Separation of menthol isomers: isomenthol (1); neoisomenthol (2); menthol (3); neomenthol (4); same conditions as previous figure

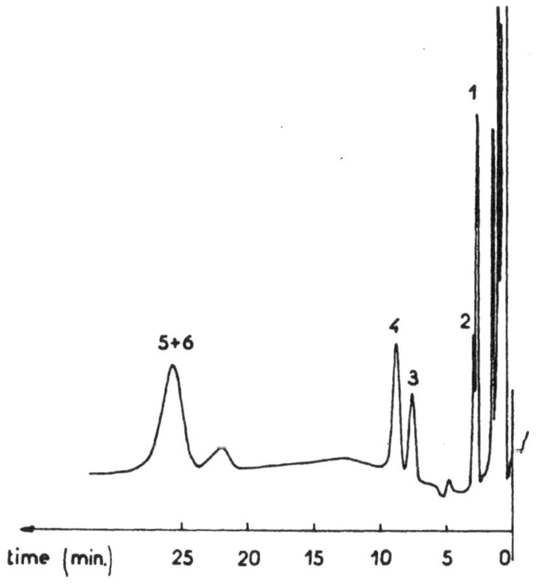

Fig. 6. Separation of a mixture of polynuclear aromatic hydrocarbons. 5 % metal-free phthalocyanine on STERLING MT. 302 °C. 2 m column length. Pe = 3 kg/cm^2. u_s = 28 cm/sec. phenanthene (1); anthracene (2); fluoranthene (3); pyrene (4); chrysene (5); naphtacene (6)

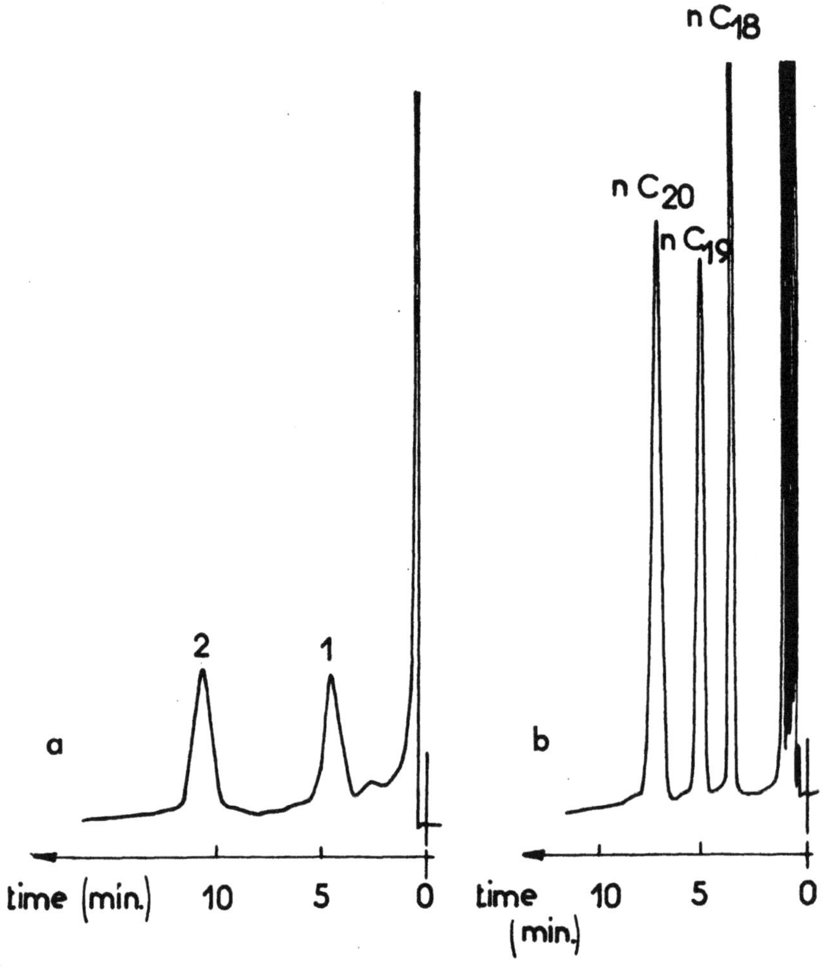

Fig. 7. a) Separation of etiocholanolone (1) and androsterone (2)

b) Separation of saturated hydrocarbons.

Same conditions as for previous figure

F. M ZADO and R. S. JUVET

Gas Chromatography, Thermodynamic of Adsorption and Solution and Nonideality of Gas Phase of Some Volatile Metal Chlorides

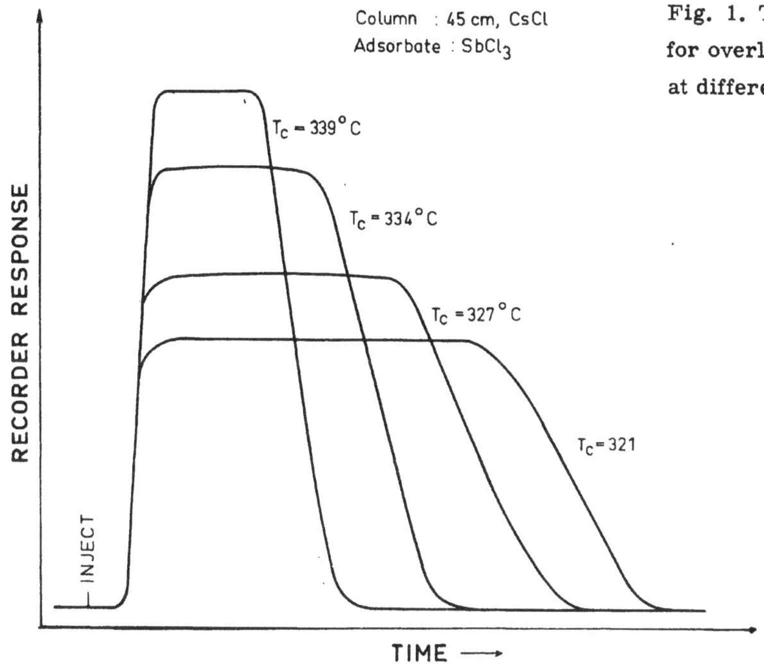

Fig. 1. Tabletop peaks obtained for overloaded CsCl column at different temperatures

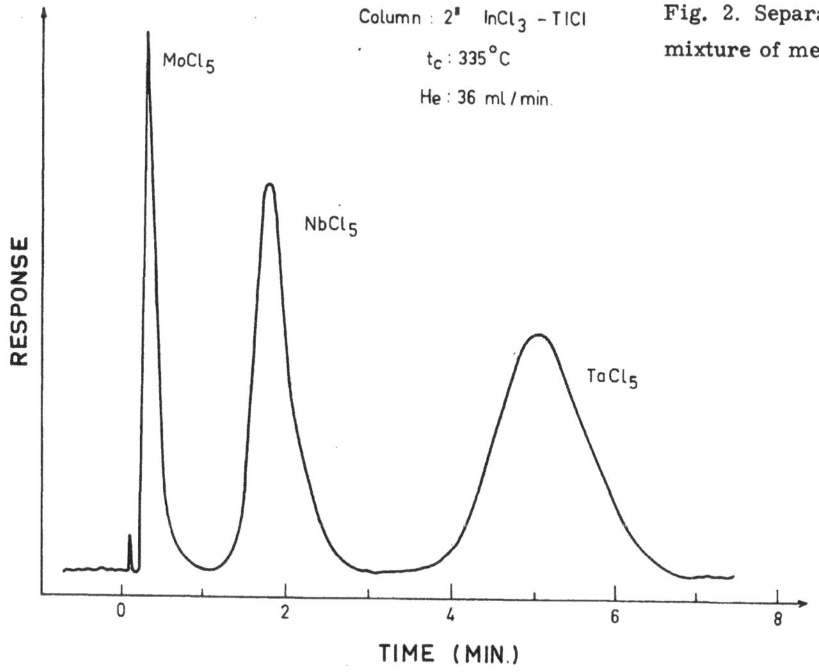

Fig. 2. Separation of ternary mixture of metal pentachlorides

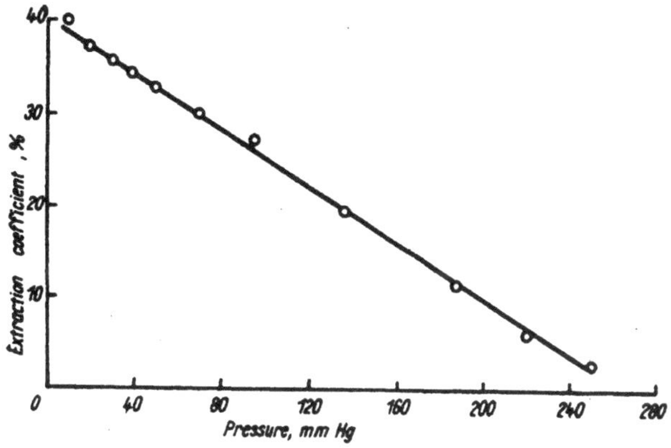

Fig. 1. Plot of extraction of helium from air vs. initial pressure in the enrichment column

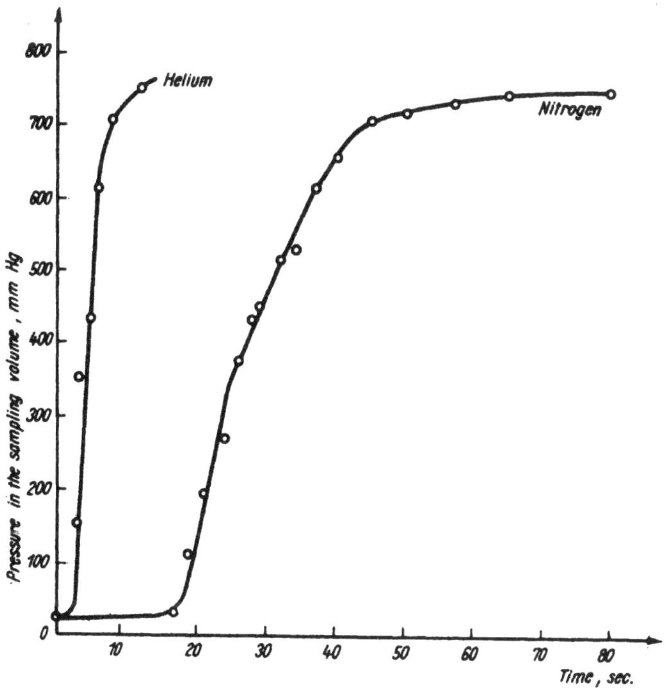

Fig. 2. Plot of pressure in the enrichment column vs. time of inflow
1 - helium, 2 - nitrogen

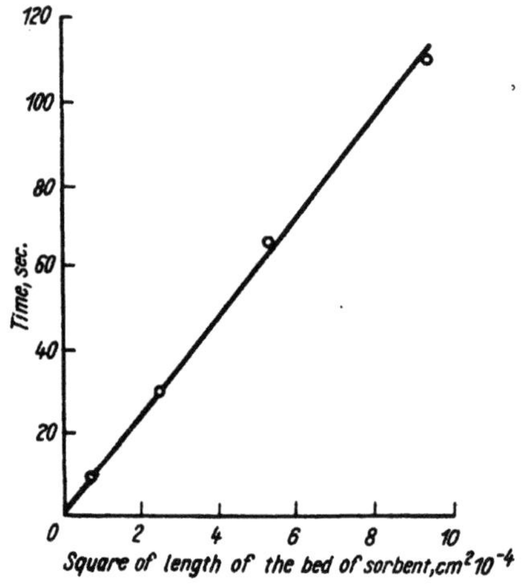

Fig. 3. Plot of time of filling of the sorbent bed vs. length of column

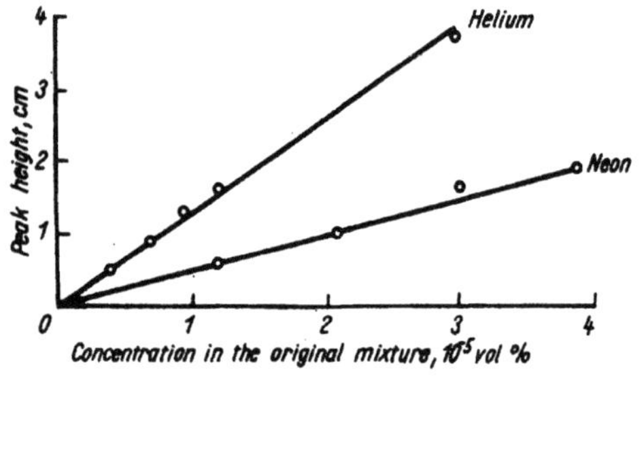

Fig. 4. Calibration curve for helium and neon, obtained in measuring trace amounts of these gases in argon

Author Index

SUBJECT INDEX

Bei Fragen zur Produktsicherheit wenden Sie sich bitte an:
If you have any questions regarding product safety,
please contact:

Walter de Gruyter GmbH
Genthiner Straße 13
10785 Berlin
productsafety@degruyterbrill.com